FUNDAMENTALS OF
LOGISTICS MANAGEMENT
European Edition

David B. Grant, Douglas M. Lambert,
James R. Stock and Lisa M. Ellram

FUNDAMENTALS OF
LOGISTICS MANAGEMENT

The *McGraw·Hill* Companies

London	Boston	Burr Ridge, IL	Dubuque, IA	Madison, WI	New York
San Francisco	St Louis	Bangkok	Bogotá	Caracas	Kuala Lumpur
Lisbon	Madrid	Mexico City	Milan	Montreal	New Delhi
Santiago	Seoul	Singapore	Sydney	Taipei	Toronto

Fundamentals of Logistics Management
David B. Grant, Douglas M. Lambert, James R. Stock, Lisa M. Ellram
ISBN-13 9780077108946
ISBN-10 0-07-710894-9

Published by McGraw-Hill Education
Shoppenhangers Road
Maidenhead
Berkshire
SL6 2QL
Telephone: 44 (0) 1628 502 500
Fax: 44 (0) 1628 770 224
Website: www.mcgraw-hill.co.uk

British Library Cataloguing in Publication Data
A catalogue record for this book is available from the British Library

Library of Congress Cataloging in Publication Data
The Library of Congress data for this book has been applied for from the Library
of Congress

Acquisitions Editor: Melanie Smith
Development Editor: Rachel Crookes
Marketing Manager: Marca Wosoba
Production Editor: Jennifer Harvey/Beverley Shields

Text design by SCW
Cover design by Ego Creative
Printed and bound in the UK by Bell and Bain

ISBN-13 9780077108946
ISBN-10 0-07-710894-9

The *McGraw-Hill* Companies

BRIEF TABLE OF CONTENTS

DETAILED TABLE OF CONTENTS

Detailed Table of Contents

PREFACE

Logistics continues to be an important business topic for several reasons: the shrinking global market in which many firms now operate; ongoing advances in computer and information technology; and a continuing emphasis on quality and customer satisfaction. This new, European edition of *Fundamentals of Logistics Management* maintains the focus of the original book by Professors Lambert, Stock and Ellram to reflect these and many other developments that have made logistics critical for business success.

This brand new edition contains features that provided much of the uniqueness and success of previous editions. The book continues to take a marketing approach to the subject of logistics by recognizing that customer satisfaction is the primary output of logistics activities. The theme of global logistics is integrated throughout this edition and is enhanced by discussion of the dichotomy that the European market represents both a common yet international environment. This edition also continues to approach the topic from a managerial perspective. Each chapter introduces basic logistics concepts in a format that is useful for management decision-making, and these concepts are examined in light of how they interrelate with other functions of the firm. Additionally, each chapter includes examples of corporate applications of these concepts to illustrate how logistics activities can be managed to properly implement the marketing concept.

Further, a number of important topics not covered in many other logistics text are also covered in this edition: order processing and management information systems, materials flow, financial control of logistics performance, logistics organizations, global logistics, decision support systems, channels of distribution, and the strategic logistics plan. Other topics covered include partnerships, green and reverse logistics, computer technology, market globalization, warehouse location, strategic planning and customer service. The goal in covering these topics in addition to the traditional activities is to provide readers with a grasp of the total picture of the logistics process.

The **Global**, **Technology** and **Creative Solutions** mini-case study boxes provide examples of current practice and issues, and include a related question for students to consider the material 'outside the box' in another context. All three types of box appear in each chapter. There are now **Logistics Challenge** scenarios and questions at the end of each chapter, except for the inventory chapter (Chapter 5), where there are appendices on inventory calculations instead, to further stimulate student thinking.

The pragmatic, applied nature of the text and its managerial orientation make it a useful reference book for present and future logistics professionals. The **end-of-chapter Questions and Problems**, and the mini-case study boxes that appear in each chapter, are structured to challenge readers' understanding of the topics covered as well as their managerial skills. They are integrative in nature and examine issues that are important to today's logistics executive. Finally, there is **Suggested Reading** at the end of each chapter, and key terms included in the **Subject Index** and **Glossary** are emboldened in the text so that they can easily be located.

The Structure of the Book

In this European edition, the developing subject of supply chain management (SCM) has been incorporated into **Chapter 1**, the first introductory chapter. SCM is a strategic business management concept encompassing the marshalling of resources to provide products and services to customers. Logistics activities comprise the functional processes that move such products and services through a firm's supply chain. This chapter integrates these two concepts and sets the scene for the rest of the text.

Chapter 2, on customer service, incorporates new material related to service performance measurement, such as key performance indicators and quality of logistics service. This material has emerged during the last decade in both logistics literature and practice, and is considered important in the logistics discipline. Chapter 3, on logistics information systems, has new material regarding logistics technology advancements such as efficient consumer response and quick response, and new technology such as radio frequency identification. These first three chapters discuss the strategic nature of logistics processes, driven by technology, that lead to the primary output of customer satisfaction.

The next series of chapters discuss the functional activities of logistics, starting with the customer order and purchase as the initial driver of the process. Chapter 4, on purchasing and procurement, encompasses all phases of the purchasing process, and introduces advances in electronic purchasing of products and services. Chapter 5 combines the topics of inventory concepts and management – however, the popular appendices concerning inventory calculations remain for this new, European edition.

Chapter 6, on materials management, has been updated regarding current thought on distribution resources planning (DRP) and the manufacturing or conversions process. Chapter 7, on transportation, discusses European positions on regulatory issues such as transport congestion and taxation, and the Working Time Directive. Chapter 8, on warehousing, incorporates new material on advances in automated warehouses and discuss technologies for designing and operating warehouses or distribution centres. Chapter 9 focuses on materials handling, packaging, and green and reverse logistics as part of the European Union's drive to reduce waste from manufacturing and materials handling and SCM systems.

Chapter 10, on organizing for logistics, discusses new topics on logistics organizations, and current aspects on logistics organizational structures drawn from European Logistics Association and Council of Supply Chain Management Professionals surveys. Chapter 11, on accounting and control for logistics, has been updated to reflect customer and product profitability, linking to discussions of performance measurement in Chapter 3.

These seven foregoing chapters present the functional logistics activities of a firm. Chapter 12 considers global aspects of these functions and methods to go global. While the European Union may be considered one large market, there are still many issues surrounding business and logistics processes and strategies among member countries, particularly the new eastern European nations that joined the EU in 2004. This chapter encompasses new material on global logistics issues facing firms doing business both within and outside the European Union.

Finally, Chapter 13 discusses implementing logistics strategy in a firm's overall corporate strategy to achieve customer satisfaction (i.e. a strategic approach related to Chapters 1–3) and operational objectives for functional logistics activities (i.e. a tactical approach related to Chapters 4–11), within a European and global context (i.e. Chapter 12). These amendments bring this new, European edition up to date with current academic thought and logistics practice within and focusing on issues unique to Europe.

Extra resources

A range of lecturer resources to accompany this book are available online: www.mcgraw-hill.co.uk/textbooks/grant. Resources include a Lecturer Manual with solutions to Logistics Challenges, PowerPoint Slides for teaching, and a Glossary for students.

AUTHOR BIOGRAPHY

Dr David B. Grant is Lecturer in Logistics at the Logistics Research Centre, School of Management and Languages, Heriot-Watt University, Edinburgh, UK. He has previously lectured at the Universities of Calgary, Lethbridge and Edinburgh.

Dr Grant holds BComm and MBA degrees from the University of Calgary and MSc by Research (with Distinction) and PhD degrees from the University of Edinburgh. His doctoral thesis investigated customer service, satisfaction and service quality in food-processing logistics and was given the James Cooper Memorial Cup PhD Award by the UK Chartered Institute of Logistics and Transport.

Dr Grant's logistics and supply chain management research interests focus on customer service and satisfaction, service quality, relationships, the integration of logistics and marketing, services marketing, research methodologies and techniques, and SME logistics. His business experience includes corporate banking, retailing, and financial and marketing consulting.

Dr Grant is a member of the Council of Supply Chain Management Professionals, the UK Logistics Research Network and the Nordic logistics research group NOFOMA. He is also a member of the UK Higher Education Academy and holds a Professional Certificate in University Teaching from the University of Edinburgh.

GUIDED TOUR

Chapter outline
Each chapter opens with an outline of the topics that are to be covered in the chapter.

Chapter objectives
A set of learning objectives precedes each chapter, summarizing what readers should learn from each chapter.

Chapter introduction
This briefly introduces key themes, and outlines the function of each chapter.

Figures and tables
Each chapter provides a number of figures and tables to help you to visualize the various economic models, and to illustrate and summarize important concepts.

Chapter summary
This briefly reviews and reinforces the main topics you will have covered in each chapter to ensure you have acquired a solid understanding of the material covered.

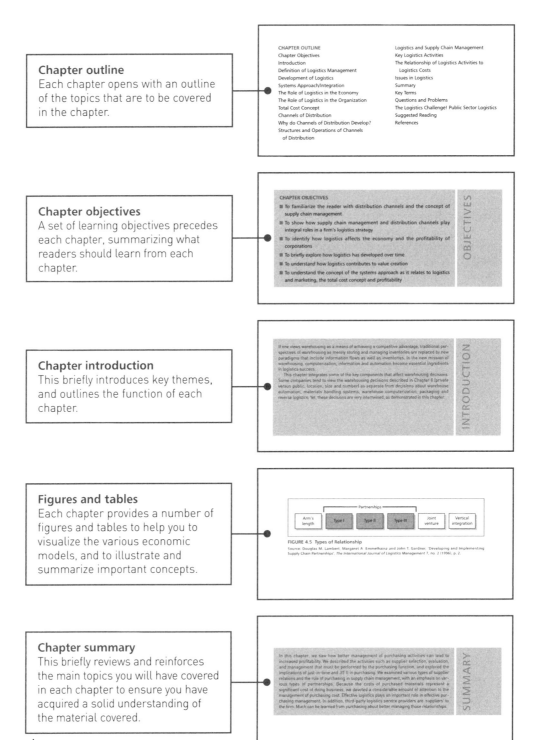

Suggested reading
A list of further sources follows each chapter to offer some pointers for continued research for projects or assessment.

QUESTIONS AND PROBLEMS

1 Describe the relationship between a firm's organizational structure and the integrated logistics management concept.
2 Coordination of the various logistics activities can be achieved in a variety of ways. Explain each of the following within the context of logistics organizational structure:
 a) process-based versus market-based versus channel-based strategies
 b) strategic versus operational
 c) centralized versus decentralized
 d) line versus staff
3 'There is no single ideal or optimal logistics organizational structure.' Do you believe this statement is accurate? Briefly present the arguments for and against such a statement.
4 How do personnel affect the degree of organizational effectiveness or productivity of a firm's logistics activity?

Questions and problems
These questions encourage you to review and apply the knowledge you have acquired from each chapter.

THE LOGISTICS CHALLENGE!

PROBLEM: PUBLIC SECTOR LOGISTICS
Most people think logistics and supply chain management are only relevant for consumer products demanded in a retail setting. However, logistics plays a major role in everyday life including public sector services such as the military, national postal services and health care. Public sector spending across Europe is under pressure, thus there is a big need to get logistics and supply chain activities right.
 One example of a public sector organization doing just that is the National Health Service (NHS) in England. It has set up its own NHS Logistics group to serve the needs of its customer base of 450 organisations such as hospitals, clinics and local surgeries. NHS Logistics has 1,250 employees and maintains seven distribution centres and a fleet of over 200 trucks that make 1,200 deliveries a day to over 10,000 delivery points. It handles over €900 million worth of products and 27 million line items per year and receives 100 percent of its customer orders electronically thanks to an Internet supplier portal initiated in late 2003.
 On the supply side NHS Logistics deals with over 750 suppliers and products are stocked on a 48 hour lead time. It issues 83 percent of its orders electronically and pays over 420,000 supplier

The Logistics Challenge!
This end-of-chapter feature is the perfect way to practise the techniques you have been taught and apply the methodology to real-world situations.

KEY TERMS

	Page
Channel power	00
Network	00
Systems approach	00
Form	00
Possession utility	00
Time utility	00
Place utility	00
Marketing concept	00

Key terms
Key terms are highlighted throughout each chapter, listed at the end of each chapter, and compiled in a Glossary at the back of the book for easy reference.

Mini-case studies
Each chapter contains three mini-case studies based around three themes: global, technology and creative solutions.

Study Skills

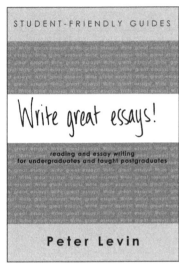

We publish guides to help you study, research, pass exams and write essays, all the way through your university studies.

Visit www.openup.co.uk/ss/ to see the full selection and get a £2 discount by entering the promotional code **study** when buying online!

Computing Skills

If you'd like to brush up your computing skills, we have a range of titles covering MS Office applications such as Word, Excel, PowerPoint, Access and more.

Get a £2 discount on these titles by entering the promotional code **app** when ordering online at www.mcgraw-hill.co.uk/app.

ACKNOWLEDGEMENTS

The year 2005 is a personal milestone as it is my fifteenth year teaching and researching the subject of logistics. Throughout this time I have referred to various books by Professors Lambert and Stock as I believe they provide a balanced and thorough view of the subject. Thus, I am pleased and honoured that I have been able to provide this European adaptation of the *Fundamentals of Logistics Management*. I have endeavoured to retain Professors Lambert, Stock and Ellram's objectives that this book is readable for both the instructor and student. Their original aim was to present instructors and students with the best textbook of this type on the market, and my aim was to continue in that vein. I believe, and hope they do too, that we have again succeeded in this quest and that I have not let the side down.

I want to acknowledge and thank everyone who has been instrumental in the development of my logistics career and in the preparation of this book. Space considerations prohibit me from naming everyone as it is a very long list, and I also fear I might leave someone off it. However, I want to specifically note some people who have been important catalysts for my own learning and development. Many thanks to my two mentors and former colleagues: Professor Stanley Paliwoda and Professor John Dawson, the latter of whom was also my PhD supervisor. You two more than any others have shaped my career and abilities. Thanks also to Professors Lambert, Stock, Martin Christopher and Herbert Kotzab for your significant influences on my views and beliefs about marketing and logistics.

Regarding this book, thanks to my Logistics Research Centre colleagues at Heriot-Watt University, Professor Alan McKinnon, Professor John Fernie and Clive Marchant, for their ongoing contributions and assistance. At McGraw-Hill, thanks to Kevin Watt for providing the idea of this adaptation and to Mark Kavanagh for seeing it through the editorial approval process. Lastly, I want to thank very much Rachel Crookes and her editorial development team for their guidance and support throughout a long and dark winter of writing to meet our ambitious deadlines.

David B. Grant
June 2005

The Publisher and Author would also like to extend their thanks to the followng for their contribution to the review and development process:

Dr Hans Voordijk, Ulf Hoglind, Susanne Hofmann, Sean Ennis, Sander de Leeuw, Ulrike Kussing, Dr Herbert Kotzab, Dick Leegwater, Christer Lindh, Chris Savage, James Stone, Chris Blake, Philip Barbonis, Mattias Gustavsson, and Willem J. H. van Groenendaal.

The Publisher and Author would also like to thank:

Touch Briefings for permission to use material from *Business Briefing: Global Purchasing & Supply Chain Strategies*; **Copenhagen Business School Press** for permission to use material from *Managing the Global Supply Chain*; **Council of Supply Chain Management Professionals** (CSCMP) for permission to use material from the CSCMP website, *Development and Implementation of Reverse Logistics Programs* and the *Journal of Business Logistics*; Philip Boyd, **Diageo** for permission to use material for and assisting with the Global Box in Chapter 4; **European Logistics Association** (ELA) for permission to use material from *Differentiation for Performance: Results of the Fifth Quinquennial European Logistics Study "Excellence in Logistics*

Acknowledgements

2003/2004", *Towards the 21st Century – Trends and Strategies in European Logistics* and *Success Factor PEOPLE in Distribution Centres*; **UK Transport Press Ltd.** for permission to use material from *Distribution Business* (now *Supply Chain Business*) and *Logistics Europe*; **Emerald Group Publishing Limited** for permission to use material from *Business Process Management Journal, International Journal of Physical Distribution & Logistics Management, Industrial Marketing Management, Journal of Business & Industrial Marketing, International Journal of Retail and Distribution Management, Supply Chain Management: An International Journal, The TQM Magazine* and *The International Journal of Logistics Management*; **European Union** (EU) and **Eurostat** for permission to use material from the EU website, *EU Transport in Figures: Statistical Pocket Book 2000* and *White Paper: European transport policy for 2010*; **Pearson Education Limited** (UK) for permission to use material from *Operations Management, Logistics and Supply Chain Management: Creating Value-Adding Networks, Purchasing and Supply Chain Management* and *Exploring Corporate Strategy*; Professor Alan C. McKinnon, **Heriot-Watt University** for permission to use material from *European Logistical and Supply Chain Trends: 1999–2005, Use of Vehicle Telematics Systems for the Collection of Key Performance Indicator Data in Road Freight Transport* and *Lorry Road User Charging: A Review of the UK Government's Proposals*; **Institute of Grocery Distribution** (IGD) for permission to use material from *Factory Gate, Open Book & Beyond* and *Retail Logistics*; **Braybrooke Press Ltd.** for permission to use material from the *Journal of General Management*; **American Marketing Association** (AMA) for permission to use material from the *Journal of Marketing*; **Kogan Page Limited** for permission to use material from *Logistics and Retail Management*; Doctor Håkan Aronsson, **Linköping University** for permission to use material from *Challenging Boundaries with Logistics – Proceedings of the XVI*[th] *Annual Conference for Nordic Researchers in Logistics*; **David Priestman** for permission to use material from *Logistics Business Magazine*; **The Chartered Institute of Logistics and Transport (UK)** for permission to use material from *Logistics & Transport Focus*; **Oxford University Press** for permission to use material from *Understanding Supply Chains*; Professors Dale S. Rogers and Ronald S. Tibben-Lembke, **University of Nevada-Reno** for permission to use material from *Going Backwards: Reverse Logistics Trends and Practices*; **John Wiley and Sons Ltd.** for permission to use material from *Supply Network Strategies* and *Inventory Control and Management*; **CEC Europe Media & Marketing Communications Ltd.** for permission to use material from *Supply Chain Europe*; **Taylor & Francis Group plc** for permission to use material from *The International Journal of Logistics: Research and Applications* (http://www.tandfco.uk); **Spice Court Publications Limited** for permission to use material from *fulfilment & e.logistics*.

CHAPTER 1
LOGISTICS AND SUPPLY CHAIN MANAGEMENT

OBJECTIVES

CHAPTER OBJECTIVES

- ■ To familiarize the reader with distribution channels and the concept of supply chain management
- ■ To show how supply chain management and distribution channels play integral roles in a firm's logistics strategy
- ■ To identify how logistics affects the economy and the profitability of corporations
- ■ To briefly explore how logistics has developed over time
- ■ To understand how logistics contributes to value creation
- ■ To understand the concept of the systems approach as it relates to logistics and marketing, the total cost concept and profitability

INTRODUCTION

Logistics is a broad, far-reaching function that has a major impact on a society's standard of living. In a modern society, we have come to expect excellent logistics services and tend to notice logistics only when there is a problem. To understand some of the implications to consumers of logistics activity, consider:

- ■ the difficulty in shopping for food, clothing and other items if logistical systems do not conveniently bring all of those items together in one place, such as a single store or a shopping mall
- ■ the challenge in locating the proper size or style of an item if logistical systems do not provide for a wide mix of products, colours, sizes and styles through the assortment process
- ■ the frustration of going to a store to purchase an advertised item, only to find out the store's shipment is late in arriving.

These are only a few of the issues often taken for granted that illustrate how logistics touches many facets of our daily lives. Because of the magnitude of the impact of logistics on society and individuals, a macro approach is taken in this initial chapter.

This chapter focuses on how logistics has developed over time, explains its relationships with supply chain management, examines the systems approach as it applies to a logistics value chain, explores the role of logistics in the economy and the firm, and examines the key interfaces of logistics with channels of distribution and other marketing activities. This chapter also shows the relationship between the systems concept and the total cost of ownership perspective. The discussion closes with a summary of key trends and current issues in logistics management.

Definition of Logistics Management

Because logistics is the topic of this textbook, it is important to establish the meaning of the term. Logistics has been called by many names, including the following:

- business logistics
- channel management
- distribution
- industrial logistics
- logistical management
- materials management
- physical distribution
- quick-response systems
- supply chain management
- supply management.

What these terms have in common is that they deal with the management of the flow of goods or materials from point of origin to point of consumption, and in some cases even to the point of disposal. The Council of Supply Chain Management Professionals (CSCMP), one of the leading professional organizations for logistics personnel, and formerly known as the Council of Logistics Management (CLM), defines *logistics management* as:

> that part of Supply Chain Management that plans, implements, and controls the efficient, effective forward and reverse flow and storage of goods, services and related information between the point of origin and the point of consumption in order to meet customers' requirements.[1]

Throughout this text, the CSCMP definition of logistics management is used. This definition includes the flow of materials and services in both the manufacturing and service sectors. The service sector includes entities such as the government, telecommunications, hospitals, banks, retailers and wholesalers. In addition, the ultimate disposal, recycling and reuse of products need to be considered because logistics is becoming increasingly responsible for issues such as removing packaging materials once a product is delivered and removing old equipment.

Logistics is not confined to manufacturing operations alone. It is relevant to all enterprises, including government institutions such as hospitals and schools, and service organizations such as retailers, banks and financial services organizations. Examples from these sectors will be used throughout the book to illustrate the relevance of logistics principles to a variety of operations.

Some of the many activities encompassed under the logistics umbrella are given in Figure 1.1, which illustrates that logistics is dependent upon natural, human, financial and information resources for inputs. Suppliers provide raw materials, which logistics manages in the form of raw materials, in-process inventory and finished goods. Management actions provide the framework for logistics activities through the process of planning, implementation and control. The outputs of the logistics system are competitive advantage, time and place utility, efficient movement to the customer, and providing a logistics service mix such that logistics becomes a proprietary asset of the organization. These outputs are made possible by the effective and efficient performance of the logistics activities shown at the bottom of Figure 1.1. Each of these activities will be explained in varying depth in this chapter and throughout the book.

Development of Logistics

Logistics activity is literally thousands of years old, dating back to the earliest forms of organized trade. As an area of study, however, it first began to gain attention in the early 1900s in the distribution of farm products as a way to support the organization's business strategy and as a way of providing time and place utility.

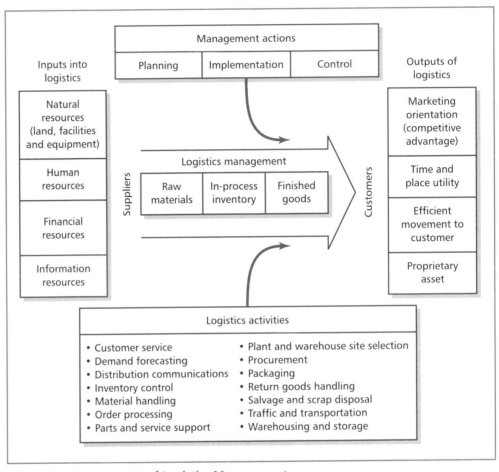

FIGURE 1.1 Components of Logistics Management

Following the clear importance of the contribution of logistics during the Second World War, logistics began to receive increased recognition and emphasis. The first dedicated logistics texts began to appear in the early 1960s, which was also when Peter Drucker, a noted business expert, author and consultant, stated that logistics was one of the last real frontiers of opportunity for organizations wishing to improve their corporate efficiency.[2] These factors combined to increase the interest in logistics.

Deregulation

To further fuel the focus on logistics, deregulation of the transportation industry in the late 1970s and early 1980s gave organizations many more options and increased the competition within and between transportation modes. As a result, carriers became more creative, flexible, customer-orientated and competitive in order to succeed. Shippers are now faced with many more transportation options. They can focus on negotiation of rates, terms and services, with their overall attention directed towards getting the best transportation buy.

Competitive Pressures

With rising interest rates and increasing energy costs during the 1970s, logistics received more attention as a major cost driver. In addition, logistics costs became a more critical issue for many organizations because of the globalization of industry. This has affected logistics in two primary ways.

First, the growth of world-class competitors from other nations has caused organizations to look for new ways to differentiate their organizations and product offerings. Logistics is a logical place to begin because domestic organizations should be able to provide much more reliable, responsive service to nearby markets than overseas competitors.

Second, as organizations increasingly buy and sell offshore, the supply chain between the organization and those it does business with becomes longer, more costly and more complex. Excellent logistics management is needed to fully leverage global opportunities.

Another factor strongly contributing to the increased emphasis and importance of logistics is a continued emphasis on cost control. A recent survey of various industry executives across 15 European countries indicates that logistics costs as a percentage of sales decreased from 12.1 per cent in 1982 to 6.1 per cent in 2003.[3] Thus, despite all the talk and emphasis on other issues – such as quality and customer service, which CEOs rated as second and third in importance – cost cutting is still considered an important factor.

Information Technology

At about this same time, information technology really began to explode. This gave organizations the ability to better monitor transaction-intensive activities such as the ordering, movement and storage of goods and materials. Combined with the availability of computerized quantitative models, this information increased the ability to manage flows and to optimize inventory levels and movements. Systems such as materials requirements planning (MRP, MRP II), distribution resource planning (DRP, DRP II) and just-in-time (JIT) allow organizations to link many materials management activities, from order processing to inventory management, ordering from the supplier, forecasting and production scheduling.

Other factors contributing to the growing interest in logistics include advances in information systems technology, an increased emphasis on customer service, growing recognition of the systems approach and total cost concept, the profit leverage from logistics, and the realization that logistics can be used as a strategic weapon in competing in the marketplace. These and other factors will be discussed throughout this book.

Channel Power

The shifting of **channel power** from manufacturers to retailers, wholesalers and distributors has also had a profound impact on logistics. When competition increases in major consumer goods industries, there is a shakeout of many suppliers and manufacturers, so that a few leading competitors remain. Those remaining are intensely competitive and offer very high-quality products. In many cases, the consumer sees all of the leading brands as substitutes for each other. Lower brand-name loyalty decreases a manufacturer's power. This increases the retailer's power because sales are determined by what is in stock, not by what particular brands are offered.

Profit Leverage

The profit leverage effect of logistics illustrates that €1.00 saved in logistics costs has a much greater impact on the organization's profitability than a €1.00 increase in sales. In most organizations, sales revenue increases are more difficult to achieve than logistics cost reductions. This is particularly true in mature markets, where price cuts are often met by the competition, and revenue in the whole industry thus declines.

5

There are many costs associated with a sale, such as the cost of goods sold and logistics-related costs. Thus, a €1.00 increase in sales does not result in a €1.00 increase in profit. If, for example, an organization's net profit margin (sales revenue less costs) is 2 per cent, the firm receives a before-tax profit of only €0.02 from each euro of sales. Yet, any euro saved in logistics does not require sales or other costs to generate the savings. Therefore, a euro saved in logistics costs is a euro increase in profit! As a result, logistics cost savings have much more leverage, euro for euro, than an increase in sales. Thus the term 'profit leverage effect of logistics' is relevant.

Systems Approach/Integration

The systems approach is a critical concept in logistics and supply chain management (SCM). Logistics is, in itself, a system; it is a **network** of related activities with the purpose of managing the orderly flow of material and personnel in addition to performing value-adding activities within the logistics channel. This is illustrated in Figure 1.2, which shows a simplified example of the network of relationships that logistics has to manage in a channel of distribution or supply chain.

The systems approach is a simplistic yet powerful paradigm for understanding interrelationships. The **systems approach** simply states that all functions or activities need to be understood in terms of how they affect, and are affected by, other elements and activities with which they interact. The idea is that if one looks at actions in isolation, one will not understand the big picture or how such actions affect, or are affected by, other activities. In essence, the sum, or outcome, of a series of activities is greater than its individual parts.

While it might be desirable to have high inventory levels in order to improve customer order fulfilment, high inventory levels increase storage costs as well as the risk of obsolescence. These unfavourable factors must be 'traded off' with the favourable aspects of a decision before arriving at a decision on inventory levels. Without considering the impact of decisions on the larger system, such as the firm or the distribution channel, suboptimization often occurs. This means that, while individual activities in that system appear to be operating well, the net result

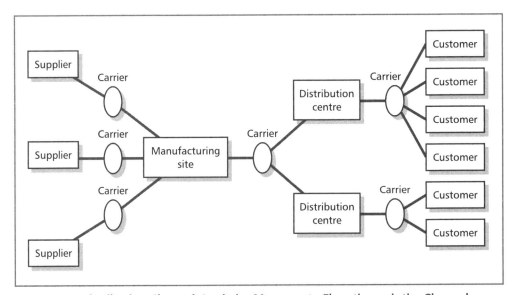

FIGURE 1.2 Distribution Channel: Logistics Manages to Flow through the Channel
Note: while the flow is primarily left to right, logistics is also responsible for returns, or movements from right to left, hence the term *reverse logistics* has developed.

on the total system is relatively poor performance. To understand the opportunities for improvement, and the implication of these opportunities, the system must be viewed as a whole.

Without understanding the channel-wide implications of logistics decisions to improve service levels, excess inventory will begin to build up at the links along the supply chain. This excess inventory will tend to increase costs throughout the channel, but it serves as a buffer to protect against the uncertainty of how other channel members will behave. Thus, the system as a whole is less efficient than it could otherwise be. To get around this issue, organizations like Hewlett-Packard's DeskJet Division have taken a systems approach to managing channel inventories.

The systems approach is at the core of the next few topics discussed. The systems approach is key to understanding the role of logistics in the economy, its role in the organization, including its interface with marketing, the total cost concept and logistics strategy.

The Role of Logistics in the Economy

Logistics plays a key role in the economy in two significant ways. First, logistics is one of the major expenditures for businesses, thereby affecting and being affected by other economic activities. In the EU, for example, European businesses are expected to spend €210 billion on logistics services in 2005, representing an annual growth of 6.5 per cent.[4]

In 1986, logistics expenditures accounted for more than double that amount, at 22 per cent of GDP.[5] If logistics expenditures were still that high in 2004, an additional €218 billion would have been spent on logistics costs in the EU. This would translate into higher prices for consumers, lower profits for businesses, or both. The result could be a lower overall standard of living and/or a smaller tax base. Thus, by improving the efficiency of logistics operations, logistics makes an important contribution to the economy as a whole.

Second, logistics supports the movement and flow of many economic transactions; it is an important activity in facilitating the sale of virtually all goods and services. To understand this role from a systems perspective, consider that if goods do not arrive on time, customers cannot buy them. If goods do not arrive in the proper place, or in the proper condition, no sale can be made. Thus all economic activity throughout the supply chain will suffer.

One of the fundamental ways that logistics adds value is by creating utility. From an economic standpoint, utility represents the value or usefulness that an item or service has in fulfilling a want or need. There are four types of utility: form, possession, time and place. The latter two, time and place utility, are intimately supported by logistics.

Form utility is the process of creating the good or service, or putting it in the proper form for the customer to use. When Volvo Car Corporation transforms parts and raw materials into a car, form utility is created. This is generally part of the production or operations process.

Possession utility is the value added to a product or service because the customer is able to take actual possession. This is made possible by credit arrangements, loans, and so on. For example, when Vauxhall Leasing extends credit to a prospective automotive customer, possession utility becomes possible.

While form and possession utility are not specifically related to logistics, neither would be possible without getting the right items needed for consumption or production to the right place at the right time and in the right condition at the right cost.[6] These 'five rights of logistics' are the essence of the two utilities provided by logistics: time and place utility.

Time utility is the value added by having an item when it is needed. This could occur within the organization, as in having all the materials and parts that are needed for manufacturing, so that the production line does not have to shut down. This occurs when the logistics function at Allied Bakeries delivers flour from one of its mills in Tilbury, Manchester or Belfast to one of its group bakeries throughout the UK, such as Glasgow, so that household bread brands such as Kingsmill and Sunblest may be produced on schedule. Or it could occur in the marketplace, as in having an item available for a customer when the customer wants it. The item does the customer no good if it is not available when it is needed.

This is closely related to **place utility**, which means having the item or service available where it is needed. If a product desired by consumers is in transit, in a warehouse or in another store, it does not create any place utility for them. Without both time and place utility, which logistics directly supports, a customer could not be satisfied.

The Role of Logistics in the Organization

In recent years, effective logistics management has been recognized as a key opportunity to improve both the profitability and competitive performance of firms. By the early 1990s, customer service took centre stage in many organizations. Even organizations that had previously adhered to the 'marketing concept' were re-examining what it meant to be customer-driven. The trend towards strong customer focus continues today.

Logistics Supports Marketing

The **marketing concept**, as mentioned above, is a philosophy which can be expressed as 'the achievement of corporate goals through meeting and exceeding customer needs better than the competition'.[7] Thus the marketing concept is a 'customer-driven' perspective which holds that a business exists to meet customer needs. The relationships between logistics and the three critical elements of the marketing concept (customer satisfaction, integrated effort/systems approach, and adequate corporate profit), are shown in Figure 1.3. Logistics plays a key role in each of these elements in several ways.

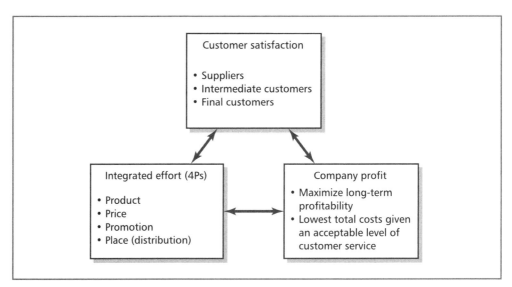

FIGURE 1.3 Marketing/Logistics Management Concept

The '4Ps' of the marketing mix require that, for a firm to be successful, any marketing effort must integrate the ideas of having the right product, at the right price, publicized with the proper promotion and available in the right place. Logistics plays a critical role particularly in support of getting the product to the right place. As discussed previously in conjunction with utility, a product or service provides customer satisfaction only if it is available to the customer when and where it is needed. Figure 1.4 summarizes the trade-offs required between and among the major elements of the marketing mix and logistics.

The organization needs, then, to use the 'systems approach' in linking the needs foreseen by marketing with production as well as logistics. Achieving customer satisfaction requires an integrated effort both internally and with suppliers and ultimate customers.

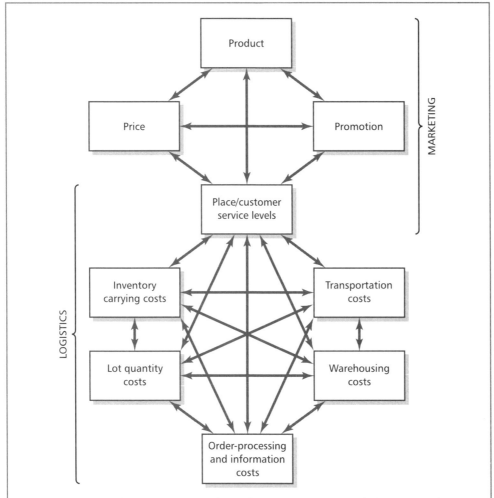

Marketing objective: allocate resources to the marketing mix to maximize the long-run profitability of the firm.
Logistics objectives: minimize total costs given the customer service objective where Total costs = Transportation costs + Warehousing costs + Order-processing and information costs + Lot quantity costs + Inventory carrying costs.

FIGURE 1.4 Cost Trade-offs Required in Marketing and Logistics

Source: adapted from Douglas M. Lambert, *The Development of an Inventory Costing Methodology: A Study of the Costs Associated with Holding Inventory*. Chicago: National Council of Physical Distribution Management, 1976, p. 7.

Also, it is important to understand that a central goal of an organization is to maximize long-term profitability or effective use of assets in the public or non-profit sectors. One of the key ways to accomplish this, as shown in Figure 1.4 and presented later, is through examining trade-offs among alternatives, thereby reducing the overall total cost of activities within a system.

To better understand Figure 1.4, the sections below briefly explore the manner in which each of the major elements of the marketing mix (the 4Ps) interact and are affected by logistics operations.

Product

Product refers to the set of utilities/characteristics that a customer receives as a result of the purchase of either a product or service. In an effort to lower price, management may decide to reduce product quality, eliminate product features, reduce the breadth of product offerings, reduce customer service or warranty support, or increase the time between model changes. However, any of these actions may reduce the attraction of the product for consumers, creating a loss of customers and thereby a reduction in long-term profits. To avoid making poor decisions, management needs to understand the trade-off and interrelationships between logistics and other marketing activities.

Price

Price is the amount of money that a customer pays for the product or service offering. Some of the items that should be factored into price include discounts for buying in quantities or for belonging to a certain class of customers, discounts for prompt payment, rebates, whether inventory is offered on consignment, and who pays delivery costs. A supplier may attempt to increase sales by reducing the price of its product, changing the terms or service offering. Unless the item in question is very price sensitive (i.e. sales change dramatically due to changes in price), such a strategy may create higher unit sales, but not enough to offset the lower price, yielding lower profit. This is particularly true in mature industries where customer demand is relatively fixed and the competition may follow the price decrease. The sales and the profitability of the entire industry suffer.

Promotion

Promotion of a product or service encompasses both personal selling and advertising. Whereas increasing advertising expenditures or the size of the direct sales force can have a positive impact on sales, there is a point of diminishing returns. A point exists where the extra money being spent does not yield sufficiently high increases in sales or profits to justify the added expense. It is important for organizations to know when they reach that point, so that they can avoid misallocating funds. A more prudent idea may be to try to use those funds more effectively, perhaps training the sales force to provide more value-added services to the customer, or make the customer more aware of the value added it currently provides through superior logistics service.

Place

Place is the key element of the marketing mix with which logistics interfaces directly. Place expenditures support the levels of customer service provided by the organization. This includes on-time delivery, high order fill rates, consistent transit times, and similar issues. Customer service is an output of the logistics system. On the other hand, when the organization performs well on all the elements of the marketing mix, customer satisfaction occurs.

For many organizations, customer service may be a key way to gain competitive advantage.[8] By adjusting customer service levels to meet what the customer desires and is willing to pay, the organization may simultaneously improve service levels and reduce cost. All of the logistics trade-offs illustrated at the bottom of Figure 1.5 must be considered in terms of their impact on customer service levels. To accomplish this analysis, the total cost concept must be used.

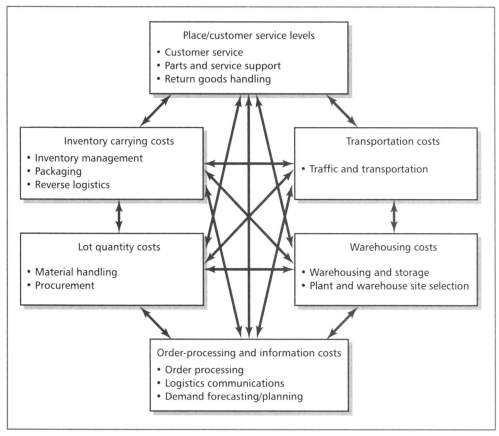

FIGURE 1.5 How Logistics Activities Drive Total Logistics Costs

Source: adapted from Douglas M. Lambert, *The Development of an Inventory Costing Methodology: A Study of the Costs Associated with Holding Inventory*. Chicago: National Council of Physical Distribution Management, 1976, p. 7.

Total Cost Concept

The **total cost concept** is the key to effectively managing logistics processes. The goal of the organization should be to reduce the *total* cost of logistics activities, rather than focusing on each activity in isolation.[9] Reducing costs in one area, such as transportation, may drive up inventory carrying costs as more inventory is required to cover longer transit times, or to balance against greater uncertainty in transit times.

Management should be concerned with the implications of decision-making on all of the costs shown in Figure 1.5. These six major cost categories cover the key logistics activities that will be discussed later in this text. Figure 1.5 illustrates how the logistics activities drive the six major logistics cost categories.

The place element is also known as distribution in recognition of channels of distribution. Channels have also been referred to as supply chains. We will explore the relationship of logistics to channels of distribution and supply chain management in the next two sections.

Channels of Distribution

In any industrialized or non-industrialized society, goods must be physically moved between the place they are produced and the place they are consumed. Except in very primitive cultures, in which each family meets its own household needs, the exchange process has become the cornerstone of economic activity. Exchange takes place when there is a discrepancy between the amount, type and timing of goods available and the goods needed. If a number of individuals or organizations within a society have a surplus of goods that someone else needs, there is a basis for exchange. **Channels** develop when many exchanges take place between producers and consumers. The alignment of firms that bring products or services to market has been called the **supply chain**, the demand chain or the value chain.

A **channel of distribution** can be defined as the 'means by which products are moved from producer to ultimate consumer'.[10] Channel functions are pervasive; they include buying, selling, transporting, storing, grading, financing, bearing market risk and providing marketing information. Any organizational unit, institution or agency that performs one or more of the marketing functions is a member of a channel of distribution.

The structure of a distribution channel is determined by the channel functions that specific organizations perform. Some channel members perform single functions – carriers transport products and public warehousers store them. Others, such as third-party logistics service providers (3PLSPs) and wholesalers, perform multiple functions. Channel structure affects (1) control over the performance of functions, (2) the speed of delivery and communication, and (3) the cost of operations.[11]

Most distribution channels are loosely structured networks of vertically aligned firms. The specific structure depends to a large extent on the nature of the product and the firm's target market. There is no 'best' channel structure for all firms producing similar products. Management must determine channel structure within the framework of the firm's corporate and marketing objectives, its operating philosophy, its strengths and weaknesses, and its infrastructure of manufacturing facilities and warehouses. If the firm has targeted multiple market segments, management may have to develop multiple channels to service these markets efficiently. For example, the BMW Group sells its BMW, Mini and Rolls-Royce automotive products through forecourt dealers under the BMW name, uses motorcycle dealers for its BMW motorcycle brand line, and sells BMW branded lifestyle products such as clothing, bicycles and children's toys on its Internet site.

Why do Channels of Distribution Develop?

The emergence of channels of distribution has been explained in terms of the following factors.[12]

- Intermediaries evolve in the process of exchange because they can increase the efficiency of the process by creating time, place and possession utility.
- Marketing agencies form channel arrangements to make transactions routine.
- Channels facilitate the searching process by consumers.
- Channel intermediaries enable the adjustment of the discrepancy of assortment by performing the functions of sorting and assorting. The discrepancy of assortment is described in more detail next.

The Discrepancy of Assortment and Sorting

Intermediaries provide possession, time and place utility. They create possession utility through the process of exchange, the result of the buying and selling functions. They provide time utility by holding inventory available for sale. And they provide place utility by physically moving goods to the market. The assortment of goods and services held by a producer and the assortment demanded by the customer often differ. The primary function of channel intermediaries is to adjust this discrepancy by performing the following 'sorting' processes.

- *Sorting out*: grouping a heterogeneous supply into relatively homogeneous separate stocks. Sorting out is typified by the grading of agricultural products or by pulling out rejects in some manufacturing operations.
- *Accumulating*: bringing similar stocks together into a larger homogeneous supply.
- *Allocation*: breaking down a homogeneous supply into smaller lots. Allocating at the whole-sale level is referred to as 'breaking bulk'. Goods received in carloads are sold in case lots. A buyer in case lots in turn sells individual units.
- *Assorting*: building up the assortment of products for use or sale in association with each other. Wholesalers build assortment of goods for retailers, and retailers build assortment for their customers.

Sorting out and accumulating predominate in the marketing of agricultural and extractive products. Allocation and assorting predominate in the marketing of finished manufactured goods. Because customers may demand a much broader assortment of goods and services than that provided by a single manufacturer, specialization develops in the exchange process to reduce distribution costs. That is, the customer's desire for **discrepancy of assortment** drives the producer to use intermediaries to reach the customer because doing so leads to improved distribution efficiency.

Structures and Operations of Channels of Distribution

The most detailed theory of channel structure was developed by Louis P. Bucklin,[13] who stated that the purpose of the channel is to provide consumers with the desired combination of its outputs (i.e. lot size, delivery time and market decentralization) at minimal cost. Consumers determine channel structure by purchasing combinations of service outputs. The best channel forms when no other group of institutions generates more profits or more consumer satisfaction per euro of product cost. Bucklin concluded that functions will shift from one channel member to another in order to achieve the most efficient and effective channel structure.

Given a desired level of output by the consumer and competitive conditions, channel institutions will arrange their functional tasks in a way that minimizes total channel costs. This shift of specific functions may lead to the addition or deletion of channel members.

In deciding when and where to use channel intermediaries, a firm is really considering the make/buy or 'outsourcing' decision. Does the organization need to develop the required skills and capabilities internally, or can it be done faster and more efficiently by a third party? Outsourcing will be discussed later in this chapter.

Postponement and Speculation

Bucklin's theory of channel structure is based on the concepts of **postponement** and **speculation**.[14] Under postponement, costs can be reduced by:

- postponing changes in the form and identity of a product to the last possible point in the marketing process
- postponing inventory location to the last possible point in time since risk and uncertainty costs increase as the product becomes more differentiated.

Postponement results in savings because it moves differentiation nearer to the time of purchase, when demand is more easily forecast. This reduces costs from risk and uncertainty. Logistics costs are reduced by sorting products in large lots in relatively undifferentiated states. Third-party service providers can support postponement.

Companies can use postponement to shift the risk of owning goods from one channel member to another. A manufacturer may refuse to produce until it receives firm orders. A middleman may postpone owning inventories by purchasing from sellers who offer faster delivery, purchasing on consignment or purchasing only when a sale has been made. Consumers may postpone ownership by buying from retail outlets that have the products in stock.

An excellent example of postponement is the mixing of paint colours at the retail store. Instead of having to forecast the exact colours that consumers will want to buy, the retailer mixes paint in any colour the consumer wishes at the time of purchase. Other examples include: the colour panels in the front of built-in kitchen appliances that enable the same unit to be in any one of a number of colours to match the kitchens; the centralization of slow-selling products in one warehouse location; and the assembly of slow-moving items only after orders have been received.

Speculation is the opposite of postponement – that is, a channel institution assumes risk rather than shifting it. Speculation can reduce marketing costs through:

- economies of large-scale production
- placement of large orders that reduce the costs of order processing and transportation
- reduction of stockouts and their associated costs
- reduction of uncertainty.

An example of speculation is the production of snow tyres. A tyre manufacturer will begin to produce snow tyres in late summer or early autumn in anticipation of seasonal demand once winter weather sets in. However, this seasonal demand is uncertain as it will partly be based on the amount and timing of snowfall and other weather conditions throughout the winter. In this situation manufacturers and retailers try to work together to establish forecasts that are as accurate as possible.

Time-to-market Pressures

To reduce the need for speculative inventories, many firms are exploring strategies of time-based competition or time compression to reduce their time to manufacture products significantly while reducing inventory, improving inventory turns and reducing the cost of ownership. Customer satisfaction should also improve. Thus 'speed' or 'velocity' can be used as a source of competitive advantage.[15] This is true in virtually all market sectors: services, manufacturing and retailing. Retailers have been leaders in the area of time-based competition, relying heavily on advanced computer systems involving bar coding and EDI to support quick response (see Chapter 3). The use of such systems is growing among carriers. However, computer systems are not enough to create speed to market; fundamental changes in operational relationships are required, such as information sharing between suppliers, manufacturers and retailers about lead-times, forecasts of sales, production and purchasing needs, shipping, new product plans and payment information.

Logistics and Supply Chain Management

The term **supply chain management** (SCM) was introduced by consultants in the early 1980s and has subsequently attracted a great deal of attention.[16] In the 1990s academics attempted to give structure to SCM.

Until recently most practitioners, consultants and academics viewed SCM as not appreciably different from the contemporary understanding of logistics management as defined by the Council of Supply Chain Management Professionals (CSCMP). According to the CSCMP's predecessor, the Council of Logistics Management (CLM), until 1998 logistics represented a supply chain orientation, from point of origin to point of consumption. So why the confusion? It is probably due to the fact that logistics has been represented as a functional silo within companies and is also a bigger concept that deals with the management of material and information flows across the entire supply chain. This is similar to the confusion over marketing as a concept and marketing as a functional area.

Thus SCM was reconceptualized from integrating logistics across the supply chain to integrating and managing key business processes across the supply chain (i.e. a more strategic approach). Based on this emerging distinction between SCM and logistics, CSCMP has now provided a separate definition of SCM.

What is a Supply Chain?

The CSCMP defines supply chain management as encompassing:

> the planning and management of all activities involved in sourcing and procurement, conversion, and all Logistics Management activities. Importantly, it also includes coordination and collaboration with channel partners, which can be suppliers, intermediaries, third-party service providers, and customers. In essence, Supply Chain Management integrates supply and demand management within and across companies.[17]

Throughout this text, the CSCMP definition of logistics management is used. CSCMP considers the scope of SCM to include:

> all of the Logistics Management activities noted . . . as well as manufacturing operations, and it drives coordination of processes and activities with and across marketing, sales, product design, finance and information technology.[18]

The Scope of Supply Chain Management

Michael Porter's value chain analysis[19] disaggregates a firm into nine value-creating and strategic activities in either a primary or support category. Inbound and outbound logistics comprise two of the primary activities. The goal of the primary activities is to generate profit margins by creating value that exceeds the cost of these activities. Support activities are seen as non-value creating, however the goal for these activities is to support the primary activities in a way that creates competitive advantage.

The value chain analysis is applied to one firm: the focal firm. However, it can be extended to the supply chain. The focal firm's immediate suppliers and customers are known as Tier 1, while immediate suppliers' suppliers and immediate customers' customers are known as Tier 2. Supply chain value comprises the collective value of many firm value chains. While logistics management is about optimizing the flows within the firm, SCM seeks to achieve trust and coordination between processes of all firms in the supply chain. Competitive advantage is consequently obtained when all firms understand the importance of subordinating their self-interests to the benefit of the supply chain as a single entity.[20]

Successful supply chain management requires a change from managing individual functions to integrating activities into key supply chain processes. Traditionally, both upstream and downstream portions of the supply chain have interacted as disconnected entities that receive sporadic flows of information over time.

Operating an integrated supply chain requires continuous information flows, which in turn help to create the best product flows. The customer remains the primary focus of the process. However, improved linkages with suppliers are necessary because controlling uncertainty in customer demand, manufacturing processes and supplier performance is critical to effective SCM. Achieving a good customer-focused system means that information must be processed with accuracy and timeliness, because quick-response systems require frequent changes in response to fluctuations in customer demand.

Optimizing the product flows cannot be accomplished without an exhaustive review of the underlying processes shown in Figure 1.6. While the specific processes identified by individual firms may vary somewhat, managers involved in SCM should consider these fundamental processes in addition to the returns channel process.

Of course, performance metrics must be changed to reflect process performance across the supply chain, and rewards and incentives must be aligned to these metrics in order to affect change.

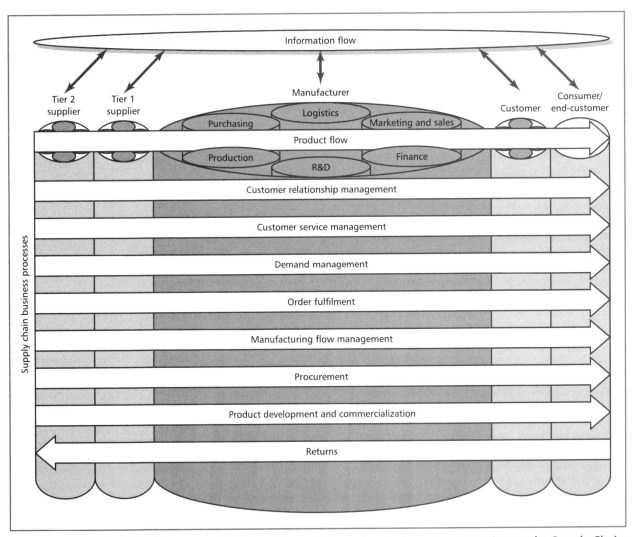

FIGURE 1.6 Supply Chain Management: Integrating and Managing Business Processes Across the Supply Chain

Source: Douglas M. Lambert, Martha C. Cooper and Janus D. Pagh, 'Supply Chain Management: Implementation Issues and Research Opportunities', *The International Journal of Logistics Management* 9, no. 2 (1998), p. 2.

Key Logistics Activities

Outlined below are the key logistics activities required to facilitate the flow of a product from point of origin to point of consumption. All these activities, listed alphabetically below, may be considered part of the overall logistics process:

- customer service and support
- demand forecasting/planning
- purchasing and procurement
- inventory management
- logistics communications and order processing
- material handling and packaging
- transportation
- facilities site selection, and warehousing and storage
- return goods handling and reverse logistics.

While all organizations may not explicitly consider these activities to be part of logistics activities, each affects the logistics process, as was shown in Figure 1.1.

Customer Service and Support

Customer service has been defined as 'a customer-oriented philosophy that integrates and manages all elements of the customer interface within a predetermined optimum cost-service mix'.[21] Customer service is the output of the logistics system, and the effective and efficient provision of the 'five rights of logistics' needed for consumption or production. Good customer service supports customer satisfaction, which is the output of the entire marketing process. Customer service and support is the topic of the next chapter.

Demand Forecasting/Planning

There are many types of demand forecast. Marketing forecasts customer demand based on promotions, pricing, competition, and so on. Manufacturing forecasts production requirements based on marketing's sales demand forecasts and current inventory levels. Logistics usually becomes involved in forecasting in terms of how much should be ordered from its suppliers (through purchasing), and how much of finished product should be transported or held in each market that the organization serves. Forecasting is a complex issue, with many interactions among functions and forecast variables. This topic will be explored in greater depth in Chapters 5 and 11.

Purchasing and Procurement

With the increase in outsourcing of goods and services, the purchasing and procurement of materials and services from outside organizations to support the firm's operations plays a more important role in the organization. Most industries spend from 40 to 60 per cent of their revenues on materials and services from sources outside of the organization. As organizations form longer-term relationships with fewer key suppliers, purchasing and procurement continue to grow in importance and contribution to the organization. This is examined in greater depth in Chapter 4.

Inventory Management

Inventory management involves trading off the level of inventory held to achieve high customer service levels with the cost of holding inventory, including capital tied up in inventory, variable storage costs and obsolescence. These costs can range from 14 to over 50 per cent of the value of inventory on an annual basis![22] With high costs for items such as high-tech merchandise,

ems that rapidly become obsolete, many organizations, including
Carrefour, are paying inventory management much more attention.
d in Chapter 5.

tions and Order Processing

oming increasingly automated, complex and rapid. Logistics interfaces
functions and organizations in its communication processes.
ccur between the organization, its suppliers and customers and various
y chain who may not be directly linked to the organization, and the major
rganization such as logistics, engineering, accounting, marketing and pro-
tion is key to the efficient functioning of any integrated system, whether it
ystem of an organization or the wider supply chain.

g entails the systems that an organization has for getting orders from customers, chec. on the status of orders and communicating to customers about them, and
actually filling an order and making it available to the customer. Because the order-processing
cycle is a key area of customer interface with the organization, it can have a big impact on a customer's perception of service and, therefore, satisfaction. Increasingly, organizations today are
turning to advanced order-processing methods such as electronic data interchange (EDI) and electronic funds transfer (EFT) to speed the process and improve accuracy and efficiency, and advanced
scanning technology such as radio frequency identifications (RFID) to track and trace products.
These topics are presented in more depth in Chapter 3, which describes information systems.

Materials Handling and Packaging

Materials handling is a broad area that encompasses virtually all aspects of all movements of
raw materials, work in process or finished goods within a plant or warehouse. Because an
organization incurs costs without adding value each time an item moves or is handled, a primary
objective of materials management is to eliminate handling wherever possible. This includes
minimizing travel distance, bottlenecks, inventory levels and loss due to waste, mishandling, pilferage and damage. Thus, by carefully analysing material flows, materials management can
save the organization significant amounts of money, as illustrated in Chapters 6 and 9.

As Chapter 9 will explain, packaging is valuable both as a form of advertising/marketing, and
for protection and storage from a logistical perspective. Packaging can convey important information to inform the consumer. Aesthetically pleasing packaging can also attract the
consumer's attention. Logistically, packaging provides protection during storage and transport.
This is especially important for long distances over multiple transportation modes such as international shipping. Packaging can ease movement and storage by being properly designed for the
warehouse configuration and materials handling equipment.

Transportation

A key logistics activity is to actually provide for the movement of materials and goods from point
of origin to point of consumption, and perhaps to its ultimate point of disposal as well.
Transportation involves selection of the mode (e.g. air, rail, water, truck or pipeline), the routing
of the shipment, assuring of compliance with regulations in the region of the country where
shipment is occurring, and selection of the carrier. It is frequently the largest single cost among
logistics activities. Transportation issues are covered in Chapter 7.

Facilities Site Selection, and Warehousing and Storage

Determining the location of the company's facilities, or plant(s) and warehouse(s), is a strategic
decision that affects not only the costs of transporting raw materials inbound and finished goods

outbound, but also customer service levels and speed of response. Issues to consider include the location of customers, suppliers, transportation services, availability and wage rates of qualified employees, governmental cooperation, and so on. The enlargement of the EU in 2004 has led to consideration of Bratislava, Budapest and Warsaw as possible new locations for logistics operations. These cities are centres of population density and economic activity but also feature lower labour and infrastructure costs.[23]

Warehousing supports time and place utility by allowing an item to be produced and held for later consumption. It can be held near the location where it will be needed, or transported later. Warehousing and storage activities relate to warehouse layout, design, ownership, automation, training of employees and related issues. All these issues are presented in Chapter 8.

Return Goods Handling and Reverse Logistics

Returns may take place because of a problem with the performance of the item or simply because the customer changed his or her mind. Return goods handling is complex because it involves moving small quantities of goods back *from* the customer rather than *to* the customer as the firm is accustomed. Many logistics systems have a difficult time handling this type of movement. Costs tend to be very high. The cost of moving a product backwards through the channel from the consumer to the producer may be as much as nine times as high as moving the same product forwards from the producer to the customer.[24] For this reason, this significant cost and service area is beginning to receive more attention.

Logistics is also involved in the removal and disposal of waste materials left over from the production, distribution or packaging processes. There could be temporary storage followed by transportation to the disposal, reuse, reprocessing or recycling location. As the concern for recycling and reusable packaging grows, this issue will increase in importance. This is of particular concern in Europe, which has very strict regulations regarding removal of packaging materials and even obsolete product due in part to limited landfill space. These topics are discussed in Chapter 9.

The Relationship of Logistics Activities to Logistics Costs

Logistics costs are driven or created by the activities that support the logistics process. Each of the major cost categories – customer service, transportation, warehousing, order processing and information, lot quantity and inventory carrying – is discussed below.

Customer Service Levels

The key cost trade-off associated with varying levels of customer service is the cost of lost sales. Monies that are spent to support customer service include the costs associated with order fulfilment, parts and service support. They also include the costs of return goods handling, which has a major impact on a customer's perception of the organization's service as well as the ultimate level of customer satisfaction.

The cost of lost sales includes not only the lost contribution of the current sale, but also potential future sales from the customer and from other customers due to word-of-mouth negative publicity from former customers. A disgruntled customer will, on average, tell nine others about his or her dissatisfaction with the product or service.[25] It is no wonder that it is extremely difficult to measure the true cost of customer service!

The best approach, then, is to determine desired levels of customer service based on customer needs, and how those needs are affected by expenditures on other areas of the marketing mix. The idea is to minimize the total cost given the customer service objectives. Because each of the other five major logistics cost elements work together to support customer service, good data are needed regarding expenditures in each category.

Transportation Costs

The activity of transporting goods drives transportation costs. Expenditures that support transportation can be viewed in many different ways, depending on the unit of analysis. Costs can be categorized by customer, product line, type of channel (such as inbound versus outbound), and so on. Costs vary considerably with volume of shipment (cube), weight of shipment, distance, and point of origin and destination. Costs and service also vary considerably with the mode of transportation chosen. These costs will be described in more depth in Chapter 7.

Warehousing Costs

Warehousing costs are created by warehousing and storage activities, and by the plant and warehouse site selection process. Included are all the costs that vary due to a change in the number or location of warehouses. Warehousing costs are explored in Chapter 8.

Order-Processing and Information Systems Costs

This category includes costs related to activities such as order processing, distribution communications and forecasting demand. Order-processing and information costs are an extremely important investment to support good customer service levels and control costs. Order-processing costs include such costs as order transmittal, order entry, processing the order, and related internal and external costs such as notifying carriers and customers of shipping information and product availability. Shippers and carriers have invested a great deal in improving their information systems to include technology such as RFID, EDI, satellite data transmission, and bar coding and scanning shipments and sales. There has also been a growth in more sophisticated information technology, such as decision support systems, artificial intelligence (AI) and expert systems. These topics are dealt with in Chapter 3.

Lot Quantity Costs

The major logistics lot quantity costs are due to procurement and production quantities. **Lot quantity costs** are purchasing- or production-related costs that vary with changes in order size or frequency and include the following.

- *Setup costs*:
  - time required to set up a line or locate a supplier and place an order
  - scrap due to setting up the production line
  - operating inefficiency as the line begins to run, or as a new supplier is brought on board.
- *Capacity lost* due to downtime during changeover of line or changeover to a new supplier.
- *Materials handling*, scheduling and expediting.
- *Price differentials* due to buying in different quantities.
- *Order costs* associated with order placement and handling.

These costs must not be viewed in isolation because they also may affect many other costs. For example, a consumer goods manufacturer that produces large production runs may get good prices from suppliers and have long, efficient production runs, but requires more storage space to handle large runs. Customer service levels may suffer as order fulfilment declines because products are produced infrequently, in large batches, and with inventory going to zero and creating stockout situations in between runs. This may increase information and order-processing costs, as customers frequently call to check on availability of back-ordered products and cancel back orders.

Transportation costs may also rise as customers are sent partial or split shipments. Inventory carrying costs will rise as large quantities of inventory are held until depleted, due to large batch sizes. The implication of one cost upon another must explicitly be considered.

Inventory Carrying Costs

The logistics activities that make up inventory carrying costs include inventory control, packaging, and salvage and scrap disposal. Inventory carrying costs are made up of many elements. For decision-making purposes, the only relevant inventory costs to consider are those that vary with the amount of inventory stored. The four major categories of inventory costs that do so are:

1 *capital cost*, or *opportunity cost*, which is the return that the company could make on the money it has tied up in inventory
2 *inventory service cost*, which includes insurance and taxes on inventory
3 *storage space cost*, which includes those warehousing space-related costs that change with the level of inventory
4 *inventory risk cost*, including obsolescence, pilferage, relocation within the inventory system and damage.

These costs will be explored in detail in Chapter 4.

Logistics Costs and Strategy

Understanding the organization's overall strategy and the key cost trade-offs in that organization are important to developing logistics strategy and competitive advantage. The primary goal of logistics in any organization is to support the organization's customer service goals in a value-adding, effective and efficient manner. Logistics costs are an important aspect of analysing alternative logistics service offerings to meet those goals. The financial performance of logistics will be discussed in Chapter 11, while Chapter 13 will explore logistics strategy.

Issues in Logistics

This section presents some of the key challenges and issues that logistics faces today and will continue to face in the future. These themes will be integrated throughout this text to provide continuity and an understanding of how these issues affect the performance and perceived importance of various logistics activities.

As the role of logistics grows and takes on greater importance in achieving the overall goals of the organization, it needs to meet the challenge and improve its performance to support those goals (see the Creative Solutions box). Some areas of opportunity include:

- greater participation in setting organizational strategy and the strategic planning process
- total quality management (TQM) and just-in-time (JIT) logistics
- the use of quick response (QR) and efficient consumer response (ECR) techniques
- identification of opportunities for using logistics as a competitive weapon/marketing strength
- better understanding of global logistics issues and improved logistics information systems
- improved understanding of and accounting for logistics costs
- greater understanding and appropriate application of technology
- greater participation of logistics professionals in work teams
- green or environmental logistics
- appropriate understanding and use of outsourcing, partnerships and strategic alliances.

Each of these issues is explored on the following pages.

CREATIVE SOLUTIONS

Transforming Logistics Infrastructure at Procter & Gamble

Procter & Gamble (P&G) manufactures and markets nearly 300 brands of consumer products, including Ariel, Pampers, Sunny Delight and Pringles. With a workforce of nearly 100,000, P&G operates in nearly 80 countries and generated net sales of almost £24.5 billion in 2003. However, its supply chain was inefficient. One under-performing measure was missed cases – availability was 96 per cent and the number of missed cases was 4 per cent – and it was delivering only about two-thirds of orders on time. The 4 per cent of missed cases amounted to seven or eight million cases a year and cost P&G an estimated £40 million.

The company has since been transforming its UK supply chain into a more efficient and collaborative operation. This process has involved implementing a range of innovative schemes to help optimize and consolidate the supply chain. P&G invested almost £23 million to upgrade its logistics infrastructure, including: focusing on supplier relationship management and improving the speed and accuracy of information flow up the supply chain; creating two distribution centres in the north and south to deliver the full range of P&G's ambient products on one truck; introducing cross-docking so that products bypass storage, saving time and resources by going straight onto store shelves; and using IT solutions such as GPS tracking and electronic proof of delivery. P&G also built a brand new automated distribution centre (DC) at its London plant and upgraded its existing DC at Skelmersdale.

In the 15 months since implementation P&G's 'first time fill rate' for customers now stands at 99 per cent and the percentage of missed cases has reduced from 4 per cent to less than 0.5 per cent. Losses – reduced inventory and fewer truck miles – have been cut out of the supply chain. P&G is already working on further innovative schemes, including joint forecasting with customers on data synchronisation and currently implementing trials with customers. One trial concerns cross-docking, where some of its products arrive at a customer's warehouse but rather than being put away in racking, are picked and sorted immediately. Two other trials with Tesco include using wheeled dollies for promotions and wheeled half pallets for laundry cleaning products. It is also developing a 'satellite pen' for signatures and is considering providing 24/7 deliveries.

Question: Discuss the cost trade-offs involved in P&G's decision to invest in its information systems.

Source: adapted from 'The Innovative Trailblazer', *Logistics Manager* 11, no. 5 (June 2004) pp. 11–13.

Strategic Planning and Participation

Activities such as logistics budgeting and control, inventory planning and positioning, and customer service have become important parts of the organization's strategic planning process. Porter's value chain analysis, introduced above, allows a firm to analyse and integrate independent aspects of corporate strategy. The strategic planning process will be discussed in Chapter 13.

Total Quality Management and Just-in-time

Total quality management (TQM) is a philosophy that should be embedded in all aspects of logistics operations. Going beyond simple 'quality control', which monitors for problems in actual performance after the fact, TQM is a philosophy that is integrated in designing logistics systems to achieve desired results, performing logistics activities and monitoring results. Total quality management involves being proactive in performing the right activity the right way the first time and continuing to perform it to the required level. In logistics, that could translate into short, predictable transit times, certain levels of in-stock availability and certain fill rates on customer orders.

The **ISO** (International Organization for Standardization) series is an internationally recognized certification programme whereby the quality processes of firms are audited to verify whether they have well-documented and effective quality processes in place. It was born in Europe in 1987 in an effort to support trade between countries and companies.[26]

Just-in-time (JIT) is an inventory management philosophy aimed at reducing waste and redundant inventory by delivering products, components or materials just when an organization needs them. JIT requires close coordination of demand needs among logistics, carriers, suppliers and manufacturing. JIT also represents a tremendous opportunity for the logistics function to contribute to the organization's success by reducing inventory while simultaneously maintaining or improving customer service levels. Thus, JIT represents an important trend in inventory management that will be discussed throughout this text. Applications of JIT principles to the retail and grocery sectors are discussed below in relation to quick response and efficient consumer response.

As will be discussed in Chapters 4 and 9, TQM and JIT have profound implications for logistics systems.

Quick Response

Quick response (QR) is a retail-sector strategy that combines a number of tactics to improve inventory management and efficiency, while speeding inventory flows. Most QR is between manufacturer and retailer only. When fully implemented, QR applies JIT principles throughout the entire supply chain, from raw material suppliers through ultimate customer demand.

The concept works by combining EDI with bar-coding technology, so that customer sales are tracked immediately. This information can be passed on to the manufacturer, which can then notify its raw material suppliers, and schedule production and deliveries as required to meet replenishment needs. This allows inventory reductions while speeding response time, lowering the number of out-of-stock products, and reducing handling and obsolescence. While QR began in the textile and apparel industry, it is now being applied by many industries in the retail sector. The grocery industry has begun an adaptation of this approach, called efficient consumer response (ECR), as discussed in the next section.

QR has had a major impact on distribution operations. Rather than 'warehousing' product, distribution centres are now charged with 'moving' the product through quickly. This frequently entails **cross-docking**, whereby the inbound product is unloaded, sorted by store and reloaded onto trucks destined for a particular store, without ever being warehoused. To further improve retail efficiency, some suppliers are shipping goods pre-hung and pre-ticketed. This concept, known as 'floor-ready merchandise', is also growing in popularity.

Efficient Consumer Response

Efficient consumer response (ECR) combines several logistics strategies in an effort to improve the competitiveness of the grocery industry by cutting waste in the supply chain. It is the grocery industry's answer to QR. ECR includes the following strategies:[27]

- widespread implementation of electronic data interchange up and down the supply chain, both between suppliers and manufacturers, manufacturers and distributors, and distributors and customers
- greater use of point-of-sale data obtained by greater and more accurate use of bar coding
- cooperative relationships between manufacturers, distributors, suppliers and customers
- continuous replenishment of inventory and flow through distribution
- improved product management and promotions.

By applying the fourth point – continuous replenishment and flow through distribution – inventory is managed on a just-in-time basis, rather than stockpiled in warehouses and distribution centres. Product is cross-docked, whereby it is unloaded at one dock, broken down into store-sized shipments, and reloaded on trucks to go directly to the stores. Thus cooperation and coordination are very important to ensure proper sequencing of truck loading and unloading, as well as the proper product mix. The belief is that the potential exists to reduce total supply chain inventory by up to 41 per cent and incur savings of 5.7 per cent based on retail prices of about €20 billion.[28]

A key feature of ECR that distinguishes it from QR is the emphasis on moving away from the grocery industry's 'deal mentality'. Cooperation is required among industry participants to move away from the heavy use of promotional strategies. Such strategies encourage grocers to 'stock-pile' or forward buy product due to promotions such as a temporary low price or 'buy two, get one free' deals. This creates excessive inventory in the supply chain, and reduces the number of times inventory turns over each year.

The ECR concept was developed in 1993 in the USA to offset some of the pressure on the grocery industry by mass merchandisers like Wal-Mart and warehouse clubs. The ECR-Europe initiative followed in 1995 and national ECR initiatives have been established in 22 European countries.

Logistics as a Competitive Weapon

Logistics may be the best source of competitive advantage for a firm because it is less easily duplicated than other elements of the marketing mix: product, price and promotion. For example, forming close, ongoing relationships with carriers or logistics service providers can help give the firm a distinct competitive advantage in terms of speed to the customer, reliability, availability or other customer service factors.

The power of logistics in achieving an organization's customer service goals and supporting customer satisfaction has received an increased amount of attention in the press. Companies that understand and utilize the potential of logistics as a competitive weapon include logistics as a key component of their strategic planning process.

Global Logistics

Many leading organizations are heavily involved in international markets through purchasing inputs to production, other importing, exporting, joint ventures, alliances, foreign subsidiaries and divisions, and other means (see the Global box). This creates a need for familiarity with global logistics and global logistics networks, as will be discussed in Chapter 12. This need is likely to continue in the future.

GLOBAL

Integrated Global Logistics for Motorcycle Racing Teams

The World Superbike (WSB) championship has 11 motorcycle races each year; 10 at various European racing circuits and one at Phillip Island in Australia. WSB organizers have developed a unique logistics strategy to enable the 13 race teams competing in the series to ship their entire operations to and from the event.

Teams prepare up to 25 shipping crates including motorcycles, spare parts, tools and riders' equipment two weeks before the race date. Each team has a weight allowance paid for by WSB organizers dependent on the status of their riders, from 900 kg per rider for a top team reducing to 500 kg per rider. Teams can ship as much freight as they wish; for example, the Fila Ducati team ships 9600 kg while the Foggy Petronas team ships 5810 kg, but they must pay for excess weight above the allowance. Ten days before the race date, 30 articulated lorries are sent throughout Europe to pick up each team's cargo as well as 4300 tyres weighing 25 tonnes from Pirelli and a total of 5000 litres of fuel from Shell, Agip, Petronas and Elf.

Two days later a Boeing 747 cargo freighter is loaded at Turin, Italy, and departs for Melbourne, Australia, 24 hours later. Two fuel stops are required during the 11,000-mile journey, in Dubai and Kuala Lumpur. Six days before the race date, 17 lorries ferry the cargo to the Phillip Island racing circuit, and the race teams unpack and begin their usual race routines. The packing up for the return journey begins on the evening of the race date, and while there is no fuel to return used tyres are recovered by Pirelli for analysis.

The 747 cargo freighter is hired from Malaysian Airlines at a cost of US$550,000. The cargo is insured for €50 million at a premium cost of €15,000. While the operation is complicated, Pierluigi Matta, who is in charge for WSB, told *Motorcycle News* it is not as difficult as it sounds as WSB has been handling the logistics for 11 years and knows each team's needs. This integrated and global approach to logistics allows WSB organizers to effectively and efficiently transport participating teams to Phillip Island and thus gives Australian motorcycle racing fans an opportunity to be part of the WSB championship.

Question: **What other types of sporting or entertainment events would face similar logistics issues as WSB?**

Source: adapted from 'Shifting the WSB Circus Down Under', *Motorcycle News* 34, no. 4 (31 March 2004), pp. 58–9.

Accounting for Logistics Costs

Implementation and utilization of the integrated logistics concept requires total cost analysis to be effective. The focus of management should be to minimize total logistics costs for a given customer service level. For this reason, it is important to understand the costs associated with the logistics trade-offs in Figure 1.5.

In general, accounting systems have not changed and adapted to accurately account for the many trade-offs inherent in logistics activity and logistics decision-making. The availability of timely, accurate and meaningful logistics information is relatively rare in practice. However, this is beginning to change as more organizations move into activity-based costing (ABC) systems to allocate costs to activities on a more accurate and meaningful basis. Much work remains to be done in this area. Some of the issues associated with the use of ABC in logistics are presented in greater depth in Chapter 11.

In addition, accounting and management support systems that are flexible in nature are needed. Logistics professionals must be able to get the information required to make decisions as they arise. Not all logistics decisions can be anticipated in advance and prepared for in a regularly scheduled logistics report. Thus, accounting systems that provide easy access to real-time data are needed to support unanticipated decisions.

Logistics as a Boundary-spanning Activity

As we have described extensively in this chapter, the logistics function and the activities performed by logistics do not exist in isolation. Logistics plays a key role in activities throughout the supply chain, both within and outside the organization. Outside the organization, logistics interfaces with customers in the order-processing, order fulfilment and delivery cycles. Logistics also interfaces with carriers, warehousers, suppliers and other third parties that play a role in the supply chain.

Within the organization, logistics interfaces with virtually every functional area in some capacity. Logistics interfaces with finance in the planning process and in the analysis of capital expenditures on investments in building and equipment to support distribution, transportation, warehousing, information technology and related issues.

Logistics interfaces with accounting in establishing logistics costs (transportation, distribution, storage) for various products, customers and distribution channels. Logistics also requires information from accounting regarding budgets and actual expenditures.

As discussed earlier, the interaction of logistics with other marketing activities is extensive. Logistics plays an instrumental role in customer satisfaction by providing high levels of customer service through good product availability, reliable service and efficient operations that keep prices competitive.

Logistics must work closely with production and operations in a number of capacities. First, logistics often receives order releases for materials from production, and it needs to ensure that the items required are ordered, transported and received on a timely basis. Storage may also need to be arranged. Logistics often manages the flow of materials or work in process within the organization. Logistics must also work with production in terms of stocking and shipping the finished product as it is available.

Logistics should be involved with research and development, product engineering, packaging engineering and related functions in the new product development process. This often occurs through logistics' participation in a new product team. It is vital for the logistics area to be represented very early in the new product development process.[29] This is critical in terms of designing the proper distribution channel, anticipating needs for inventory build-up, ensuring the availability of materials for production, and properly configuring the packaging for maximum efficiency and production within the distribution channel.

An increasing number of organizations are using the team approach to facilitate communications, create buy-in from multiple functions and to anticipate problems. Logistics should be an active participant in teams that deal with issues affecting the supply chain. These issues will be discussed in more detail in Chapter 10.

Logistics Technology

Part of an organization's ability to use logistics as a competitive weapon is based on its ability to assess and adjust actual logistics performance in real time. This means the ability to monitor customer demands and inventory levels as they occur, to act in a timely manner to prevent stockouts, and to communicate potential problems to customers. This requires excellent, integrated logistics information systems. These systems impact all of the logistics activities presented earlier, and must be integrated with and take into account marketing and production activities. Such systems must also be integrated with other members of the supply chain, to provide accurate information throughout the channel from the earliest supplier through to the ultimate customer (see the Technology box).

There has been a proliferation of technological developments in areas that support logistics. As discussed above, there have been major technological developments in the information systems area: RFID, EDI, bar coding, point-of-sale data and satellite data transmissions are just a few examples. In addition, improvements in automated warehousing capabilities should be integrated into logistics plans for upgrading technology. Technology is having a profound effect on the way that logistics personnel interface with other functional areas, creating the ability to access more timely, accurate information. Combining information technology with automated warehousing reduces inherent human variability, creating an opportunity to improve customer service.

Integrating Systems at Electrolux

Logistics information systems may link a variety of information technologies, as is the case with AB Electrolux, the Swedish producer of appliances and equipment. Electrolux uses an integrated pan-European IT set-up to communicate with its supply chain partners in over a thousand operational flows of 200–500,000 units per year. Electrolux uses 'information to replace inventory' to underscore various supply chain replenishment systems. Under 'reactive replenishment' customers place orders with local distributors and factories replenish national inventories. Electrolux also provides 'proactive replenishment', which involves sharing information including forecasts with its customer base. The company also provides make-to-order with direct deliveries and vendor managed inventory (VMI) for customers to ensure adequate availability of Electrolux 'white goods' appliances.

Question: Why is information important in today's global economy generally, and for logistics activities specifically?

Source: adapted from Walter Koch, 'Face to Face', *Logistics Europe* 10, no. 8 (November 2002), p. 60.

Increasing Skill Requirements

As suggested above, the demands on logistics professionals are increasing. As logisticians become increasingly involved in setting corporate strategy and other aspects of the strategic planning process, different skill sets are required in quality issues, global logistics and improving relationships with third-party providers.

Green or Environmental Logistics

Environmental issues have been an area of growing concern and attention for businesses on a global scale. Transportation and disposal of hazardous materials are frequently regulated and controlled. In Europe, organizations are increasingly required to remove and dispose of packaging materials used for their products. These issues complicate the job of logistics, increasing costs and limiting options. Organizations are continually looking at reducing, reusing and reapplying packaging materials, by-products of production and obsolete items. Companies are substituting items that are more readily recyclable. These activities are covered by the term **Green logistics**.

Outsourcing, Partnering, Third-party Logistics (3PL) and Fourth-party Logistics (4PL)

During the 1980s, many organizations began to recognize that they could not effectively and efficiently 'do it all' themselves and still remain competitive. They began to look to third-party specialists to perform activities that were not a part of their 'core competency'. This activity is known as **outsourcing**, in which an organization hires an outside organization to provide a good or service that it had traditionally provided itself, because this third party is an 'expert' in efficiently providing this good or service, while the organization itself may not be.

Outsourcing has been an area of growing interest and activity since the early 1990s. Approximately €64 billion, or 30.5 per cent, of the €210 billion logistics market in Europe is expected to be outsourced in 2005, and there are significant opportunities to outsource additional logistics services.[30] Logistics outsourcing often involves third-party warehouses and use of public or contract transportation carriers.

Outsourcing offers the opportunity for organizations to use the best 3PL service providers available to meet their needs. Outsourcing may involve a partnership relationship or be ad hoc, on a transaction-to-transaction basis. Traditionally, such relationships have been arm's length, with each party concerned only for its own welfare. However, the concept of 4PL sees a logistics service provider undertaking the SCM and assembly of finished goods on behalf of customers.[31]

Managers in many firms are accepting the concept of partnering or establishing close, long-term working relationships with suppliers of goods or services, customers and third-party providers. The most closely integrated partnerships are often referred to as **strategic alliances**. For a partnership to be a strategic alliance, it must be strategic in nature and must directly support one of the organization's distinctive competencies. Strategic alliances are rare in practice.

We have briefly summarized some of the current and future issues facing logistics professionals. These issues are recurring themes that will appear throughout this text.

SUMMARY

In this chapter, we introduced the concept of logistics, and described its development and relevance to channels of distribution, SCM, and the organization and economy as a whole. The concepts of value chain analysis and the systems approach were introduced and related to the role of logistics and its interface with marketing and other functions. The key role of logistics in customer service was emphasized. The systems approach was also related to the total cost concept and the principle of trade-offs as it relates to both the performance of logistics activity and the costs associated with such activity. The key logistics costs identified were customer service, inventory carrying costs, transportation, warehousing, order-processing/information systems costs, and lot quantity costs.

We also examined the issue of logistics strategy, and the role of logistics in corporate strategy. This chapter closed with a summary of future challenges for logistics professionals. These challenges range from playing an active role in the strategic planning process to improving accounting information, information technology, and other types of technology and practices, such as TQM, JIT, QR and ECR. The changing nature of logistics relationships, from team participation to forming partnering relationships with suppliers, was also explored.

Many opportunities and challenges will face the logistics function in the future. We will begin to describe these in more depth in Chapter 2, which focuses on customer service.

KEY TERMS

chapter 1 Logistics and Supply Chain Management

A full Glossary can be found at the back of the book.

QUESTIONS AND PROBLEMS

1 Define supply chain management, including an identification of upstream and downstream supply chain elements.
2 How is logistics related to the marketing effort? Be sure to discuss customer service/customer satisfaction, integration of efforts, and cost and performance outputs.
3 What are the different types of utility? How does logistics directly or indirectly affect each one?
4 Why has logistics recently been receiving more attention as a strategic function of the organization?
5 What is meant by the profit leverage affect of logistics?
6 What are the greatest cost savings opportunities for logistics?
7 Why do channels of distribution develop?
8 Give an example of:
   (a) a firm that uses postponement
   (b) a firm that uses speculation in the channel of distribution.
9 Discuss the key challenges facing logistics today. What do you see as the greatest area of opportunity for logistics? Why?
10 How has the role and performance of logistics been enhanced by the growth of technology, particularly information technology? What do you see as key trends in the future?

THE LOGISTICS CHALLENGE!

PROBLEM: PUBLIC SECTOR LOGISTICS

Most people think logistics and supply chain management are only relevant for consumer products demanded in a retail setting. However, logistics plays a major role in everyday life including public sector services such as the military, national postal services and health care. Public sector spending across Europe is under pressure; thus there is a big need to get logistics and supply chain activities right.

One example of a public sector organization doing just that is the National Health Service (NHS) in England. It has set up its own NHS Logistics group to serve the needs of its customer base of 450 organizations such as hospitals, clinics and local surgeries. NHS Logistics has 1350 employees and maintains seven distribution centres and a fleet of over 200 trucks that make 1200 deliveries a day to over 10,000 delivery points. It handles over €900 million worth of products and 27 million line items per year and receives 100 per cent of its customer orders electronically thanks to an Internet supplier portal initiated in late 2003.

On the supply side NHS Logistics deals with over 750 suppliers and products are stocked on a 48 hour lead time. It issues 83 per cent of its orders electronically and pays over 420,000 supplier invoices per year. However, the biggest challenge, according to NHS chief executive Barry Mellor, has been getting people to think about logistics and supply chain issues. To that end NHS Logistics has been instrumental in getting customers involved with technology such as hand-held bar code scanners used by nurses in hospital wards to improve stock management and establishing an online NHS Supply Chain Knowledge Centre that allows authorized persons to log on and gain access to examples of best practice, case studies and descriptions of projects. The results of all these efforts speak for themselves; NHS Logistics maintains a consistent 98 per cent service level for product availability and delivery to a half-hour 'window' of agreed time by its customers.

But what about other public sector organizations? The challenge here is: in a dynamic public sector environment that is driven by a demand for better financial performance, i.e. reduced costs and value for money, how can they also effectively manage their logistics and supply chains? What can they do to smooth out product and service flows to meet customers' needs in this environment at lower cost without seriously compromising their service standards?

What Is Your Solution?

Source: adapted from Chris Lewis, 'In the Public Interest,' *Logistics Europe* 12, no. 2 (March 2004), pp. 28–32.

SUGGESTED READING

BOOKS

Christopher, Martin, *Logistics and Supply Chain Management: Creating Value-Adding Networks*, 3rd edn. Harlow, UK: FT Prentice Hall, 2005.

Lambert, Douglas M., *The Development of an Inventory Costing Methodology: A Study of the Costs Associated with Holding Inventory*. Chicago: National Council of Physical Distribution Management, 1976.

Michael E. Porter, *Competitive Advantage: Creating and Sustaining Superior Performance*. New York: Free Press, 1985.

Stern, Louis W., Adel I. El-Ansary and Anne T. Coughlan, *Marketing Channels*, 5th edn. Englewood Cliffs, NJ: Prentice Hall, 1996.

Stock, James R., *Development and Implementation of Reverse Logistics Programs*. Oak Brook, IL: Council of Logistics Management, 1998.

Valarie A. Zeithaml and Mary Jo Bitner, *Services Marketing*, 3rd edn. New York: McGraw-Hill Irwin, 2003.

JOURNALS

Alvarado, Ursula Y. and Herbert Kotzab, 'Supply Chain Management: The Integration of Logistics in Marketing', *Industrial Marketing Management* 30 (2001), pp. 183–98.

Bechtel, Christian and Jayanth Jayaram, 'Supply Chain Management: A Strategic Perspective', *The International Journal of Logistics Management* 8, no. 1 (1997), pp. 15–34.

Cavinato, Joseph L., 'Identifying Interfirm Total Cost Advantages for Supply Chain Competitiveness', *International Journal of Purchasing and Materials Management* 27, no. 4 (Fall 1991), pp. 10–15.

Clendenin, John A., 'Closing the Supply Chain Loop: Reengineering the Returns Channel Process', *The International Journal of Logistics Management* 8, no. 1 (1997), pp. 75–85.

Cooper, Martha, Douglas M. Lambert and Janus D. Pagh, 'Supply Chain Management: More than a New Name for Logistics', *The International Journal of Logistics Management* 8, no. 1 (1997), pp. 1–14.

Ellram, Lisa M. and Martha C. Cooper, 'The Relationship Between Supply Chain Management and Keiretsu', *The International Journal of Logistics Management* 4, no. 1 (1993), pp. 1–12.

Fernie, John, Frances Pfab and Clive Marchant, 'Retail Grocery Logistics in the UK', *The International Journal of Logistics Management* 11, no. 2 (2000), pp. 83–90.

Giunipero, Larry C. and Richard R. Brand, 'Purchasing's Role in Supply Chain Management', *The International Journal of Logistics Management* 7, no. 1 (1996), pp. 29–38.

La Londe, Bernard J. and Terrance L. Pohlen, 'Issues in Supply Chain Costing', *The International Journal of Logistics Management* 7, no. 1 (1991), pp. 1–12.

Lambert, Douglas M., Martha Cooper and Janus D. Pagh, 'Supply Chain Management: Implementation Issues and Research Opportunities', *The International Journal of Logistics Management* 9, no. 2 (1998), pp. 1–19.

Mentzer, John T., 'Managing Channel Relations in the 21st Century', *Journal of Business Logistics* 14, no. 1 (1993), pp. 27–41.

Stock, James R., 'Logistics Thought and Practice: A Perspective', *International Journal of Physical Distribution & Logistics Management* 20, no. 1 (1990), pp. 3–6.

REFERENCES

[1] Definition on the CSCMP's website, http://clm1.org.

[2] Peter F. Drucker, 'The Economy's Dark Continent', *Fortune* (April 1962), pp. 103, 265–70.

[3] European Logistics Association and A.T. Kearney Management Consultants, *Differentiation for Performance: Results of the Fifth Quinquennial European Logistics Study 'Excellence in Logistics 2003/2004'*. Hamburg: Deutcher Verkehrs-Verlag GmbH, 2004, p. 11.

[4] 'An Industry on the Up', *Logistics Europe* 9, no. 7 (October 2001), p. 4.

[5] Alan C. McKinnon, *Physical Distribution Systems*. London: Routledge, 1989, p. 13.

[6] Christine Harland, Louise Knight and Paul Cousins, 'Supply Chain Relationships', in *Understanding Supply Chains*, Steven New and Roy Westbrook, eds. Oxford: Oxford University Press, 2004, p. 213. By permission of Oxford University Press.

[7] David Jobber, *Principles and Practice of Marketing*, 4th edn. Maidenhead, UK: McGraw-Hill, 2004, p. 5.

[8] Joseph B. Fuller, James O'Conor and Richard Rawlinson, 'Tailored Logistics: The Next Advantage', *Harvard Business Review* 71, no. 3 (May–June 1993), pp. 87–98.

[9] This section draws heavily on work by Douglas M. Lambert, *The Development of an Inventory Costing Methodology: A Study of the Costs Associated with Holding Inventory*. Chicago: National Council of Physical Distribution Management, 1976, pp. 5–15, 59–67.

[10] Jobber, *Principles and Practice of Marketing*, p. 634.

[11] Louis W. Stern, 'Channel Control and Interorganization Management', in *Marketing and Economic Development*, Peter D. Bennett, ed. Chicago: American Marketing Association, 1965, pp. 655–65.

[12] Wroe Alderson, 'Factors Governing the Development of Marketing Channels', in *Marketing Channels for Manufactured Products*, R.M. Clewett, ed. Burr Ridge, IL: Richard D. Irwin, 1954, pp. 8–16.

[13] Louis P. Bucklin, *A Theory of Distribution Channel Structure*. Berkeley: University of California, Institute of Business and Economic Research, 1966.

[14] Louis P. Bucklin, 'Postponement, Speculation and the Structure of Distribution Channels', *Journal of Marketing Research* 2, no. 1 (February 1965), pp. 26–31.

[15] Martin Christopher, *Logistics and Supply Chain Management: Creating Value-adding Networks*, 3rd edn. Harlow, UK: FT Prentice Hall, 1998, pp. 31–2.

[16] This section draws heavily on work by Martha C. Cooper, Douglas M. Lambert and Janues D. Pagh, 'Supply Chain Management: More than a New Name for Logistics', *The International Journal of Logistics Management* 8, no. 1 (1997), pp. 1–14; and Douglas M. Lambert, Martha C. Cooper and Janues D. Pagh, 'Supply Chain Management: Implementation Issues and Research Opportunities', *The International Journal of Logistics Management* 9, no. 2 (1998), pp. 1–19.

[17] Definition on the CSCMP's website, http://clm1.org.

[18] Ibid.

[19] Michael E. Porter, *Competitive Advantage: Creating and Sustaining Superior Performance*. New York: Free Press, 1985.

[20] Christopher, *Logistics and Supply Chain Management*, pp. 6–13.

[21] Bernard J. La Londe and Paul H. Zinszer, *Customer Service: Meaning and Measurement*. Chicago: National Council of Physical Distribution Management, 1976, p. iv.

[22] Lambert, *The Development of an Inventory Costing Methodology*, pp. 104–24.

[23] Simon Lloyd, 'Europe's DC Hot Spots', *Logistics Europe* 12, no. 3 (April 2004), pp. 38–42.

[24] James R. Stock, *Development and Implementation of Reverse Logistics Programs*. Oak Brook, IL: Council of Logistics Management, 1998.

[25] Jobber, *Principles and Practice of Marketing*, p. 781.

[26] Nigel Slack, Stuart Chambers and Robert Johnson, *Operations Management*, 4th edn. Harlow, UK: FT Prentice Hall, 2004, pp. 722–3.

[27] Herbert Kotzab, 'Improving Supply Chain Performance by Efficient Consumer Response? A Critical Comparison of Existing ECR Approaches', *Journal of Business & Industrial Marketing*, 14, no. 5/6 (1999), pp. 364–77.

[28] Ibid., p. 371.

[29] Edward A. Morash, Cornelia Dröge and Shawnee Vickery, 'Boundary Spanning Interfaces Between Logistics, Marketing, Production and New Product Development', *International Journal of Physical Distribution & Logistics Management* 28, no. 6 (1996), pp. 43–62.

[30] 'An Industry on the Up', *Logistics Europe*, p. 4.

[31] Cecilia Cabodi, 'The Fourth Way', *Logistics Europe* 12, no. 3 (April 2004), pp. 24–8.

CHAPTER 2
CUSTOMER SERVICE

CHAPTER OBJECTIVES

- ■ To define customer service
- ■ To show the central role that customer service plays in an organization's marketing and logistics efforts
- ■ To show how to calculate cost/revenue trade-offs
- ■ To illustrate how to conduct a customer service audit
- ■ To identify opportunities for improving customer service performance

In times of tough competition when many organizations offer similar products in terms of price, features and quality, customer service differentiation can provide an organization with a distinct advantage over the competition. Customer service represents the output of the logistics system as well as the 'place' component of the organization's marketing mix. Customer service performance is a measure of how well the logistics system functions in creating time and place utility, with a focus on external customers.

The level of service provided to functions, such as marketing and production, affects the organization's ability to serve the needs of customers and will determine how well these functions communicate and interact with logistics on a day-to-day basis. The level of customer service provided to customers determines whether the organization will retain existing customers and how many new customers it will attract.

In virtually every industry today, from computers to clothing to cars, customers have a wide variety of choices. A company cannot afford to offend its customers. The customer service level that an organization provides has a direct impact on its market share, its total logistics costs and, ultimately, its overall profitability.

To illustrate, a key to corporate profitability is to successfully attract and retain customers. However, it has been estimated that average customer turnover is between 10 and 30 per cent for companies.[1] If customer turnover could be reduced by 5 per cent, bottom-line profitability could possibly increase significantly, perhaps by 60–95 per cent annually.[2]

For these reasons, it is of the utmost importance that customer service be an integral part of the design and operation of all logistics systems.

Customer Service Defined

The definition of **customer service** varies across organizations. Suppliers and their customers can view the concept of customer service quite differently. In a broad sense, customer service is the measure of how well the logistics system is performing in providing time and place utility for a product or service. This includes activities such as the ease of checking stock, placing an order and post-sale support of the item.

Customer service is often confused with the concept of customer satisfaction. In contrast to customer service, **customer satisfaction** represents the customer's overall assessment of all elements of the marketing mix: product, price, promotion and place (the so-called 4Ps). Thus, customer satisfaction is a broader concept that encompasses customer service. A thorough description of customer satisfaction can be found in many introductory marketing textbooks.[3]

In most organizations, customer service is defined in one or more ways, including (1) an activity or function to be managed, such as order processing or handling of customer complaints; (2) actual performance on particular parameters, such as the ability to ship complete orders for 98 per cent of orders received within a 24-hour period; or (3) part of an overall corporate philosophy, rather than simply an activity or performance measure.[4] If an organization views customer service as a philosophy, it will likely have a formal customer service function and various performance measures.[5]

A Recent View of Customer Service

Customer service can be defined as:

> . . . a process which takes place between the buyer, seller, and third party. The process results in a value added to the product or service exchanged. This value added in the exchange process might be short term as in a single transaction or longer term as in a contractual relationship. The value added is also shared, in that each of the parties to the transaction or contract are better off at the completion of the transaction than it was before the transaction took place. Thus, in a process view: Customer service is a process for providing significant value-added benefits to the supply chain in a cost-effective way.[6]

Successful implementation of the marketing concept requires that companies both win and retain customers. Too often, the emphasis is on winning new customers and gaining new accounts. But this is an extremely short-sighted approach for an ongoing business concern. Many business strategy textbooks used to say that 'the objective of a firm is to make a profit', but this attitude is shifting. The objective of the firm is still to make a profit, but before that can take place, the firm needs to establish service policies and programmes that will satisfy customers' needs and deliver them in a cost-efficient manner – that is, customer service.

Companies like Marriott, KLM and Nokia have gained significant advantage in the marketplace by meeting and sometimes exceeding a customer's service expectations; below are details of how these firms address customer service issues.

- Marriott hotels' reputation for superior customer service rises out of a long tradition to provide 'Good Food and Good Service at a Fair Price', 'Do Whatever it Takes to Take Care of the Customer', pay extraordinary attention to detail, take pride in their physical surroundings and use their creativity to find new ways to meet the needs of customers.[7]
- Royal Dutch Airlines, KLM, is a signatory to the European Airline Passenger Service Commitment of 2002. KLM's commitments are based on four concepts: responsibility to deliver what it promises; transparency to make flight information understandable and accessible; efficiency of ticketing, check-in and baggage services; and prompt and clear care and assistance if service is disrupted.[8]
- Nokia, as a high-technology manufacturer, focuses on quality for its customer service. Its quality programme is continuous and has three main aspects: quality in processes for better productivity and innovation; quality in products, which is equated to customer experiences and perceptions; and quality in management to balance values-based leadership and fact-based management.[9]

Customer Retention and Service Quality

It is very expensive to both win and lose customers. Keeping customers should be a paramount concern. Determining what customers need in terms of service levels and delivering upon those needs in a cost-effective and efficient manner should be a key concern of the logistics function. A goal should be to 'do it right the first time', to prevent complaints from ever occurring. Many studies indicate that for every customer that complains, many others simply choose to stop doing business with the organization and usually tell their friends and associates about their negative experience.

However, the complainers have much to offer in terms of potential learning. They may alert the organization to a widespread problem which, if addressed, could reduce future complaints and help retain those 'non-complaining' customers who otherwise would simply have walked away. In addition, if handled well, complaining customers actually become more loyal and are more likely to do business with that organization again in the future.

Quality in customer service, then, from initial dealings with the customer to proper handling of problems, is critical in achieving high levels of customer service. This in turn contributes to high levels of customer satisfaction. This topic is discussed further below.

Elements of Customer Service

The elements of customer service can be classified into three groups: pre-transaction, transaction and post-transaction elements. These groups are linked to the definitions of marketing that incorporate the notion of market transactions – before, during, and after the sale.[10] This conceptualization is depicted in Figure 2.1.

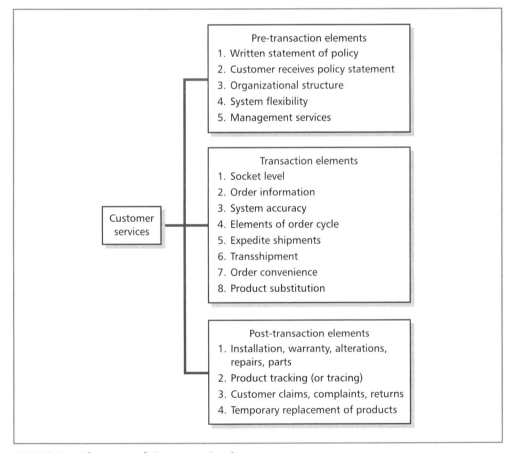

FIGURE 2.1 Elements of Customer Service

Source: based on Bernard J. La Londe and Paul H. Zinszer, *Customer Service: Meaning and Measurement*. Chicago: National Council of Physical Distribution Management, 1976, p. 281.

Pre-transaction Elements

The pre-transaction elements of customer service tend to be related to the organization's policies regarding customer service, and can have a significant impact on customers' perceptions of the organization and their overall satisfaction. These elements are not all directly related to logistics. They must be formulated and in place before the organization can consistently implement and execute its customer service activities. Pre-transaction elements include the following.

1 *A written statement of customer service policy*. This policy would define service standards, which should be tied to customers' needs. It should include metrics for tracking service performance and the frequency of reporting actual performance, and be measurable and actionable.

2 *Customers provided with a written statement of policy*. A written statement lets the customer know what to expect and helps to safeguard against unreasonable expectations. It should provide the customer with information about how to respond if expected service levels are not achieved by the firm.

3 *Organizational structure*. The organization structure best suited to ensure the achievement of customer service goals varies across organizations, but the senior logistics executive should be positioned at a high level and have high visibility within the firm. The structure should facilitate both internal and external communication of policies, performance and corrective actions as needed. Customers should have easy access to individuals within the organization who can satisfy their needs and answer their questions. Imagine the frustration felt by a customer who has experienced a problem with product delivery or performance, who telephones the selling organization only to be put on hold, and then transferred from one representative to another, continually having to re-explain his or her entire problem. The customer may never call that organization again for anything.

4 *System flexibility*. Flexibility and contingency plans should be built into the system, and should allow the organization to respond successfully to unforeseen events such as labour strikes, material shortages, and natural disasters such as hurricanes or flooding.

5 *Management services*. Offering help with merchandising, improving inventory management and ordering are examples of some of the services an organization may provide to its customers. These may be provided in the form of training manuals, seminars or one-to-one consultations. The services may be free of charge or fee based.

All of these pre-transaction elements may be experienced by the customer outside of the normal order cycle. Decisions relating to the pre-transaction elements tend to be relatively stable, long-term decisions that are changed infrequently. This provides some stability for the customer in terms of expectations.

Transaction Elements

Transaction elements are the elements that are *normally* considered to be associated with customer service, and include the following.

1 *Stockout level*. The stockout level measures product availability. Stockouts should be monitored by product and customer in order to better track potential problems. When stockouts occur, the organization should endeavour to maintain customer goodwill by offering a suitable substitute, drop-shipping from another location to the customer if possible, or expediting the shipment once the out-of-stock item arrives.

2 *Order information availability*. Customers' expectations regarding access to all types of information related to their orders have increased dramatically because of the availability of relatively inexpensive computing power. This includes information on inventory status, order status, expected or actual shipping date, and back-order status. Tracking back-order performance is important because customers pay close attention to problems and exceptions to

delivery. Back-orders should be tracked by customer and by product type, so that recurring problems become visible and can be addressed in a timely fashion.

3 *System accuracy.* In addition to the ability to rapidly obtain a wide variety of data, customers expect that the information they receive about order status and stock levels will be accurate. Inaccuracies should be noted and corrected as quickly as possible. Continuing problems require major corrective action and a high level of attention. Errors are costly to correct for customers and suppliers in terms of time delays and paperwork created.

4 *Consistency of order cycle or lead-time.* The order cycle or lead-time is the total time from customer initiation of the order through to receipt of the product or service by the customer. Thus, if a salesperson obtains an order from a customer and holds it for five days before entering the order, that adds five days to the order cycle time, even though those five days were invisible to the distribution centre. Elements of the order cycle include placing the order, order entry if separate from placement, order processing, order picking and packing for shipment, transit time and the actual delivery process. Customers tend to be more concerned with the consistency of lead-times than with absolute lead time, so it is important to monitor actual performance in this regard and take corrective action if needed. However, with the increased emphasis on time-based competition, reducing total cycle time has received greater attention. This topic will be discussed later in this chapter.

5 *Special handling of shipments.* Special handling of shipments relates to any order that cannot be managed through the normal delivery system. This could happen because it needs to be expedited or has unique shipping requirements. The costs of such shipments are considerably higher than standard shipments. However, the cost of a lost customer could be higher still. The company should determine which customers or situations warrant special treatment and which do not.

6 *Transshipment.* Transshipment refers to shipping products between various distribution locations to avoid stockouts and possibly obsolescence. For companies with multiple locations, some sort of policy should be in place concerning transshipments as opposed to back-ordering or drop-shipping directly to a customer from more than one location.

7 *Order convenience.* Order convenience refers to how easy it is for a customer to place an order. Customers prefer suppliers that are user-friendly. If forms are confusing, terms are not standardized or the waiting time on hold on the telephone is long, customers may experience dissatisfaction. Order-related problems should be monitored and identified by talking directly with customers. Problems should be noted and corrected.

8 *Product substitution.* Product substitution occurs when the product that the customer ordered is not available, but is replaced by a different size of the same item or a different product that will perform just as well or better. Figure 2.2 illustrates that if a product currently has a 70 per cent service level and one acceptable substitute that also has a 70 per cent service level, a manufacturer can effectively raise its service level for that product to 91 per cent. If the product has two acceptable substitutes, the in-stock availability becomes 97 per cent. Thus, the ability to provide a customer with acceptable substitutes can significantly improve the firm's service level.

The manufacturer should work with its customers to develop product substitution policies and should keep its customers informed of those policies. It is always a good idea to check with the customer before substituting one product for another. For example, if a custom furniture manufacturer orders one-litre cans of furniture stain and the distributor is out of stock, the distributor may offer five-litre units in their place. This may not be suitable because the furniture manufacturer may use only two litres for each job and does not want to have partially used five-litre containers of stain. However, if the distributor offers 500-millilitre cans at the same price per litre, this may be a perfectly acceptable substitute to the customer.

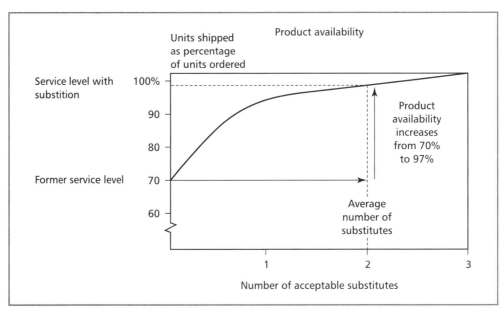

FIGURE 2.2 Impact of Product Substitution on Customer Service Levels

The transaction elements of customer service often receive the most attention, because they are the most immediate and apparent to the customer. For example, in a recent survey of over 1200 UK food-processing companies, respondents considered that a product arriving undamaged and according to specification was more important than the product quality itself.[11]

Post-transaction Elements

The post-transaction elements of customer service support the product or service after the customer has received it. Historically, this has tended to be the most neglected of the three groups of customer service elements, in part because a relatively small proportion of customers complain about poor service. However, retaining and satisfying current customers can be much more profitable than finding new customers. Post-transaction elements include the following.

1 *Installation, warranty, repairs and service parts.* These elements should be an important consideration in almost all purchases, especially purchases of capital equipment where such costs tend to far outweigh the cost of the purchased item itself.[12] These elements should receive the same attention and scrutiny as transaction elements.

2 *Product tracking.* Product tracking, also referred to as product tracing, is an important customer service element. For example, in order to inform consumers of potential problems, firms must be able to re-call potentially dangerous products from the market once the potential hazard has been identified.

3 *Customer complaints, claims and returns.* To resolve customer complaints, an accurate online information system is needed to process the data from the customer, monitor trends and provide the customer with the most current information available. Logistics systems are designed to move products to customers, so the cost of non-routine handling, particularly of small shipments such as customer returns, tends to be high. Customer returns go through the logistics process in reverse, hence the term **reverse logistics** (see Chapter 9). Corporate policies should be established to handle these complaints as efficiently and effectively as possible.

4 *Product replacement*. Depending on the item, having back-up product temporarily available when the item is being serviced can be critical. For example, some automobile dealerships provide loan cars to their customers at no charge while their cars are being serviced. This minimizes inconvenience and may create a more loyal customer.

Importance of Customer Service for Gaining Strategic Advantage

Customer service is the output of the logistics system and is the key interface between the marketing and logistics functions, supporting the 'place' element of the marketing mix. But even more important, customer service plays a significant role in developing and maintaining customer loyalty and ongoing satisfaction.

The product, pricing and promotion elements of the marketing mix create value added for customers. However, when the performance of competitors is similar on these attributes, it is customer service that really brings the customer back.

Products and prices are relatively easy for competitors to duplicate. Promotional efforts can also be matched by competitors, with the possible exception of a well-trained and motivated sales force. The satisfactory service encounter, or favourable complaint resolution, is one important way that the organization can really distinguish itself in the eyes of the customer. Thus logistics can play a key role in contributing to the organization's competitive advantage by providing excellent customer service.

TECHNOLOGY

Techniques for Home Delivery

Home delivery, and the specialist urban logistics services it uses, is a burgeoning business with recent market growth of 7.3 per cent to a total value of almost €45 billion. Therefore, the need to get it right is more urgent than ever before if those involved are keen to take a share of this buoyant market and fulfil growing customer demand.

Technology is at the forefront when it comes to getting urban delivery right. For instance, track and trace systems are increasingly being used to enable customers to log on to the retailer's/carrier's website to find out the status of their orders. Customers are provided with a number of points of reference, including: order received and being processed, order being picked and packed in the warehouse, order dispatched, and with the carrier for delivery. By providing this information, customers are given an improved service and the number of enquiries made to the call centre as to the whereabouts of orders are reduced. Track and trace technology, such as tapograph systems and satellite technology (General Packet Radio Service, better known as GPRS), are also beneficial from an operations point of view as monitoring the carriers' progress allows for a more efficient and effective delivery service.

Often carriers will be away from the warehouse for a week, delivering products to customers over a large geographical area, and so without the use of some form of tracking technology the head office is blind to which products have been delivered and to whom. By having electronic information consistently fed back to head office, it is made aware of which customers have or have not received their goods, which customers were not at home to receive the goods, as well as other issues, such as damaged goods. This allows head office to immediately work to rectify any problems and the call centre can be pre-warned should customers call up with a query or complaint.

Cost-effective telecommunications devices, such as portable Bluetooth units, allow automatic, efficient transmissions back to the head office computer system from a cradle in the carrier's vehicle. Personal digital assistants (PDAs) are also increasingly being used by carriers so that on delivering goods, customer signatures can be obtained electronically with the use of a pen-like device. The signature and date and time stamp are then uploaded to head office as confirmation that the job has been completed and so that head office knows where the carrier is up to in its delivery schedule.

But in order to make a success of home delivery, a number of factors need to be considered, including identifying the proposition from the outset, using technology to support home delivery, ensuring staff reflect the brand and taking into account returns. However, all of these are unlikely to result in success if the home delivery service is not integrated with the rest of the business – an integrated approach to multi-channel retail is the key to getting home delivery right. One company employing that sort of expertise is Amtrak Express parcels. The company recently took on a new client, the major UK electrical consumer goods retailer Comet.

Amtrak delivers electrical items, working alongside existing carrier Parcelforce and Comet's own fleet, which handles larger goods. Comet makes 13 million sales a year and with home deliveries increasing, peak volumes reach 7000 parcels a night. To enable this service Amtrak has made significant investment in core IT systems and new technology, utilizing wireless bar code scanning systems and the web. The automation has helped in reducing overheads, but is primarily aimed at giving Amtrak class-leading service levels. With the new 'DayTrak' courier service and established 'Collections Service', customers can track their bookings using real-time information via the web.

Comet stocks much of its one-man delivery items like phones, PDAs and small vacuum cleaners at its central England distribution centre. Evening collections are arranged by the nearby Amtrak office in Nottingham with trunking direct to the Amtrak hub near Birmingham. Amtrak then delivers next morning through its 100-strong depot network. Amtrak has become the first UK parcel carrier to offer evening deliveries as part of a standard service, adding to Saturday and unattended delivery services geared for the urban delivery sector. The sort of service offered to Comet has boosted Amtrak's reputation and in 2004 the company achieved a 4 per cent growth in turnover to £70 million and an increase of £6 million in operating profit.

Question: **What other applications would benefit from Amtrak's technology?**

Source: adapted from 'All Homeward Bound', *Logistics Business Magazine* (November/December 2004), pp. 10–14.

How to Establish a Customer Service Strategy

An organization's entire marketing effort can be neutralized by poorly conceived or badly executed customer service policies. Yet customer service is often a neglected element of the marketing mix. As a result, customer service standards tend to be based on industry norms, historical practices or management's judgement of what the customer wants, rather than what the customer really desires. Management often treats all customers the same, not recognizing that different customers want different levels and types of service.[13]

It is essential that a firm establish customer service policies based on customer requirements and supportive of the overall marketing strategy. What is the point of manufacturing a great product, pricing it competitively and promoting it well, if it is not readily available to the consumer? At the same time, customer service policies should be cost efficient, contributing favourably to the firm's overall profitability.

One popular method for setting customer service levels is to benchmark competitors' customer service performance. While it may be interesting to see what the competition is doing, this information has limited usefulness. In terms of what the customer requires, how does the firm know if the competition is focusing on the right customer service elements? Therefore, competitive benchmarking alone is insufficient.

Competitive benchmarking should be performed in conjunction with customer surveys that measure the importance of various customer service elements. Opportunities to close differences between customer requirements and the firm's performance can be identified. The firm can then target the primary customers of specific competitors while protecting its own key accounts from potential competitor inroads.

A number of methods have been suggested for establishing customer service strategies. The following five are of greatest value:

1 determining customer service levels based on customer reactions to stockouts at the retail level
2 cost/revenue trade-offs
3 determining service quality differences between customers' expectations and a firm's performance
4 ABC analysis of customer service
5 customer service audits.

Each of these techniques is discussed below.

Customer Reactions to Stockouts

Most manufacturers do not sell exclusively to end users. Instead, they sell to wholesalers or other intermediaries who sell to the final customer. For this reason, it may be difficult for a manufacturer to assess the impact of stockouts on end users. For example, an out-of-stock situation at the manufacturer's warehouse does not necessarily mean an out-of-stock product at the retail level. One way to establish the desirable level of customer service at the retail level is to determine consumers' response to stockouts, which can include substituting another size of the same brand, switching brands or perhaps going to a different store to buy the items. For most products, consumers will switch stores only if they believe that the product they desire is superior to or considerably less expensive than the available substitutes.

To see how stockouts have a different effect at various levels of the channel of distribution, we can examine the infant formula industry. Most infant formula manufacturers do not advertise their products on national television, and generally limit the amount they spend on consumer-directed media advertising. They also limit the use of price promotion. Instead, they spend their marketing budget on sample products to give to doctors and hospitals, who in turn give the product samples to new mothers. New mothers are often told not to switch brands because the baby develops a preference and may not adapt well to another brand. In addition, most mothers assume that their doctor would give them only the recommended products as samples. So, when the mother goes to the store to buy infant formula and the product is out of stock, she will go to a different store to find it rather than risk switching products.

Understanding behaviour at different levels in the channel is critical in formulating customer service strategies.[14] The penalty for being out of stock at a particular retail store is relatively low for the manufacturer of infant formula because the vast majority of customers will just switch stores.

However, the penalty of running out of stock at a particular hospital is very high if, for example, it causes the doctor to switch from Mead Johnson's brand, Enfamil, say, to Abbott-Ross Laboratories' brand, Similac. Mead Johnson will likely lose all the potential future business from mothers who give birth at that hospital and continue to feed their baby with Similac rather than Enfamil. The customer service implications are clear: hospitals and doctors require a very high level of customer service, which may mean in-stock availability above 99 per cent, and very short lead times of 24 to 48 hours.

The retailer is also likely to lose the sale if it is out of the customer's preferred brand of infant formula. The inventory position on an item such as infant formula, which would actually cause the customer to switch stores in case of a stockout, must be monitored closely by the retailer. Frequent stockouts on such an item can cause customers to switch stores permanently.

When the manufacturer is aware of the implications of stockouts at the retail level, it can make adjustments in order cycle times, fill rates, transportation options and other strategies that will result in higher levels of product availability in retail stores.

For some items, customers may be more willing to wait, even expecting to place a special order rather than have the item available in stock. For example, consumers in the UK buying sofas and other large furniture items wait up to 12 weeks for their purchases to be made and delivered to their homes.

For some products, consumers who face an out-of-stock situation on the particular stock-keeping unit (SKU) they desire will choose to switch brands or sizes. A **stockkeeping unit (SKU)** is an individual product that differs from other products in some way. The difference could be in size, colour, scent, flavour or some other relevant characteristic. If a customer switches to another size of the same product, the customer service level should be measured relative to the desired SKU and all substitutes. If a customer switches brands, then that manufacturer has lost a sale today; in addition, the customer may come to prefer the new brand, so that significant future sales are lost as well. The value of these lost sales is difficult to determine. As long as some sort of acceptable substitute is available to the customer, retailers are less concerned with this type of stockout because they have still made the sale.

Recent research has found that an average out-of-stock (OOS) rate for fast moving consumer goods (FMCG) retailers across the world is 8.3 per cent, or an average on-shelf availability (OSA) of 91.7 per cent. Five consumer responses to OOS were found to be: buy the item at another store (31 per cent), substitute a different brand (26 per cent), substitute the same brand (19 per cent), delay their purchase until the item becomes available (15 per cent), and do not purchase any item (9 per cent).[15] The implication of these findings is that 55 per cent of consumers will not purchase an item at the retail store while 50 per cent of consumers will substitute or not purchase the manufacturer's item.

If there are unplanned decreases in customer service levels because of labour strikes, materials shortages or other factors, the company can look at sales 'before', 'during' and 'after' such events to assess the impact of various levels of customer service on retail sales. This assessment is only relevant if the competitor's product was still available during the same period.

Cost/Revenue Trade-offs

The total of logistics expenditures such as carrying inventory, transportation and information/order processing can be viewed as the company's expenditures on customer service. Figure 2.3 illustrates the cost trade-offs and considerations required to implement an integrated logistics management concept. The objective is to provide the organization with the lowest total logistics costs, given a specific customer service level. While Figure 2.3 shows logistics issues as trade-offs, in some cases simultaneous improvement may occur in multiple areas, and the organization reduces its total cost while providing improved customer service. This is only possible by taking the perspective of the total system in the long run.

For example, if a major department store chain wishes to increase its retail in-stock levels to 98 per cent, point-of-sale (POS) data that track actual sales by store and by SKU might be used. Thus, it has to invest in information technology such as in-store scanners of bar codes at each cash register, and software to compile and analyse the data in addition to generating meaningful management reports. To maximize its leverage, the discount store chain also might want to invest in an **electronic data interchange (EDI)** system to provide rapid, two-way communication

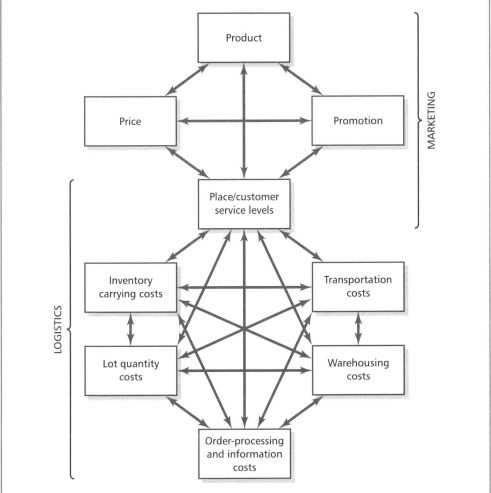

Marketing objective: allocate resources to the marketing mix to maximize the long-run profitability of the firm.
Logistics objective: minimize total costs given the customer service objective where: Total costs = Transportation costs + Warehousing costs + Order-processing and information costs + Lot quantity costs + Inventory carrying costs.

FIGURE 2.3 Cost Trade-offs Required in a Logistics System
Source: adapted from Douglas M. Lambert, *The Development of an Inventory Costing Methodology: A Study of the Costs Associated with Holding Inventory*. Chicago: National Council of Physical Distribution Management, 1976, p. 7.

with its suppliers. Hypothetically, this could cost the chain roughly €100,000 per store. Thus management appears to be making a trade-off: by investing in information technology, the store is increasing its costs to improve customer service levels.

If each euro of additional sales revenue costs the company 60 cents in product costs, plus variable logistics and marketing expenses, the contribution margin is 40 per cent. For each euro of sales revenue, what sales need to be generated to break even? We can calculate the additional sales required to offset this investment by dividing the €100,000 investment by the 40 per cent contribution margin. Thus, the company needs to increase sales by €250,000 per store, on average, to break even on this investment. If sales increase more than that, they will be ahead on this investment, achieving a positive return, rather than 'trading off'.

This decision would need to be evaluated considering how likely it is that each store would increase sales by €250,000 over some specified time period. If current sales are €10 million per store per year, it would seem much more feasible to recover this investment in a timely manner than if sales are currently €2 million per store per year.

Service Quality and Customers' Expectations

The functional areas of logistics such as transportation, warehousing and inventory management were introduced in Chapter 1. The characteristics of these functional areas reflect the nature of services as opposed to products or goods, i.e. these functional areas are generally intangible, inconsistent, perishable and inseparable relative to the products and goods they act upon in their role as logistics activities.[16]

Customers evaluate services differently due to these characteristics. One popular method to examine such evaluations is the service quality or 'gaps' model.[17] In this model, customers develop prior expectations of a service based on several criteria, such as previous experience, word-of-mouth (WOM) recommendations, or advertising and communication by the service provider. Once the customers 'experience' the service they compare their perceptions of that experience to their expectations. If their perceptions meet or exceed their expectations then they are satisfied; conversely, if perceptions do not meet expectations then they are dissatisfied. The difference between expectations and perceptions forms the major 'gap' that is of concern to firms.

Figure 2.4 shows the service quality model that includes the customer and firm's positions. The expectations and perceptions 'gap' is affected by four other 'gaps' related to the firm's customer service and service quality activities that are for the most part invisible to the customer. First, the firm must understand the customer's expectations for the service. Second, the firm must then turn the customer's expectations into tangible service specifications. Third, the firm must actually provide the service according to those specifications and, last, the firm must communicate its intentions and actions to the customer.

Using the service quality model forces a firm to examine what customer service and service quality it provides to customers in a customer-centric framework.

ABC Analysis/Pareto's Law

In Chapter 1, we used the abbreviation ABC for activity-based costing. Here, ABC analysis is used to denote a tool for classifying items or activities according to their relative importance. This concept will also be discussed in greater depth in conjunction with inventory management in Chapter 5.

The logic behind ABC classification is that some customers and products are more beneficial to a firm than others: beneficial in terms of profitability, sales revenues, segment growth rates or other factors deemed important by corporate management. Using profitability as an example, the most profitable customer–product combinations should receive the most attention and, hence, higher customer service levels. Profitability should be measured according to a product's contribution towards fixed costs and profits.

Like ABC analysis, **Pareto's law** notes that many situations are dominated by relatively few critical elements. For example 80 per cent of the bottlenecks in the logistics system may be caused by the failure of one carrier. This concept is also commonly referred to as the **80/20 rule**.

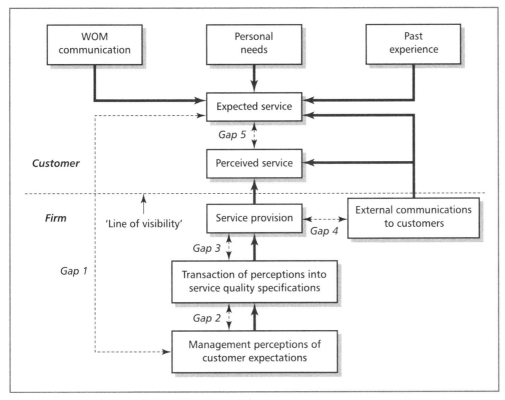

FIGURE 2.4 Service Quality or 'Gaps' Model

Source: adapted from Ari Parasuraman, Valarie A. Zeithaml and Leonard L. Berry, 'A Conceptual Model of Service Quality and its Implications for Future Research', *Journal of Marketing* 49 (Fall 1985), pp. 41–50. Reproduced with permission of the American Marketing Association (AMA).

Table 2.1 illustrates how the importance of customers can be combined with the importance of products to establish customer service levels that are the most beneficial to a firm. This matrix can be interpreted as follows (using profitability as the relevant factor).

Customer classification	Product			
	A	B	C	D
I	1	2	6	10
II	3	4	7	12
III	5	8	13	16
IV	9	14	15	19
V	11	17	18	20

TABLE 2.1 A Customer–product Contribution Matrix

Source: adapted from Bernard J. La Londe and Paul H. Zinszer, *Customer Service: Meaning and Measurement*. Chicago: National Council of Physical Distribution Management, 1976, p. 181.

1   Products in category A are the most profitable, followed by B, C and D. The products in category A usually represent a small percentage of the firm's product line. The products in category D, on the other hand, are the least profitable and probably make up about 80 per cent of the firm's product line.

2   Customers in category I are the most profitable, but are few in number, maybe 5 to 10. Those in category V are the least profitable, accounting for the majority of the firm's customers.

3   The most profitable customer–product combination occurs when customers in category I buy products in category A. The next most profitable combines B products with category I customers, then category II customers with A products, and so on. Management will use some logical approach in ranking the various customer and product combinations. An example of one such approach is illustrated in Table 2.1.

An organization can use the data in Table 2.1 in setting customer service policies, as illustrated in Table 2.2. For example, the standards for customer–product combinations in priority range 1–5 may be 100 per cent in-stock product availability, 48 hours' order cycle time and 99 per cent of all orders shipped complete (i.e. no partial shipments).

Priority range	In-stock availability standard (%)	Order cycle time standard (hours)	Order completeness standard (%)
1–5	100	48	99
6–10	95	72	97
11–15	90	96	95
16–20	85	120	93

TABLE 2.2 Making the Customer–product Contribution Matrix Operational

Source: adapted from Bernard J. La Londe and Paul H. Zinszer, *Customer Service: Meaning and Measurement*. Chicago: National Council of Physical Distribution Management, 1976, p. 182.

Again, if profitability is the relevant measure of customer and product importance, this method recognizes the need to provide the most profitable customers with the highest service levels in order to encourage customer loyalty and thus repeat business. Those less profitable accounts can be made more attractive to the firm by reducing the service levels, which makes them less costly to service and therefore more profitable.

A lower level of customer service does not mean that the service provided is less consistent. In other words, whatever the service level, 100 per cent consistency of service is provided whenever possible. Consistency is important to all customers, irrespective of size or type. However, the important issue is that it is usually less expensive for a firm to deliver lower levels of customer service (e.g. lead time) with high consistency than it is to provide higher levels of service with low consistency; for example, a 72-hour order cycle time with high consistency is less expensive to the provider firm than a 48-hour order cycle time with low consistency.

As a generic example, it may be possible for a firm to develop several strategies that allow it to reduce its logistics costs while providing an acceptable level of customer service. Generally, the longer the order cycle time, the less inventory a firm must carry. As we will see later in this book, less inventory increases a firm's profits through reductions in inventory carrying costs. Also, it may be possible to obtain lower rates when transportation carriers are given more time to deliver products. As long as consistent service is provided to customers, most firms can plan ahead, knowing that order cycle times will be longer.

The principle is illustrated by Gillette, a global producer of razors and other personal care products. Gillette requires that small, **less-than-truckload (LTL)** customer orders be received by a certain day and time to be processed that week. These orders are then pooled and shipped with other orders destined for a particular geographical area. This is referred to as scheduled delivery.

The benefit of this approach is that Gillette can fill a truck, which lowers its transportation cost. It also reduces the absolute transit time and the variability in transit time, because the truck is going to one general area and dropping off products, or delivering to a regional carrier in the market for local delivery rather than having many stops in dispersed areas. If customers miss the order deadline, they are given the option of waiting until the next week or having their order shipped LTL. Most opt to wait, for the cost and service reasons mentioned above.

The key to developing a customer–product contribution matrix that meets the needs of both the customer and the firm is knowing how customers define service, identifying which service components are most important, and determining how much customer service to provide. Customer service audits are often conducted to obtain this information prior to a firm establishing any kind of policy relating to customer–product profitability and customer service levels.

The Customer Service Audit

A customer service audit is used as a means of evaluating the level of service a company is providing and as a benchmark for assessing the impact of changes in customer service policies. The objectives of the audit are to (1) identify critical customer service elements, (2) identify how performance of those elements is controlled, and (3) assess the quality and capabilities of the internal information system.

The audit typically includes four distinct stages:

1 external customer service audit
2 internal customer service audit
3 identifying opportunities and methods for improvement
4 establishing customer service levels.

Each of these stages is discussed below. While the stages tend to occur sequentially, there is some overlap between them.

The External Customer Service Audit

The external audit should be the starting point in an overall customer service audit, one that examines both internal and external factors. The key goals of the external audit are to:

- identify the elements of customer service that customers believe are important in their buying decision
- determine the customer's perception of the service being offered by the firm and each major competitor.

The first step is to determine which service elements the customer perceives as important. This should be accomplished using an interview format with a sample of the firm's customers. Interviewing can provide insights into customer service issues of which the firm may otherwise be unaware. Some of the key customer service elements to a retailer assessing a manufacturer might be the consistency of order cycle time, absolute length of the order cycle, whether the supplier uses EDI, the number of orders shipped complete, back-order policies of the firm, billing procedures and backhaul policies. Because many service variables differ by industry, it is important to survey customers in order to establish their service requirements.

It is beneficial to involve the marketing function in the external customer service audit for a number of reasons. First, marketing often has the major decision-making authority in terms of making customer service trade-offs within the marketing mix. Also, the marketing function can provide useful insights into a better understanding of customer needs and the incorporation of the most relevant issues in the design of any instruments used to collect customer data. A high level of involvement will create buy-in, so that more support will be provided by the marketing function when it later implements the audit findings.

Once the important customer service elements have been identified, the next major step is to develop a questionnaire to gain feedback from a statistically valid and representative group of the firm's customers.

The questionnaire is used to determine the relative importance of various customer service elements, other marketing mix elements, and measures of the perceived performance of the firm and its major competitors on each element. Ideally, the organization would like to perform very well on those elements that customers and potential customers evaluate as the most important. This enables management to develop strategies by customer segments while considering the strengths and weaknesses of specific competitors.

It is important that the questionnaire ask about customers' relative market share with each supplier, as well as their overall perception/satisfaction with their supplier(s). This will allow the firm to examine the relationship between its sales and its performance as perceived by customers.

The questionnaire also should explore customers' expected levels of performance on key issues now and in the future. Demographic data will allow the firm to assess perceived performance differences according to geographical region, customer type and other relevant dimensions.

For best results and validity, the questionnaire should be pretested with a small sample of customers to ensure that no critical issues have been missed and that customers are able to understand and answer the questions.

The results obtained from the customer service survey can reveal both opportunities and potential problems. The variables that receive the highest importance ratings from customers should be the focus of analysis and action. For example, if a firm scores significantly higher than competitors on key ratings, it could use those findings in its promotional mix – and perhaps generate increased sales revenues. If an organization scores significantly lower than competitors on important service variables or big gaps exist between desired performance and actual performance, it faces potentially significant problems. Without some corrective action, market share could erode if competitors take advantage of a firm's weaknesses.

It would appear that the variables that receive the highest ratings should play the greatest role in share of business. However, this may not always be the case, for a number of reasons:

■ all of an industry's major suppliers may be performing at 'threshold' levels or at approximately equal levels, which makes it difficult to distinguish among suppliers
■ variables for which there are significant variances in vendor performance may be better predictors of market share than the variables described above
■ customers may rate a variable as extremely important, but there may be few or no suppliers providing satisfactory levels of service for that variable; such variables offer opportunities to provide differential service in the marketplace
■ a variable may be rated low in importance with a low variance in response; in addition, there may be no single supplier providing adequate service levels; therefore, customers do not recognize the advantages of superior service for that variable; if one supplier improved performance, it could lead to gains in market share.[18]

The organization should look simultaneously at the importance of various elements and its position relative to competing suppliers. If customers perceive the supplier's performance to be poor on some attributes relative to management's beliefs on organization performance, management should determine whether it is measuring the firm's performance the same way that customers are and, if not, adjust measurements to align with customer measures.

If the actual performance is better than customers believe it to be, management should determine how to educate customers and to inform them of actual performance. This might include providing the sales force with monthly or quarterly performance reports by customer that salespeople would review with each of their accounts.

The internal customer service audit can be conducted while the external customer service audit is occurring.

The Internal Customer Service Audit

The internal customer service audit reviews the firm's current service practices. This provides a benchmark for assessing the impact of changes in customer service levels. As such, the internal customer service audit should address the following issues.

- How is customer service currently measured within the firm?
- What are the units of service measurement?
- What are the service performance standards or objectives?
- What is the current level of attainment: results versus objectives?
- How are these measures derived from the firm's information and/or the order-processing systems?
- What is the internal customer service reporting system?
- How do each of the functional areas of the business (e.g. logistics, marketing) perceive customer service?
- What are the relationships between these functional areas in terms of communication and control?[19]

The major goal of the internal audit is to measure gaps between the firm's service practices and customer requirements. Customers' perceptions of current service levels should also be determined, because they may perceive service as worse than it really is. If that is the case, customers' perceptions should be the focus of change through education and promotion, rather than changing the firm's service levels.

Another key area to assess in the internal customer service audit is the communication flows from the customer to the company, and communication flows within the company, including the measurement and reporting of service performance. Communication is a major factor in determining how well customer service-related issues are understood. Without excellent internal communications, customer service tends to be reactive and problem focused, rather than proactive.[20]

Communication between the customer and the organization relates primarily to the order–ship–receive cycle. The seven major issues are order entry, post-order entry inquiry/change, delivery, post-delivery reports of any shipment-related problems, billing, postbilling discrepancies, and payment-related issues. The audit can help determine the effectiveness of the communications.

Management interviews are an important source of information. Interviews should be conducted with managers responsible for all logistics activities and activities with which logistics interacts, such as accounting/finance, sales/marketing and production. The interviews should examine:

- definition of responsibilities
- size and organizational structure
- decision-making authority and process
- performance measurements and results
- definition of customer service
- management's perception of how customers define customer service
- company plans to alter or improve customer service

- intrafunctional communications
- interfunctional communications
- communications with key contacts such as consumers, customers transportation carriers and suppliers

In addition, management should give its assessment of customer service measurement and reporting. This should include not only assessment of how current systems measure performance, but also how the firm interfaces with customers on service-related issues.

Identifying Potential Solutions

The external service audit enables management to identify problems with the firm's customer service and marketing strategies. Used in combination with the internal service audit, it may help management adjust these strategies and vary them by segment in order to increase profitability. But if management wants to use such information to develop customer service and marketing strategies for optimal profitability, it must use these data to benchmark against competitors.

The most meaningful competitive benchmarking occurs when customer evaluations of competitors' performance are compared with each other and with customers' evaluations of the importance of supplier attributes. Once management has used this type of analysis to determine opportunities for gaining a competitive advantage, every effort should be made to identify best practice; that is, the most cost-effective use of technology and systems regardless of the industry in which it has been successfully implemented. Non-competitors are much more likely to share their knowledge and, through such contacts, it is possible to uncover potential opportunities.

A methodology for competitive benchmarking can be demonstrated from the data contained in Table 2.3. The analysis involves a comparison of the performance of the major manufacturers in the office furniture industry. The first step is to generate a table with evaluations of the level of importance for each of the variables and the performance evaluations of all firms within the industry. The next step is to compare the importance score of each service attribute to customer evaluations of each manufacturer's performance.

Table 2.3 shows that one of the most important attributes identified by customers, 'ability to meet promised delivery date', received a score of 6.4 (out of 7.0) in overall importance. Manufacturer 1's perceived performance of 5.9 is significantly less than the mean importance score of 6.4 as well as the perceived performance of manufacturer 4 (6.6). Therefore, manufacturer 1 must improve its performance to meet customer requirements and achieve competitive parity.

The variable 'advance notice on shipping delays', rated 6.1 in overall importance, presents a different situation. None of the manufacturers was perceived to be meeting customer expectations. Therefore, if manufacturer 1 wished to improve its performance in this area, it could use this as a source of performance differentiation compared with that of its competitors, and perhaps gain a competitive advantage.

On the other hand, the variable 'free inward WATS [Wide-Area Telecommunications Service] telephone lines for placing orders with manufacturers' was ranked as 43rd in overall importance. No manufacturer received a high evaluation for its performance on this service. This indicated that customers did not perceive the advantages of this attribute because the service was not presently available from any manufacturer. If a competitor were to change its order entry procedures to allow customers the ability to telephone their orders without cost, then two things could happen: (1) customers would likely change their opinion of the advantages of this capability and consequently increase their perceptions of the importance of free WATS service; (2) the supplier that first introduced this service could achieve a definite, long-term competitive advantage.

Rank	Variable number	Variable description	Overall importance: all dealers		Mfr 1		Mfr 2		Mfr 3		Mfr 4		Mfr 5		Mfr 6	
			Mean	SD	Mean	SD	Mean	SD	Mean	SD	Mean	SD	Mean	SD	Mean	SD
1	9	Ability of manufacturer to meet promised delivery date (on-time shipments)	6.4	0.8	5.9	1.0	4.1	1.6	4.7	1.6	6.6	0.6	3.7	1.8	3.3	1.6
2	39	Accuracy in filling orders (correct product is shipped)	6.4	0.8	5.6	1.1	4.7	1.4	5.0	1.3	5.8	1.1	5.1	1.2	4.4	1.5
3	90	Competitiveness of price	6.3	1.0	5.1	1.2	4.9	1.4	4.5	1.5	5.4	1.3	4.4	1.5	3.6	1.8
4	40	Advance notice on shipping delays	6.1	0.9	4.6	1.9	3.0	1.6	3.7	1.7	5.1	1.7	3.0	1.7	3.1	1.7
5	94	Special pricing discounts available on contract/project quotes	6.1	1.1	5.4	1.3	4.0	1.7	4.1	1.6	6.0	1.2	4.7	1.5	4.5	1.8
6	3	Overall manufacturing and design quality of product relative to the price range involved	6.0	0.9	6.0	1.0	5.3	1.3	5.1	1.2	6.5	0.8	5.2	1.3	4.8	1.5
7	16	Updated and current price data, specifications, and promotion materials provided by manufacturer	6.0	0.9	5.7	1.3	4.1	1.5	4.8	1.4	6.3	0.9	4.9	1.7	4.3	1.9
8	47	Timely response to requests for assistance from manufacturer's sales representative	6.0	0.9	5.2	1.7	4.6	1.6	4.4	1.6	5.4	1.6	4.2	2.0	4.3	1.7
9	14	Order cycle consistency (small variability is promised versus actual delivery; that is, vendor consistently meets expected date)	6.0	0.9	5.8	1.0	4.1	1.5	4.8	1.4	6.3	0.9	3.6	1.7	4.4	1.7
10	4b	Length of promised order cycle (lead times [from order submission to delivery] for base line/in-stock ['quick ship'] product	6.0	1.0	6.1	1.1	4.5	1.4	4.9	1.5	6.2	1.1	4.3	1.7	3.7	2.0

TABLE 2.3 Importance and Performance of Office Furniture Manufacturers on Selected Customer Service Attributes

Rank	Variable number	Variable description	Overall importance: all dealers		Mfr 1		Mfr 2		Mfr 3		Mfr 4		Mfr 5		Mfr 6	
			Mean	SD	Mean	SD	Mean	SD	Mean	SD	Mean	SD	Mean	SD	Mean	SD
11	54	Accuracy of manufacturer in forecasting and committing to estimated shipping dates on contract/project orders	6.0	1.0	5.5	1.2	4.0	1.6	4.3	1.4	6.3	1.1	3.8	1.7	3.5	1.6
12	49a	Completeness of order (% of line items eventually shipped complete) – made-to-order product (contract orders)	6.0	1.0	5.5	1.2	4.3	1.2	4.7	1.3	6.0	1.1	4.4	1.4	4.0	1.6
43	45	Free WATS line provided for entering orders with manufacturer	5.3	1.5	3.6	2.5	4.8	2.0	3.4	2.6	3.5	2.6	2.0	1.5	3.8	1.9
50	33a	Price range of product line offering (e.g. low, medium, high price levels) for major vendor	5.0	1.3	4.4	1.5	4.6	1.6	5.1	1.5	5.2	1.4	4.3	1.6	3.9	1.6
101	77	Store layout planning assistance from manufacturer	2.9	1.6	4.2	1.7	3.0	1.5	3.4	1.6	4.7	1.6	3.0	1.4	3.4	1.2

Dealer evaluations of manufacturers

TABLE 2.3 Continued

Note: mean (average score) based on a scale of 1 (not important) through 7 (very important).

Source: adapted from Jay U. Sterling and Douglas M. Lambert, 'Customer Service Research: Past, Present and Future', *International Journal of Physical Distribution and Materials Management* 19, no. 2 (1989), p. 19.

Indeed, this is precisely what happened in the office furniture industry. One of the major manufacturers implemented an online, interactive order entry system utilizing free inward WATS service. Within three years, this capability became the norm for all of the major firms in the industry. Therefore, when looking for ways to improve customer service, it is equally critical to look at both important and relatively unimportant services because conditions change over time.

This highlights the notion that customers may not really know what they want because it has never been offered to them. By improving on performance of current parameters, companies are sentenced to 'keeping up with the competition'. Yet, history shows that the real winners are those who see an opportunity first and stake out a leadership position, taking on the risk of investing in the future before it arrives.[21] It is the unexploited, unserved market that holds real growth opportunities.

Establishing Customer Service Levels

The final steps in the audit process are the actual establishment of service performance standards and the ongoing measurement of performance. Management must set target service levels for segments based on factors such as the type of customer, geographic area, channel of distribution and product line. It must communicate this information to all employees responsible for implementing service policies, while also developing compensation programmes that encourage employees to achieve the firm's customer service objectives. Formal reports that document performance are a necessity.

Finally, management must repeat the process periodically to ensure that the firm's customer service policies and programmes reflect current customer needs. The collection of customer information over time is the most useful element in guiding overall corporate strategy and the specific strategies of the various functional areas within the firm.

Developing and Reporting Customer Service Standards

Once management has determined which elements of customer service are most important, it must develop standards of performance. These standards of performance are also known as key performance indicators (KPIs) if they are part of a firm's core business offerings and competitive advantage. Designated employees should regularly report results to the appropriate levels of management. Customer service performance can be measured and controlled by:

- establishing quantitative standards of performance for each service element
- measuring actual performance for each service element
- analysing variance between actual services provided and the standard
- taking corrective action as needed to bring actual performance into line.[22]

Customer cooperation is essential for the company to obtain information about speed, dependability and condition of the delivered product. To be effective, customers must be convinced that service measurement and monitoring will help improve future service.

Figure 2.5 contains a number of possible measures of service performance. These measures are by no means exhaustive and are also related to the transactional elements; they do not include many measures related to ongoing relationships between firms and their customers.[23] The emphasis any firm places on individual elements must be based on what customers believe is important. Service elements such as inventory availability, meeting delivery dates, order status, order tracing and back-order status require good communication between firms and their customers.

Order processing offers significant potential for improving customer service because many companies have not kept pace with technological developments in that field. Consider the possibilities for improved communications if customers can phone their orders to customer service

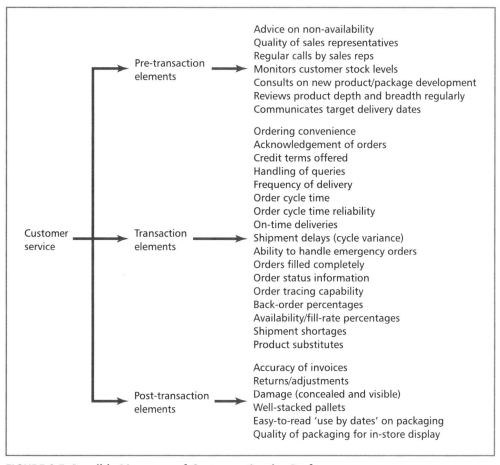

FIGURE 2.5 Possible Measures of Customer Service Performance

representatives who input orders on their own computer terminals. Immediate information on inventory availability can be provided and product substitution can be arranged when a stockout occurs. Customers can also be given target delivery dates for their orders.

Figure 2.6 gives examples of customer service standards. The standards chosen should be those that best reflect what customers actually require, rather than what management thinks they need. Designated employees should measure and compare service performance to the standard, and report this information to management on a regular and timely basis.

The firm's order-processing and accounting information systems can provide much of the information necessary for developing a customer–product contribution matrix and meaningful customer service management reports. We will discuss some of these important interfaces in Chapter 3.

Impediments to an Effective Customer Service Strategy

Many companies have ineffective or inconsistent customer service strategies, policies or programmes. Sometimes even the best of firms may have difficulty in overcoming the various barriers or impediments that can hinder the implementation of successful customer service processes.

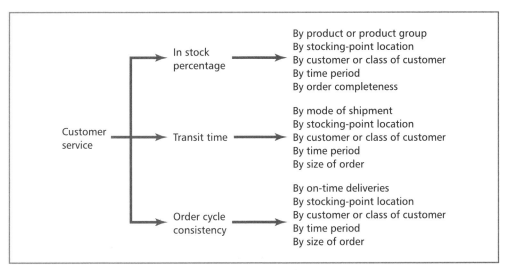

FIGURE 2.6 Examples of Customer Service Standards

Failing to target specific market segments based on the services they require can be a costly mistake. Management sometimes hesitates to offer different levels of customer service for fear of violating laws on preferential treatment. Service differentials are often viewed much like price differentials that must be cost justified. However, most firms do not have the necessary cost information to do so. Nevertheless, management can segment markets based on customers' evaluations of the importance of marketing services and can obtain the necessary financial data to determine the costs of serving such markets through a variety of research techniques.

Salespeople can create unrealistic customer service expectations by promising faster delivery of orders to 'make the sale'. But most customers value reliability and consistency in filling orders more than speed of delivery. Consequently, attempting to decrease the order cycle on an ad hoc basis typically increases transportation costs resulting from the need to expedite shipments. Order-assembly costs can rise because of the disruption of normal work flows that occur in 'rush' situations. Also, the so-called domino effect might occur. When salespeople override customer service policies on shipping dates, lead times, shipping points, modes of transportation and units of sale, they disrupt the orders of other customers and cause an increase in logistics costs throughout the system.[24]

A firm's customer service standards and performance expectations are affected substantially by the competitive environment and perceived traditional industry practices. Consequently, it is vital that management understands industry norms, expectations and the costs required to provide high levels of customer service.

Evidence suggests, however, that many firms do not measure the cost-effectiveness of service levels and have no accurate way of determining competitive service levels. Information is fed back into the company through a sales organization that is frequently concerned with raising service levels or through industry anecdotes and outraged customers. The net result of this information feedback is that firms may overreact to imprecise cues from the marketplace or even from within their own firms.[25]

Considering the vast sums of money firms spend on research and development and advertising, it makes little sense for a company not to adequately research the levels of customer service necessary for profitable long-range business development.

Global Customer Service Issues

The global perspective focuses on seeking common market demands worldwide, rather than dividing up world markets and treating them as separate entities with very different product needs.[26] On the other hand, different parts of the world have different service needs related to information availability, order completeness, expected lead times, and so on. The local congestion, infrastructure, communications and time differences may make it impossible to achieve high levels of customer service. In addition, management styles in different international or global markets may be different than those prevalent in the firm's 'home' environment.[27] The service provided by market should match local customer needs and expectations to the greatest degree possible.

For example, Coca-Cola provides very different types of service in Japan than in Europe. Coca-Cola delivery drivers in Japan focus on providing merchandising in supermarkets, help in processing bills in small 'mom and pop' operations, and respond to signals from communication systems in vending machines so that time is not wasted delivering to full machines.[28] This creates the most efficient and effective customer service policy, rather than simply duplicating domestic patterns worldwide. The latter strategy could be both ineffective and expensive.

 GLOBAL

Surviving and Thriving in an Era of Unprecedented Change: A Japanese Case Study

Japan's wholesaling industry is under extreme pressure, pushing companies like Ryoshoku Limited, one of the country's largest food wholesalers with over 40 million transactions per month, to reinvent themselves. Economic conditions are driving retail price discounting, large-format retail stores are emerging, discount chains are growing more popular, and market power is shifting to consumers and retailers. These factors are turning wholesalers' attention to meeting demands for lower supply chain costs.

Further complicating the picture, large retailers have increased the amount of direct trade between themselves and large manufacturers, in some cases squeezing out the wholesaler altogether. Moreover, large retailers have invested in point-of-sale data capture and analysis capabilities, reducing their reliance on wholesalers for these services. To meet these challenges, and to maintain and increase market share, leading wholesalers are strengthening and expanding value-added services, particularly emphasizing information services, retail support and logistics.

Faced with these changes and prospects for continued distribution system restructuring in Japan, Ryoshoku has positioned itself to thrive in these uncertain times. Aiming to bridge the gap between manufacturers and retailers, it developed market-based transfer/consolidation centres to provide its retail customers with frequent and efficient full-truckload, multiple-vendor deliveries to stores. This combined delivery service was designed to offer the same flexible small-lot, just-in-time service demanded by stores without all the inefficiencies, let alone the traffic congestion and pollution problems, of unconsolidated deliveries. Ryoshoku invested in a distribution/logistics information network called TOMAS from Fujitsu, and nine regional distribution centres to provide flexible unit picking and order assembly services.

To provide these sophisticated and constantly evolving logistics services, Ryoshoku set up an independent logistics organization that has equal status with the marketing, sales and information systems groups. Leveraging the company's strong culture of innovation and technology, the new logistics organization also pursued efficient consumer response (ECR)-based partnerships along its supply chain and expanded information systems links to retailer storefronts and suppliers' operations.

Question: **Would outsourcing have been a better alternative for Ryoshoku rather than doing its logistics in-house or with its new logistics subsidiary?**

Sources: adapted from David Frentzel, 'Surviving and Thriving in an Era of Unprecedented Change: Case Studies of Four Agile Japanese Companies', *Logistics!*, a publication of Mercer Management Consulting (Fall 1995), pp. 11–12; and case study, 'Developing an Informational Management System as a Competitive Advantage', http://www.fujitsu.com/ (2005).

Improving Customer Service Performance

The levels of customer service a firm achieves can often be improved through one or more of the following actions: (1) thoroughly researching customer needs, (2) setting service levels that make realistic trade-offs between revenues and expenses, (3) making use of the latest technology in order-processing systems, and (4) measuring and evaluating the performance of individual logistics activities.

An effective customer service strategy *must* be based on an understanding of how customers define service. The internal and external customer service audits previously discussed were utilized to obtain customer inputs into service strategies, plans and programmes. As described above, Coca-Cola has identified the distinctive logistics needs of its customers.

Once the firm has determined its customers' view of service, management must select a customer service strategy that advances the firm's objectives for long-term profits, return on investment or other relevant measures of performance. The optimum level of customer service is the one that obtains and retains the most profitable customers.

Order-processing systems can have a major impact on customer service levels and perceptions (see Chapter 3). The primary benefit of automating order-processing systems is to reduce the order cycle time. Given that most customers prefer a consistent delivery cycle to a shorter one, it is usually unnecessary – even unwise – to reduce the order cycle time for customers. But by using the additional time internally for planning, the company can achieve savings in transportation, warehousing, inventory carrying costs, production planning and purchasing.

Automation improves customer service by providing the following benefits to the customer:
- better product availability
- more accurate invoices
- the ability to lower safety stock levels and their associated inventory carrying costs
- improved access to information on order status.

In short, automated order-processing systems enhance the firm's ability to perform all of the transaction and post-transaction elements of customer service.

Finally, the development of an effective customer service programme requires the establishment of customer service standards that do the following:
- reflect the customer's point of view
- provide an operational and objective measure of service performance
- provide management with cues for corrective action.[29]

Management should also measure and evaluate the impact of individual logistics activities – transportation, warehousing, inventory management, production planning, purchasing and order processing – on customer service. Designated employees should report achievements regularly to the appropriate levels of management. Management should compare actual performance to standards and take corrective action when performance is inadequate. For management to be successful and efficient, a firm needs timely information. It is also necessary to hold individuals accountable for their performance because information alone does not guarantee improved decision-making.

The success of a firm is no longer based exclusively on selling products; instead it is the value-added services provided that can create a differential and sustainable competitive advantage. Logistics can be an important source of such service-based advantage.

SUMMARY

This chapter opened with a definition of customer service. Although the importance of the individual elements of customer service varies from company to company, we reviewed the common elements that are of concern to most companies. We also saw the necessity for a customer service strategy consistent with corporate and marketing strategies. The successful implementation of the integrated logistics management concept depends on management's knowledge of the costs associated with different system designs, and of the relationship between system design and customer service levels. We saw how management can obtain better knowledge of the costs and revenues associated with different levels of customer service, and how it can implement cost/service trade-offs.

The customer service audit is a method of determining the existing service levels, determining how performance is measured and reported, and appraising the impact of changes in customer service policy. Firms should conduct both internal and external service audits. Surveys are one means of finding out what management and customers view as important aspects of customer service.

Although customer service may represent the best opportunity for a firm to achieve a sustainable competitive advantage, many firms implement customer service strategies that are simply duplicates of those implemented by their major competitors. The audit framework represented in this chapter can be used by management to collect and analyse customer and competitive information.

We saw that there are some common roadblocks to an effective customer service strategy, as well as some ways to improve performance. In the next chapter, we will present the influence of information technology on the efficiency and effectiveness of the logistics function.

CREATIVE SOLUTIONS

Zara Designs its Whole Operation to Fit the Customer

In the last 30 years fashion has changed from an elite accessory of the super-rich to a mass-market product. Since the mid-1990s the department stores that traditionally dominated this broader market have started to lose ground to specialist clothing chains offering the latest designs at competitive prices. One of the most successful specialist players is Zara, a subsidiary of the Spanish Inditex Group. Since 1975 the Inditex Group has become an international fashion business with 2082 stores in 56 countries and will open its first Zara store in Indonesia in 2005. It sells 90 million garments a year and turnover in 2003 was just under €4.6 billion. The fact that this position has been achieved without a formal marketing department and with minimal advertising spend has prompted reams of journalistic commentary and academic study. Zara's success is based on its unique customer focus, or what theorists term 'market orientation'. The customer and whole market focus permeates every level of the Zara operation. Typical company statements are 'Our customers are the basis and the reason of our group's existence'; 'The customer is our inspiration' and 'Our own and our suppliers' production will be able to focus on trend changes happening inside each season.' It is this last statement that reveals what market orientation really means for day-to-day working.

The traditional view of fashion products is that they are durable articles with a seasonal sell-by period. The Zara concept views fashion products as disposable with a maximum three- to four-week sell-by period. Zara's efforts are therefore focused on reducing the time between design and sale, which means that its production cycle is entirely different from fashion-sector norms. The fashion sector traditionally moves along a clearly segmented time line from trade fairs through design and shows to production and sale. For example, goods sold in quarter 4 will have been designed in quarter 1 and produced in quarter 2. In contrast, Zara views design and production as processes that are 'live' throughout the entire seasonal cycle as the company responds to market information. At the same time, the focus on market changes means that 85 per cent of products are manufactured in the season they are sold.

The Zara concept means that the company is able to work to incredibly short lead-times. The average time from design to delivery is just two weeks and new stock is delivered to all Zara's stores at least twice a week. Products responding to changing market trends are brought into stores in a continuous stream. The stores are therefore always fully stocked with successful lines and there is little chance of Zara being left with large stocks of failed lines. Buying and producing lines late in the traditional cycle is also cost-effective because suppliers will offer low prices to clear materials before the season ends. The Zara concept also means that customers behave differently in its stores. The continuous stream of new products means that customers will keep coming back throughout the season to see what's new. At the same time, the short shelf life of Zara products means that customers know they will have to buy them straight away or never see them again.

Zara brings an average of 10,000 products to market each year. Information gathering drives this phenomenal offering. The design team works throughout the season, studying everything from what clothes are worn in hit TV series to how clubbers dress. 'Product-shop teams' check product sales and store trends every day, and this information is cross-checked against stores' twice-weekly orders. The information is fed to purchasing, design and production functions. Unsuccessful products are taken off the market immediately; and stores can place only small orders to avoid building up stocks.

Zara has repeated its resounding success in country after country and has never closed a single store. Industry insiders say that wherever Zara is located its sales are the leading sales per square metre. The secret of Zara's success has been to ignore traditional fashion-sector behaviours yet be customer- and market-responsive with its service and logistics activities.

Question: Discuss Zara's efforts and success in the context of quick response (QR), introduced in Chapter 1.

Source: adapted from 'Zara Creates a Ready to Wear Business', *Strategic Direction* 19, no. 11 (2003), pp. 24–6.

Fundamentals of Logistics Management

KEY TERMS

	Page
Customer service	35
Customer satisfaction	35
Reverse logistics	40
Stockkeeping unit (SKU)	44
Electronic data interchange (EDI)	44
Pareto's law or 80/20 rule	46
Less-than-truckload (LTL)	49

A full Glossary can be found at the back of the book.

QUESTIONS AND PROBLEMS

1 Customer service can be defined as an activity, a performance measure or a corporate philosophy. What are the advantages and disadvantages of each of these types of definition? How would you define customer service?
2 Explain the importance of the pre-transaction, transaction and post-transaction elements of customer service.
3 Explain why customer service should be integrated with other components of the marketing mix when management develops the firm's marketing strategy.
4 Explain how ABC analysis can be used to improve the efficiency of the customer service activity.
5 Why is the customer service audit important when establishing a firm's customer service strategy?
6 Why does automation of the order-processing system represent such an attractive opportunity for improving customer service? How is this service improvement accomplished?
7 What are some ways that management can improve the firm's customer service performance?
8 Why is it important to use pre-transaction, transaction and post-transaction customer service elements to identify and develop customer service measures? Discuss specific examples of measures in each category.

THE LOGISTICS CHALLENGE!

PROBLEM: ONLINE GROCERY SHOPPING FULFILMENT

Customer service effectiveness by online grocery retailers has recently been challenged. Research has uncovered four fundamental difficulties in online retailing: limited online sales potential; high cost of delivery; a selection-variety trade-off; and existing, entrenched competition. In UK online grocery retailing, the selection-variety trade-off is the only major difficulty in this marketplace.

However, the UK's biggest supermarkets were under investigation in late 2004 after claims that they overcharged online customers and failed to deliver the requested goods. The UK Office of Fair Trading (OFT) scrutinized British retailers, including Sainsbury's, Tesco and Asda after customers complained that they receive an inferior service from supermarket websites.

One of the most common grievances was the delivery of substitute items. Shoppers reported that groceries ordered were often out of stock and supermarkets sent a replacement, which was often inadequate. Shoppers also complained that they are sometimes overcharged. The actual price of an item could be higher than the advertised 'guide price' and based on what the local shop is charging, it was alleged.

In addition, items were missed off delivery orders because they were not available, but consumers were still required to pay the full delivery costs, which can be up to £5.99. Customers also complained that supermarkets used their websites to offload food close to its sell-by or expiry date. Non-food items, such as books, DVDs, clothes and electrical goods, could cost more online than in shops.

A spokeswoman for the OFT said, 'We have held confidential discussions with supermarkets, but this is an ongoing investigation.' The OFT refused to name the supermarkets but has been in touch with Tesco, Sainsbury's and Asda.

A spokesman for Asda said, 'We have had one meeting with the OFT, but unlike other online stores we sell our goods at one price.'

Lucy Neville Rolfe, group corporate affairs director and company secretary at Tesco, said: 'With a Tesco.com grocery shop your food is picked off the shelves of your local store, so whether it is a book you are buying or a carton of milk you will be paying the same price as you would in store. The OFT contacted the industry several weeks ago on a confidential basis on a fairly technical issue and we will be responding to them.'

A survey for Which?, formerly the Consumers' Association, found that of 1500 online shoppers, nearly nine out of ten of those using the Sainsbury's website said that their orders 'generally' contained replacement items. A spokeswoman for Sainsbury's said, 'We are in consultation with the OFT . . . It would be inappropriate to comment further.'

Part of the problem may be the way online customer distribution is structured. It is usually store-based. Orders are taken centrally, but then passed on to individual stores, where they are met from the shop's own shelves. Inevitably, not all 30,000 or so lines are available, especially in smaller stores.

By contrast, the Ocado system used for Waitrose's online grocery shoppers is warehouse-based. All orders are fulfilled from a single large warehouse in Hatfield near London. Product availability is much higher and customers get the products they order. The scope for abuse is also reduced. Other supermarkets could consider following the Ocado approach. However, based on the experiences of failed online grocery retailers who had warehouse-based fulfilment, such as Webvan in the USA, the Ocado approach is not a perfect solution either.

Could the supermarkets improve their online customer service and also their image by adopting one of the five methods for establishing customer service strategies discussed in the chapter? Which one(s) would you recommend and why?

What Is Your Solution?

Source: adapted from Helen Nugent, 'Online Shoppers Complain of Second-rate Service', *The Times*, (13 December 2004), p. 7.

SUGGESTED READING

BOOKS

Christopher, Martin, *Logistics and Supply Chain Management: Creating Value-Adding Networks*, 3rd edn. Harlow, UK: FT Prentice Hall, 2005.

Christopher, Martin, Adrian Payne and David Ballantyne, *Relationship Marketing: Creating Stakeholder Value*. London, UK: Butterworth-Heinemann, 2002.

Christopher, Martin and Helen Peck, *Marketing Logistics*, 2nd edn. London, UK: Butterworth-Heinemann, 2003.

Grant, David B., 'A Quarter-Century of Logistics Customer Service Research: Where Are We Now?', in Edward Sweeney, John Mee, Bernd Huber, Brian Fynes and Pietro Evangelista (eds), *Enhancing Competitive Advantage through Supply Chain Innovation – Proceedings of the 9th Annual Logistics Research Network Conference*. Dublin: National Institute for Transport and Logistics (2004), pp. 201–13.

La Londe, Bernard J. and Martha C. Cooper, *Partnerships in Providing Customer Service: A Third-Party Perspective*. Oak Brook, IL: Council of Logistics Management, 1989.

La Londe, Bernard J., Martha C. Cooper and Thomas G. Noordewier, *Customer Service: A Management Perspective*. Oak Brook, IL: Council of Logistics Management, 1988.

JOURNALS

Caplice, Chris and Yosef Sheffi, 'A Review and Evaluation of Logistics Metrics', *The International Journal of Logistics Management* 5, no. 2 (1994), pp. 11–28.

Caplice, Chris and Yosef Sheffi, 'A Review and Evaluation of Logistics Performance Measurement Systems', *The International Journal of Logistics Management* 6, no. 1 (1995), pp. 61–74.

Chow, Garland, Trevor D. Heaver and Lennart E. Henriksson, 'Logistics Performance: Definition and Measurement', *International Journal of Physical Distribution and Logistics Management* 24, no. 1 (1994), pp. 17–28.

'A Compendium of Research in Customer Service', *International Journal of Physical Distribution and Logistics Management* 24, no. 4 (1994), pp. 1–68.

Fuller, Joseph, James O'Conor and Richard Rawlinson, 'Tailored Logistics: The Next Advantage', *Harvard Business Review* 94, no. 3 (May–June 1994), pp. 87–94.

Grant, David B., 'UK and US Management Styles in Logistics: Different Strokes for Different Folks?', *International Journal of Logistics: Research and Applications* 7, no. 3 (2004), pp. 181–97.

Harrington, Lisa, 'Logistics Unlocks Customer Satisfaction', *Transportation and Distribution* 36, no. 5 (May 1995), pp. 41–4.

Mentzer, John T., Daniel J. Flint and John L. Kent, 'Developing a Logistics Service Quality Scale', *Journal of Business Logistics* 20, no. 1 (1999), pp. 9–32.

Mentzer, John T., Daniel J. Flint and G. Tomas M. Hult, 'Logistics Service Quality as a Segment-Customized Process', *Journal of Marketing* 65 (October 2001), pp. 82–104.

Parasuraman, Ari, Valarie A. Zeithaml and Leonard L. Berry, 'A Conceptual Model of Service Quality and its Implications for Future Research', *Journal of Marketing* 49 (Fall 1985), pp. 41–50.

Sharma, Arun, Dhruv Grewal and Michael Levy, 'The Customer Satisfaction/Logistics Interface', *Journal of Business Logistics* 16, no. 2 (1995), pp. 1–21.

Sterling, Jay U., 'Establishing Customer Service Strategies within the Marketing Mix', *Journal of Business Logistics* 8, no. 1 (1987), pp. 1–30.

Sterling, Jay U. and Douglas M. Lambert, 'Customer Service Research: Past, Present and Future', *International Journal of Physical Distribution and Materials Management* 19, no. 2 (1989), pp. 3–23.

REFERENCES

[1] Toby B. Gooley, 'How Logistics Drive Customer Service', *Traffic Management* 35, no. 1 (January 1996), p. 46.

[2] Ibid.

[3] See, for example, David Jobber, *Principles and Practice of Marketing*, 4th edn. Maidenhead, UK: McGraw-Hill, 2004.

[4] Bernard J. La Londe and Paul H. Zinszer, *Customer Service: Meaning and Measurement*. Chicago: National Council of Physical Distribution Management, 1976, pp. 156–9.

[5] For an overview of how customer service has been viewed during the previous 25 years, see 'A Compendium of Research in Customer Service', *International Journal of Physical Distribution and Logistics Management* 24, no. 4 (1994), pp. 1–68; and David B. Grant, 'A Quarter-Century of Logistics Customer Service Research: Where Are We Now?', in Edward Sweeney, John Mee, Bernd Huber, Brian Fynes and Pietro Evangelista (eds), *Enhancing Competitive Advantage through Supply Chain Innovation – Proceedings of the 9th Annual Logistics Research Network Conference*. Dublin: National Institute for Transport and Logistics, 2004, pp. 201–13.

[6] Bernard J. La Londe, Martha C. Cooper and Thomas G. Noordewier, *Customer Service: A Management Perspective*. Chicago: Council of Logistics Management, 1988, p. 5.

[7] Marriott International, Inc., http://marriott.com/, (2005).

[8] KLM's Airline Passenger Service Commitment, http://www.klm.com/, (2005).

[9] Nokia, http://www.nokia.com/, (2005).

[10] La Londe and Zinszer, *Customer Service*, pp. 272–82.

[11] David B. Grant, 'UK and US Management Styles in Logistics: Different Strokes for Different Folks?', *The International Journal of Logistics: Research and Applications* 7, no. 3 (2004), pp. 181–97.

[12] Lisa M. Ellram, *Total Cost Modeling in Purchasing*. Tempe, AZ: Center for Advanced Purchasing Studies, 1994, p.10.

[13] Joseph Fuller, James O'Conor and Richard Rawlinson, 'Tailored Logistics: The Next Advantage', *Harvard Business Review* 94, no. 3 (May–June 1994), pp. 87–94.

[14] See Larry W. Emmelhainz, Margaret A. Emmelhainz and James R. Stock, 'Logistics Implications of Retail Stockouts', *Journal of Business Logistics* 12, no. 2 (1991), pp. 129–42; and Daniel Corsten and Thomas Gruen, 'Desperately Seeking Shelf Availability: An Examination of the Extent, the Causes, and the Efforts to Address Retail Out-of-stocks', *International Journal of Retail and Distribution Management* 31, no. 12 (2003), pp. 605–17.

[15] Corsten and Gruen, 'Desperately Seeking Shelf Availability', pp. 605–6.

[16] Jobber, *Principles and Practice of Marketing*, pp. 792–6.

[17] Ari Parasuraman, Valarie A. Zeithaml and Leonard L. Berry, 'A Conceptual Model of Service Quality and its Implications for Future Research', *Journal of Marketing* 49 (Fall 1985), pp. 41–50.

[18] Jay U. Sterling and Douglas M. Lambert, 'Establishing Customer Service Strategies within the Marketing Mix', *Journal of Business Logistics* 8, no. 1 (1987), pp. 1–30.

[19] Ibid., p. 22.

[20] La Londe and Zinszer, *Customer Service*, p. 168.

[21] Gary Hamel and C.K. Prahalad, 'Seeing the Future First', *Fortune*, 5 September 1994, pp. 64–70.

[22] Fuller, O'Conor and Rawlinson, 'Tailored Logistics' pp. 87–94.

[23] Grant, 'UK and US Management Styles in Logistics: Different Strokes for Different Folks?', pp. 181–97.

[24] Douglas M. Lambert, James R. Stock and Jay U. Sterling, 'A Gap Analysis of Buyer and Seller Perceptions of the Importance of Marketing Mix Attributes', in *Enhancing Knowledge Development in Marketing*, ed. William Bearden *et al*. Chicago: American Marketing Association, 1990, p. 208.

[25] La Londe, Cooper and Noordewier, *Customer Service*, p. 29.

[26] Martin Christopher, 'Customer Service Strategies for International Markets', *Annual Proceedings of the Council of Logistics Management*. Oak Brook, IL: Council of Logistics Management, 1989, p. 327.

[27] Grant, 'UK and US Management Styles in Logistics: Different Strokes for Different Folks?', pp. 181–97.

[28] Fuller, O'Conor and Rawlinson, 'Tailored Logistics', p. 88.

[29] La Londe and Zinszer, *Customer Service*, p. 180.

CHAPTER 3
LOGISTICS INFORMATION SYSTEMS AND TECHNOLOGY

CHAPTER OBJECTIVES

■ To provide an overview of the ways that computers can be used in logistics operations

■ To show how the order-processing system can influence the performance of logistics activities

■ To show how the order-processing system can form the core of logistics information systems at both a tactical and strategic level, supporting customer service goals

■ To discuss the role of information technology in supporting time-based competition

■ To identify uses of advanced information technologies, such as decision support systems (DSSs), artificial intelligence (AI) and expert systems (ES) in logistics

Computer and information technology has been utilized to support logistics for many years. It grew rapidly with the introduction of microcomputers in the early 1980s. Information technology is seen as the key factor that will affect the growth and development of logistics.

The order-processing system is the nerve centre of the logistics system. A customer order serves as the communications message that sets the logistics process in motion. The speed and quality of the information flows have a direct impact on the cost and efficiency of the entire operation. Slow and erratic communications can lead to lost customers or excessive transportation, inventory and warehousing costs, as well as possible manufacturing inefficiencies caused by frequent production line changes. The order-processing and information system forms the foundation for the logistics and corporate management information systems. It is an area that offers considerable potential for improving logistics performance.

Organizations of all types are utilizing computers to support logistics activities. This is especially true for companies thought to be on the 'leading edge' – that is, leaders in their industry. Such firms are heavy users of computers in order entry, order processing, finished goods inventory control, performance measurement, freight audit/payment and warehousing. Several studies of logistics trends in Europe have cited logistics information systems and electronic commerce (e-commerce) as growth areas and keys to competitiveness.[1]

Going beyond 'transaction processing and tracking', **decision support systems (DSSs)** are computer-based and support the executive decision-making process. The DSS is an integrative system of subsystems that has the purpose of providing information to aid a decision-maker in making better choices than would otherwise be possible.

To support time-based competition, organizations are increasingly using information technologies as a source of competitive advantage. Systems such as quick response (QR), just-in-time (JIT), efficient consumer response (ECR) and customer relationship manage-

chapter 3 Logistics Information Systems and Technology

ment (CRM) are integrating a number of information-based technologies in an effort to reduce order cycle times, speed responsiveness and lower supply chain inventory.

In addition, more sophisticated applications of information technology, such as decision support systems, artificial intelligence and expert systems, are being used directly to support decision-making in logistics. This chapter will begin with the customer order cycle, which is at the heart of logistics information systems.

Customer Order Cycle

The **customer order cycle** includes all of the elapsed time from the customer's placement of the order to the receipt of the product in an acceptable condition and its placement in the customer's inventory. The typical order cycle consists of the following components: (1) order preparation and transmittal, (2) order receipt and order entry, (3) order processing, (4) warehousing picking and packing, (5) order transportation, and (6) customer delivery and unloading.

Figure 3.1 illustrates the flow associated with the order cycle. In this example taken from the customer's point of view, the total order cycle is 13 days. However, many manufacturers make the mistake of measuring and controlling only the portion of the order cycle that is *internal* to their firm. That is, they monitor only the elapsed time from receipt of the customer order until it is shipped. The shortcomings of this approach are obvious.

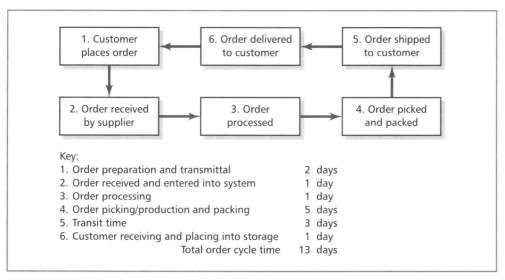

FIGURE 3.1 Total Order Cycle: A Customer's Perspective

In the example presented in Figure 3.1, the portion of the total order cycle that is internal to the manufacturer (steps 2, 3 and 4) amounts to only seven of the 13 days. This ratio is not unusual for companies that do not have an automated order entry and processing system. Improving the efficiency of the seven-day portion of the order cycle that is 'controlled' by the manufacturer may be costly compared to eliminating a day from the six days not directly under the manufacturer's control. For example, it may be possible to reduce transit time by as much as one day by monitoring carrier performance and switching business to carriers with faster and more consistent transit times.

However, a change in the method of order placement and order entry may have the potential for the most significant reduction in order cycle time. An advanced order-processing system could reduce the total order cycle by as much as two days. In addition, the improved information flows could enable management to execute the warehousing and transportation more efficiently, reducing the order cycle by another one or two days.

In Figure 3.1, we treated the performance of order cycle components as though no variability occurred. Figure 3.2 illustrates the variability that may occur for each component of the order cycle and for the total. For this illustration, we assume that each of the variable time patterns follows a normal statistical distribution. However, other statistical distributions may actually be experienced. In our example, the actual order cycle could range from a low of 4.5 days to as many as 21.5 days, with 13 days as the most likely length. Variability in order cycle time is costly to the manufacturer's customer; the customer must carry extra inventory to cover for possible delays or lose sales as a result of stockouts.

Return to the example in Figure 3.2. If the average order cycle time is 13 days but can be as long as 21.5 days, the customer must maintain additional inventory equivalent to approximately

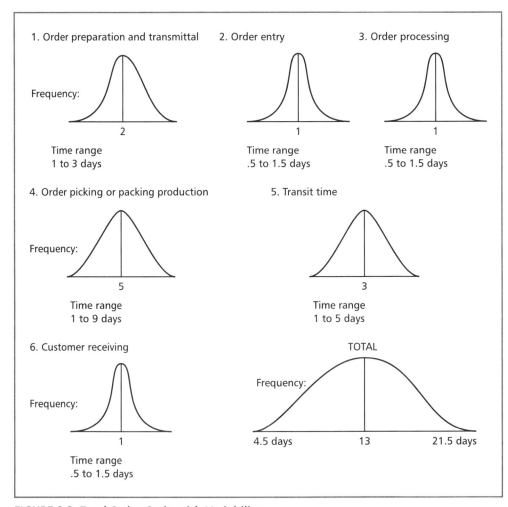

FIGURE 3.2 Total Order Cycle with Variability

eight days' sales just to cover variability in lead-time. If daily sales equal 20 units and the company's economic order quantity is 260 units (a 13-day supply) the average cycle stock is 130 units, one-half the order quantity. The additional inventory required to cover the order cycle variability of eight days is 160 units. Without even considering demand uncertainty, average inventory will increase from 130 units to 290 units because of the variability in the order cycle.

Which has the greatest impact on the customer's inventory: a five-day reduction in the order cycle or a five-day reduction in order cycle variability? Table 3.1 helps illustrate this scenario.

A. Situation 1: Base case

Daily sales = 20 units

Order 13-day supply of inventory (20 units × 13 days = 260 units)

$$\frac{260 \text{ order quantity}}{2} = 130 \text{ units average cycle stock}$$

20 units per day × 8 days order cycle variability = 160 units safety stock

Average inventory = Average cycle stock + Safety stock
= 290 units

B. Situation 2: Reduce order cycle by 5 days to 8 days

8 days × 20 units daily sales = 160 unit order quantity

$$\frac{160 \text{ order quantity}}{2} = 80 \text{ units average cycle stock}$$

20 units per day × 8 days order cycle variability = 160 units safety stock

Average inventory = Average cycle stock + Safety stock
= 240 units

C. Situation 3: Reduce safety stock by 5 days due to reduction in variability to 3 days

$$\frac{260 \text{ order quantity}}{2} = 130 \text{ units average cycle stock}$$

20 units per day × 3 days order cycle variability = 60 units safety stock

Average inventory = Average cycle stock + Average safety stock
= 190 units

TABLE 3.1 Illustration of the Impact of Reduction in Order Cycle and Order Variability on Average Inventory

If the customer continued to order the economic order quantity of 260 units, a five-day reduction in the order cycle would result in little or no change in inventories, as demonstrated in Table 3.1A. The customer simply waits five days longer before placing an order.

Table 3.1B shows that even if the customer orders 160 units every time instead of 260, making the average cycle stock 80 units instead of 130, safety stock of 160 units is still required to cover the eight days of variability. The result would be a reduction in total average inventory of 50 units, from 290 to 240 units.

However, Table 3.1C demonstrates that a five-day reduction in order cycle variability would reduce safety stock by 100 units and result in an average inventory of 190 units. This example should make clear why order cycle *consistency* is preferred over fast delivery. Gaining competitive advantage based on reduced order cycle time is fruitless without consistent performance. Calculation of safety stock is described in Chapter 5.

In the next section, we will examine how customer orders enter the order-processing function, and the typical path taken by a customer's orders.

How do Customer Orders Enter the Firm's Order-processing Function?

A customer may place an order in a number of ways. Historically, customers wrote orders by hand and gave them to salespeople, mailed them to the supplier or telephoned them to the manufacturer's order clerk, who then wrote it up. Today, one way for a customer to order is by telephone to a supplier's customer service representative, who is equipped with a computer terminal networked to the supplier's database, or to order by computer directly into the supplier's system.

This type of system allows the customer service representative to determine if the ordered products are available in inventory, and to deduct orders automatically from inventory so that items are not promised to another customer. This improves customer service because if there is a stockout on the item, the representative can inform the customer of product availability and perhaps arrange product substitution while the customer is still on the telephone. In addition, this type of system almost completely eliminates the first two days of the order cycle shown in Figure 3.1.

However, electronic methods, such as an electronic terminal with information transmitted by telephone lines, and computer-to-computer hookups such as electronic data interchange (EDI), are more commonplace today. These methods support the maximum speed and accuracy in order transmittal and order entry. Generally, rapid forms of order transmittal require an initial investment in equipment hardware and software. However, management can use the time saved in order transmittal to reduce inventories and realize opportunities in transportation consolidation, offsetting the investment.

There is a direct trade-off between inventory carrying costs and communications costs. In many channels of distribution, significant potential exists for using advanced order processing to improve logistics performance. However, the more sophisticated the communications system, the more vulnerable the company becomes to any internal or external communications malfunctions. With advanced order-processing systems and lower inventory levels, safety stocks are substantially reduced, leaving the customer with minimal protection against stockouts that result from any variability in the order cycle time. In addition, customers expect immediate information regarding availability and shipping, and become frustrated when systems go down.

The Path of a Customer's Order

When studying a firm's order-processing system, it is important to understand the information flow that begins when a customer places an order. Figure 3.3 represents one interpretation of the path that a customer's order might take. In the first step, shown at upper left, the customer recognizes the need for certain products and transmits an order to the supplier.

Once the order enters the order-processing system, various checks are made to determine if (1) the desired product is available in inventory in the quantities ordered, (2) the customer's credit is satisfactory to accept the order, and (3) the product is scheduled for production if not currently in inventory. If these activities are performed manually, a great amount of time may be required, which can slow down (i.e. lengthen) the order cycle. The norm is that these activities are performed by computer in a minimal amount of time; often these activities can be performed simultaneously with other order cycle activities. The inventory file is then updated, product is back-ordered if necessary, and production is issued a report showing the inventory balance.

Management can also use the information on daily sales as an input to its sales-forecasting package. Order processing next provides information to accounting for invoicing, acknowledge-

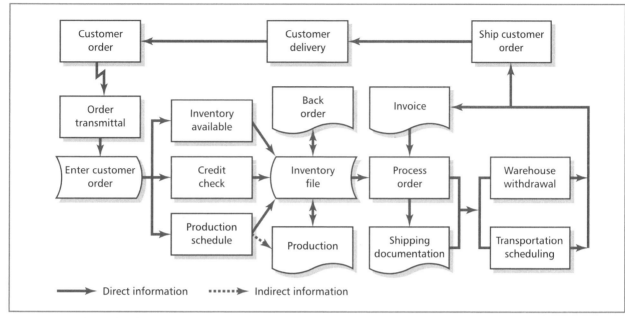

FIGURE 3.3 The Path of a Customer's Order

ment of the order to send to the customer, picking and packing instructions to enable warehouse withdrawal of the product, and shipping documentation. When the product has been pulled from warehouse inventory and transportation has been scheduled, accounting is notified so that invoicing may proceed. All of these processes can be automated seamlessly to reduce additional input of data, and avoid the errors, paper shuffling and non-value added of manual effort.

The primary function of the order-processing system is to provide a communication network that links the customer and the manufacturer. In general, greater inconsistency is associated with slower methods of order transmittal. Manual methods of order transmittal require more handling by individuals; consequently, there is greater chance of a communication error. Management can evaluate methods of order transmittal on the basis of speed, cost, consistency and accuracy. As shown in Table 3.2, order transmittal should be as direct as possible; orders transmitted electronically instead of manually minimize the risk of human error.

Level	Type of system	Speed	Cost to implement/ maintain	Consistency	Accuracy
1	Manual	Slow	Low	Poor	Low
2	Phone in to customer service rep with a CRT	Intermediate	Intermediate	Good	Intermediate
3	Direct electronic linkage	Rapid	Investment high; operating cost low	Excellent	High

TABLE 3.2 Characteristics of Various Order-processing Systems

The order-processing system can communicate useful sales information to marketing (for market analysis and forecasting), to finance (for cash-flow planning), and to logistics or production. Finally, the order-processing system provides information to those employees who assign orders to warehouses, clear customer credit, update inventory files, prepare warehouse picking instructions, and prepare shipping instructions and the associated documentation. In advanced systems, many of these activities are computerized.

Advanced Order-processing Systems

No component of the logistics function has benefited more from electronic and computer technology than order entry and processing. Some advanced systems are so sophisticated that orders are generated automatically when stock reaches the reorder point.

At the level of advanced order-processing systems, shown as the third level in Table 3.2, customers and salespeople are increasingly transmitting orders to distribution centres or corporate headquarters by means of EDI, e-mail or the Internet. The order is entered directly from a computer or personal digital assistant (PDA) into the selling company's computer system.

Deviations from standard procedure, such as products on promotion, special pricing arrangements and allocations, are usually already in the computer system. The system can match order quantity against a list of minimum shipment quantities to ensure that the order meets the necessary specifications, and may provide ongoing feedback on the order's details and status while the customer is connected to it. When the order meets all criteria for accuracy and completeness, it is released for processing.

The Creative Solutions box describes the implementation of an advanced order system at a new company formed by Shell and BASF in 2001, Basell Polyolefins Company.

 CREATIVE SOLUTIONS

Supply Chain Technology Integration in the Chemical Processing Industry

Basell Polyolefins Company is the world's largest producer of polypropylene, a leading supplier of polyethylene and advanced polyolefin products, and a global leader in the development and licensing of polypropylene and polyethylene processes and catalysts. Basell was created in 2000 from subsidiaries of Shell and BASF, and its corporate centre is located in Hoofddorp, the Netherlands. The company also maintains regional offices in Wilmington, Delaware, in the USA, Brussels, Mainz in Germany, Sao Paulo in Brazil, and Hong Kong. Basell serves more than 4000 customers in over 120 countries and its largest market segments are flexible packaging, which includes food and soft drinks, consumer products packaging, the automotive industry and the consumer goods industry.

Basell is the global leader in the production of polypropylene, with a capacity of more than 8000 kilo tonnes, including its joint ventures. It is also the seventh largest producer of polyethylene in the world and number one in Europe, with capacity of more than 2500 kilo tonnes, including its joint ventures. Basell employs 6600 people worldwide to generate annual turnover of about €6.7 billion.

In such a commoditized market, Basell faces large challenges in differentiating itself and really adding value in its supply chain. Martin Feuerhahn, senior vice president supply chain – Europe, says that the company 'recognized from the beginning that only a fully integrated supply chain would allow it to effectively and efficiently deploy its multiple resources'. At Basell, such deployment meant fully integrated order-to-cash and demand-to-supply processes supported by

SAP's R3 software system. SAP's APO software is also used for Basell's forecasting, planning, plan execution and product sourcing optimization.

Feuerhahn believes that supply chain integration – for example, with customers through vendor managed inventory (VMI) or with logistics service providers via its Internet tool 'Basell Connect' – is an important prerequisite for an adaptable and responsive supply chain.

The speed with which Basell created one unified company with aligned processes across Europe and sharing a common database is impressive. The SAP systems implementation has provided real visibility over the entire supply chain across different countries and sectors. Connectivity with Basell's supply base, customer base and transport operations already sees over 20 per cent of orders being entered electronically, and this figure is growing.

Question: **What other industry sectors could follow Basell's strategy to enhance their processes and logistics?**

Source: adapted from Sam Tulip, 'The European Supply Chain Excellence Awards 2003', *Logistics Europe* 11, no. 10 (November 2003), p. 47.

Inside Sales/Telemarketing

Inside sales through **telemarketing** is an extension of the advanced order-processing systems we have discussed. It enables the firm to maintain contact with existing customers who are not large enough to justify frequent sales visits, increase contact with large, profitable customers, and efficiently explore new market opportunities.

Customer contacts made by telephone from an inside sales group can achieve the desired market coverage in an economical, cost-effective manner. In addition, the use of data terminals for direct order input integrates inside sales with logistics operations. One of the major cost advantages of inside sales/telemarketing comes from the efficiencies of associated logistics.

One method of improving efficiency is to place all small customers on scheduled deliveries in order to allow for consolidation, thereby reducing transportation and other logistics costs. The expenditure for inside sales must be justified based on projected sales increases, and improved profitability and visibility with customers.

Electronic Data Interchange[2]

Electronic data interchange (EDI) is the electronic, computer-to-computer transfer of standard business documents between organizations. EDI transmissions allow a document to be directly processed and acted upon by the receiving organization. Depending on the sophistication of the system, there may be no human intervention at the receiving end. EDI specifically replaces more traditional transmission of documents, such as mail, telephone and even fax, and may go well beyond simple replacement, providing a great deal of additional information, as discussed later.

There are a couple of key points to note about the definition of EDI given above. First, the transfer is computer to computer, which means that fax transmissions do not qualify. Also, the transmission is of standard business documents/forms. Some of the purchasing-related documents that are currently being transmitted by EDI include purchase orders, material releases, invoices, electronic funds transfer (EFT) for payments, shipping notices and status reports. Thus e-mail and sending information over the Internet, which is non-standard, free-form data, does not fit the definition of EDI. There are, however, secure EDI networks established on the Internet.

This section on EDI will discuss the issue of standards, the various types of system available, EDI benefits and potential problems, EDI implementation, and legal issues.

EDI Standards

For EDI to function properly, computer language compatibility is required. First, the users must have common communication standards. This means that documents are transmitted at a certain speed over particular equipment, and the receiver must be able to accept that speed from that equipment. But this is not enough. In addition, the users must share a common language or message standard, or have conversion capabilities. This means that EDI trading partners must have a common definition of words, codes and symbols, and a common format and order of transmission.

One issue is the multitude of EDI protocols in use today. Some are unique systems created by and for a particular company. Some standards have been adopted within a certain industry. The American National Standards Institute (ANSI) has proposed the use of the ANSI X12 standard, for use in the USA, while the Trade Electronic Data Interchange System (TEDIS) is the standard adopted for use in the EU.

While EDI standards such as X12 and TEDIS are designed to allow communications between different computers, they are only message formats. They do not specify how communications will take place between the trading partners and another networking service is needed to transmit messages between them.

Such services also have protocols, such as the Odette file transfer protocol (OFTP), to provide a standardized way for separate computer systems to communicate with each other. OFTP is a high-level protocol and is applied to the way messages are sent rather than the physical communications line across which they travel.

The physical connections are also standardized and are covered by other protocols, such as X.25 and TCP/IP. X.25 is a packet switched data network protocol, which defines an international recommendation for data exchange as well as control information between a user device or host and a network node. TCP/IP, or Transmission Control Protocol/Internet Protocol, is a suite of communication protocols used to connect hosts on the Internet.[3]

Many industry associations have established their own standards for EDI, which are to be used between members of firms within their industry. Examples of this include, but are not limited to, the grocery, automotive, retail, warehousing, chemical and wholesale drug industries.

Types of EDI System

Several types and variations of EDI system are in use today. The main types of system are proprietary systems, **value-added networks (VANs)** and industry associations (as mentioned above). The difference between a proprietary system and a VAN is illustrated in Figure 3.4 and explained below.

Proprietary Systems

Proprietary systems, also known as one-to-many systems, are aptly named, because they involve an EDI system that is owned, managed and maintained by a single company. That company buys from, and is directly connected with, a number of suppliers. This situation works best when the company that owns the system is relatively large and powerful, and can readily persuade key suppliers to become part of the network. The EDI system can also be used over the Internet, which is discussed below.

The advantage to the system owner is control. The disadvantages are that it may be expensive to establish and maintain internally, and suppliers may not want to be part of the system because it is unique and may require a dedicated terminal.

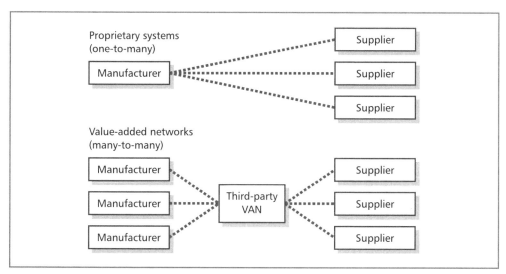

FIGURE 3.4 Typical EDI Configurations

Source: GE Information Service, as reported in Lisa H. Harrington, 'The ABCs of EDI', *Traffic Management* 29, no. 8 (August 1990), p. 51.

Value-added Networks

Value-added networks, also known as VANs, third-party networks or *many-to-many systems*, appear to be the most popular choice for EDI systems. Under VANs, all of the EDI transmissions go through a third-party firm, which acts as a central clearinghouse.

For example, a buying firm sends a number of purchase orders (POs), which go to different suppliers through the VAN. The VAN sorts the POs by supplier and transmits them to the proper supplier. The real 'value added' comes in when buyers and suppliers use incompatible communication and/or message standards. The VAN then performs translation 'invisibly', so that the user does not need to worry about system compatibility with its trading partners. This represents a big advantage over one-to-many systems.

In addition, the users do not need expertise in EDI standards and issues, as many VANs provide *turnkey*, off-the-shelf systems. This can lower start-up costs and reduce start-up lead-time.

Furthermore, a value-added network generally utilizes a 'mailbox' feature. With the mailbox, orders and other documents are not transmitted automatically to the receiver when they arrive in the network. Instead, the receiver 'picks up' the documents whenever it chooses. This may be at a regular time several times a day, to allow those sending the documents to plan accordingly. This gives the receiver flexibility, particularly if orders are placed or released to be filled at certain times. The user's system does not need to be cluttered with information that will not be acted upon immediately.

Yet another advantage of a VAN is that it can receive from and transmit to one-to-many systems. This means that the supplier who has a customer or customers that use proprietary systems does not need to have a dedicated terminal or direct linkage for each system. This capability of a VAN can increase the acceptability of networking with a customer who uses a proprietary system.

The Impact of the Internet on VANs

Using EDI over the Internet is rapidly becoming a reality, with annual growth rates in excess of 50 per cent in the late 1990s.[4] After initial software purchase and systems set-up, EDI over the Internet is virtually 'free' versus VAN transmission.

EDI over the Internet uses various protocols such as OFTP or HTTP, which is the hypertext transfer protocol we are all familiar with when using the World Wide Web. A key issue for companies using the Internet is security. A secure website uses a secured socket layer (SSL) that functions by using a private 'key' to encrypt data that is transferred over the SSL.

Major corporations, such as Unisys, are currently using the Internet for EDI on a regular basis. The revolution has begun – it is likely just a matter of time before the Internet completely replaces VANs.

Benefits of EDI Implementation

It should be obvious from the above discussion that electronic data interchange is a complex system. However, once in place, EDI tends to be a very easy system with which to interface and communicate.

The potential benefits of EDI are many (see Table 3.3). Most of these benefits are self-explanatory. The reduction in clerical work is a major benefit, reducing paperwork, increasing accuracy and speed, and allowing purchasing to shift its attention to more strategic issues.

The above improvements should also bring about a reduction in costs. One expert estimates that EDI can reduce the cost of processing a purchase order by 80 per cent.[5] Other firms claim that they have been able to reduce their inventory dramatically owing to improved inventory accuracy and reduced order cycle time.

- Reduced paperwork to be created and filed
- Improved accuracy due to a reduction in manual processing
- Increased speed of order transmission and other data
- Reduced clerical/administrative effort in data entry, filing, mailing and related tasks
- Opportunity for proactive contribution by purchasing because less time is spent on 'clerical tasks'
- Reduced costs of order placement and related processing and handling
- Improved information availability due to speed of acknowledgements and shipment advices
- Reduced workload and improved accuracy of other departments through linking EDI with other systems, such as bar-coding inventory and electronic funds transfers (EFTs)
- Reduced inventory due to improved accuracy and reduced order cycle time

---

TABLE 3.3 EDI Benefits

Electronic Mail and the Internet

A variation of EDI, **electronic mail (e-mail)**, has also become an important form of data transmission:

> Electronic mail involves electronic transmission of a variety of data. . . . EDI *always* involves one computer in contact with at least one other, usually transmitting specific documents such as invoices, waybills, or purchase orders.[6]

The e-mail market is a multibillion-euro business that has exhibited tremendous growth during the past decade. E-mail usually occurs over the Internet. Many individuals and organizations subscribe to online Internet services, such as AOL, Yahoo! and MSN. This allows them access to many data sources and services (some for a fee), as well as allowing them to send e-mail to Internet users at other organizations throughout the world. Many companies solicit bids and interact regularly with suppliers via e-mail.

One of the major reasons for the growth in e-mail, in addition to the speed and accuracy of its data transmission, is cost savings.

Integrating Order Processing and a Company's Logistics Management Information System

The order-processing system sets many logistics activities in motion, such as:

- determining the transportation mode, carrier and loading sequence
- inventory assignment and preparation of picking and packing lists
- warehouse picking and packing
- updating the inventory file; subtracting actual products picked
- automatically printing replenishment lists
- preparing shipping documents (a bill of lading if using a common carrier)
- shipping the product to the customer.

Other computerized order-processing applications include maintaining inventory levels, and preparing productivity reports, financial reports and special management reports.

Processing an order requires the flow of information from one department to another, as well as the referencing or accessing of several files or databases, such as customer credit status, inventory availability and transportation schedules. The information system may be fully automated or manual; most are somewhere in between.

Depending on the sophistication of the order-processing system and the corporate management information system (MIS), the quality and speed of the information flow will vary, affecting the manufacturer's ability to provide fast and consistent order cycle times and to achieve transportation consolidations and the lowest possible inventory levels.

Generally, manual systems are very slow, inconsistent and error prone. Information delays occur frequently. A manual system seriously restricts a company's ability to implement integrated logistics management, specifically to reduce total costs while maintaining or improving customer service. Some common problems include the inability to detect pricing errors, access timely credit information or determine inventory availability. Lost sales and higher costs combine to reduce the manufacturer's profitability.

Indeed, timely and accurate information has value. Information delays lengthen the order cycle. Automating and integrating the order process frees time and reduces the likelihood of information delays. Automation helps managers integrate the logistics system and allows them to reduce costs through reductions in inventory and freight rates. The communications network is clearly a key factor in achieving least total cost logistics.

Basic Need for Information

A logistics management information system is necessary to provide management with the ability to perform a variety of tasks, including the following:

- penetrate new markets
- make changes in packaging design
- choose between common, contract or private carriage

- increase or decrease inventories
- determine the profitability of customers
- establish profitable customer service levels
- choose between public and private warehousing
- determine the number of field warehouses and the extent to which the order-processing system should be automated.

To make these *strategic decisions*, management must know how costs and revenues will change given the alternatives being considered.

Once management has made a decision, it must evaluate performance on a routine basis to determine (1) if the system is operating under control and at a level consistent with original profit expectations, and (2) if current operating costs justify an examination of alternative systems. This is referred to as *operational decision-making*. The order-processing system can be a primary source of information for both strategic and operational decision-making.

An advanced order-processing system is capable of providing a wealth of information to various departments within the organization. Terminals for data access can be made available to logistics, production and sales/marketing. The system can provide a wide variety of reports on a regularly scheduled basis and status reports on request. It can also accommodate requests for a variety of data including customer order history, order status, and market and inventory position.

Designing the Information System

The design of a logistics management information system should begin with a survey of the needs of both customers, and a determination of standards of performance for meeting these needs. Next, customer needs must be matched with the current abilities of the firm, and current operations must be surveyed to identify areas that will require monitoring and improvement.

It is important at this stage to interview various levels of management. In this way, the organization can determine what strategic and operational decisions are made, and what information is needed for decision-making and in what form. Table 3.4 illustrates the various types of strategic and operational decision that management must make within each of the functions of logistics.

The next stage is to survey current data-processing capabilities to determine what changes must be made. Finally, a common database must be created and management reports designed, considering the costs and benefits of each. A good system design must support the management uses previously described and must have the capability of moving information from locations where it is collected to the appropriate levels of management.

Data for a logistics information system can come from many sources. The most significant sources of data for the common database are (1) the order-processing system, (2) company records, (3) industry/external data, (4) management data and (5) operating data. The type of information most commonly provided by each of these sources is shown in Figure 3.5.

Usually, the database contains computerized data files, such as the freight payment system, transportation history, inventory status, open orders, deleted orders, and standard costs for various logistics, marketing and manufacturing activities. The computerized information system must be capable of (1) data retrieval, (2) data processing, (3) data analysis and (4) report generation.

Data retrieval is simply the capability of recalling data such as freight rates (in their raw form) rapidly and conveniently. **Data processing** is the capability to transform the data to a more useful form by relatively simple and straightforward conversion. Examples of data-processing capability include preparation of warehousing picking instructions, preparation of bills of lading and printing purchase orders.

Decision type	Customer service	Transportation	Warehousing	Order processing	Inventory
Strategic	Setting customer service levels	Selecting transportation models	Determination of number of warehouses and locations	Extent of mechanization	Replenishment systems
		Freight consolidation programme	Public vs private warehousing		
		Common carriers vs private trucking	Public vs private warehousing		
Operational	Service level measurements	Rate freight bills	Picking	Order tracking	Forecasting
		Freight bill auditing	Packing	Order validation	Inventory tracking
		Claims administration	Stores measurement	Credit checking	Carrying – cost measurements
		Vehicle scheduling	Warehouse stock transfer	Invoice reconciliation	Inventory turns
		Rate negotiation	Staffing	Performance measurements	
		Shipment planning	Warehousing layout and design		
		Railcar management	Selection of materials-handling equipment		
		Shipment routing and scheduling	Performance measurements		
		Carrier selection			
		Performance measurements			

TABLE 3.4 Typical Strategic and Operational Decisions by Logistics Function
Source: American Telephone and Telegraph Company, *Business Marketing*, Market Management Division.

Data analysis refers to taking the data from orders and providing management with information for strategic and operational decision-making. A number of mathematical and statistical models are available to aid a firm's management, including linear programming and simulation models. **Linear programming** is probably the most widely used strategic and operational planning tool in

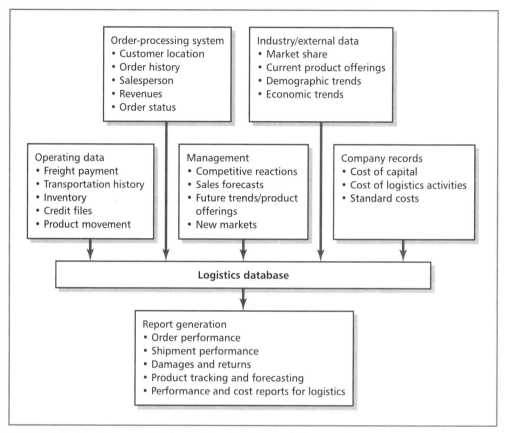

FIGURE 3.5 Key Sources of Information for the Logistics Database

logistics management. It is an optimization technique that subjects various possible solutions to constraints that are identified by management.

Simulation is a technique used to provide a model of a situation so that management can determine how the system is likely to change through the use of alternative strategies. The model is tested using known facts. Although simulation does not provide an optimal solution, the technique allows management to determine satisfactory solutions from a range of alternatives. A number of simulation models are available for purchase if the firm does not have the resources to develop its own.

The last feature of an information system is **report generation**. Typical reports that can be generated from a logistics management information system include order performance reports, inventory management reports, shipment performance reports, damage reports, transportation administration reports, system configuration reports (which may contain the results of data analysis from mathematical and statistical models) and cost reports for logistics. The company LOGiCOM effectively used a new integrated logistics information system to upgrade its existing, or 'legacy', ERP system (see the Technology box).

An Integrated ERP Solution for LOGiCOM

The information technology (IT) sector argues that if a firm integrates its ERP, SCM and CRM systems it will have the best of all possible situations. However, a survey by Siebel Systems of its customers found that, on average, 50 systems need to be integrated and 51 per cent of customers need to integrate in excess of 50 systems. This is the real world, with lots of systems and, critically, a large number of existing or 'legacy' systems requiring integration. The reasons for IT system integration can be a merger or acquisition, change of business strategy, company expansion, or the addition of suppliers and customers. But the underlying issue is always the need to manage a complex flow of information from a variety of disparate and incompatible systems.

LOGiCOM, an independent pan-European provider of comprehensive logistics services for IT service parts, faced such an issue in 2001 when it was spun off from its parent company Fujitsu Services (formerly ICL). LOGiCOM serves the needs of original equipment manufacturers (OEMs) and multi-vendor service providers by coordinating all supply-chain activities required to source, manage and deliver parts to field service engineers or end-consumers. Currently, the company has premises in more than 60 locations across Europe, providing a 365-day, 24-hour operation and distributing to 83 countries worldwide. It handles 50,000 part requests every month and maintains a multi-vendor inventory valued at £50 million.

LOGiCOM rejected the costly option of customizing its legacy ERP system in favour of an integration strategy using an integration broker to link new best-of-breed applications into the framework. Instead, LOGiCOM selected SeeBeyond's e*Gate Integrator as opposed to other solutions available on the market. Implementation began in March 2001. Cap Gemini Ernst & Young was responsible for developing the integration solution, while SeeBeyond's consultants ensured the overall design and architecture were appropriate and that the Integrator was configured and installed correctly. Two interfaces were developed: the parts for service (PAFOS) interface, which operates between LOGiCOM's ERP system and multiple external service delivery systems, and the carrier interface, which has a similar relationship with external carrier systems.

LOGiCOM achieved integration with a major new OEM customer and its maintenance company in July 2001, followed in October 2001 by integration with one of LOGiCOM's main third-party carriers, UPS. When the OEM needs spare parts for one of its own customers, it sends a parts request message to LOGiCOM over the Internet using the Integrator. This message is routed to the ERP system to automatically allocate stock to the request. The allocated stock is picked and added to a shipping manifest that is automatically transmitted to UPS by the Integrator. The carrier picks up the part at the warehouse and delivers it to the OEM's customer. During transit, the carrier sends frequent track and trace information to LOGiCOM, to automatically update the part's request status and feed this information to the OEM through the Integrator. A similar approach is used to manage the collection of failing parts when required by the customer. Normally, all the processes are completed within the time periods necessary to achieve the customer's service-level agreement, demonstrating the benefit of real-time supply chain integration.

The LOGiCOM situation illustrates one way to solve an issue of 'legacy' integration in the short term to protect an application investment in the long term. Other benefits of IT integration strategy come from the flexibility a company can portray to potential trading partners and resultant revenue opportunities from those partnerships.

Question: **What do you think are the implications for its suppliers and customers of a firm changing its legacy IT systems?**

Source: adapted from Ian Howells, 'Integrating IT Systems: Legacy No Problem', *Logistics & Transport Focus* 5, no. 2 (March 2003) pp. 37–40.

Financial Considerations

Of course, it will be necessary to justify an advanced order-processing system in terms of cost–benefit analysis. The costs of developing the system, *start-up costs*, can be justified by comparing the present value of improvement in cash flows associated with the new system to the initial investment. In most cases, cash flow will improve by changing to an advanced order-processing system if the volume of orders processed is large. In smaller operations, however, this may not be true if the proposed system is more than the company needs.

Generally, the fixed costs associated with an advanced order-processing system are higher than those incurred by a manual system. However, the variable costs per order are significantly less with the advanced system. We will expand upon this type of cost analysis in Chapter 11, which deals specifically with the calculation of logistics cost savings.

Using Logistics Information Systems to Support Time-based Competition

As discussed in Chapter 2 in conjunction with customer service, customers are becoming increasingly demanding about their expectations of suppliers. Customers want consistent delivery times, consistent order cycles, and excellent communications regarding in-stock availability and expected shipment arrival. In short, customers are demanding integrated logistics systems supported by integrated logistics information systems.

These applications are aided by integrating a number of technologies, such as bar coding, EDI and point-of-sale (POS) data gathering and transmission, and electronic funds transfer (EFT). Bar coding, POS technology and management of quality are described below. These technologies and techniques can be linked to support quick response (QR), efficient consumer response (ECR) or customer relationship management (CRM), considerably reducing the total order cycle time.

Bar Coding

Bar codes can be seen on virtually all types of consumer packaged goods today. A **bar code** is a sequence of parallel bars of various widths, with varying amounts of space between them. The pattern and spacing of the bars convey information such as letters, numbers and special characters. The bar codes are optically read by 'scanning' them with a beam of light. The information contained in the bars is read directly into a computer, or stored and downloaded into the computer system at a later time.

Although there are several forms of bar code, the most popular standard in Europe is the European Article Numbering (EAN) Uniform Code Council (UCC), with over one million firms operating under that system. To obtain a bar code, a firm must register with EAN International to receive an assigned and unique bar code prefix to use with all its products. The firm must then allocate a unique number to each product.[7]

Bar coding can be useful in logistics applications, particularly in track and trace situations. Receiving can also be automated, which further contributes to cycle time reduction and data accuracy. These data can automatically be used by the accounts payable department for generating checks and reconciling invoices with purchase orders and receiving. Thus, bar coding represents a logical extension of the organization's information systems and a linkage with EDI.

Bar codes are accurate and efficient. Since 1995 bar codes have proved their worth as an enabling technology contributing to manufacturing cost reduction, quality improvement, cycle-time reduction and improved profitability.

Bar code technology is a passive format (i.e. the firm must seek out the item and scan the bar code). A successor technology, radio frequency identification (RFID), is an active format that is just being introduced into logistics around the world (see the Global box).

RFID: The Next Revolution!

Radio frequency identification (RFID) is the latest technology being developed in the middle of this decade to improve logistics and supply chain management. RFID systems are 'active' and use radio frequencies to provide automatic identification and location of electronic data-carrying devices or 'tags' attached to items.

RFID tags consist of a semiconductor chip with memory-processing capability and a transmitter connected to an antenna. The memory can be configured to read only, write once, read many times or read/write.

The tags can be detected by radio frequencies at a remote distance from a reader without the necessity of contact or line of sight. A network or computer information system can process the data received for final application, such as materials handling, inventory replenishment, asset and product tracking, anti-counterfeiting or safety and security functions.

When compared to other auto-identification systems such as bar codes RFID technology offers additional benefits such as fewer labour requirements, reading speed and multiple reading, read/write capacity and higher security levels.

The cost of RFID tags is, at present, a barrier but costs should fall once large-scale implementation gets under way. Standards are also slow in being developed, however the Auto-ID Center at the Massachusetts Institute of Technology is working towards global harmonization.

There are also some technical issues. One issue pertains to differences in radio-communication UHF radiation limits between Europe and the USA. European limits are 0.5 watts while US limits are 4 watts. The European limits being 100 times slower than US limits means the readable range in Europe would also be 100 times lower than in the USA. Also, an issue in the retailing environment relates to the risk of tags being misread due to metal interference.

Nonetheless, RFID appears to be here to stay. In the United States, Wal-Mart informed its top 100 suppliers in June 2003 that it required RFID-tagged pallets and cases by 2005. In the United Kingdom both Marks & Spencer and Tesco have piloted RFID tags at the case level in their food retail stores and are also looking to roll out wider RFID implementation during 2005.

Question: What other uses can you foresee for RFID technology?

Sources: adapted from Richard Wilding and Tiago Delgado, 'The Story So Far: RFID Demystified', *Logistics & Transport Focus* 6, no. 3 (April 2004), pp. 26–31; 'RFID Demystified: Supply Chain Applications', *Logistics & Transport Focus* 6, no. 4 (May 2004), pp. 42–8; and 'RFID Demystified: Company Case Studies', *Logistics & Transport Focus* 6, no. 5 (June 2004), pp. 32–42.

Point-of-sale Data

Point-of-sale (POS) data gathering is simply the scanning of bar codes of items sold, generally at the retail level. The data may be transmitted to the relevant supplier, who can replenish the inventory based on sales. This allows the retailer to 'skip' order placement and to transfer responsibility directly to the supplier. In other cases, the retailer may prefer to intervene and use POS data to place the order itself.

Quick response (QR) and efficient consumer response (ECR) integrate the above technologies in an effort to speed time to market, thereby supporting time-based competition while reducing inventories and improving or maintaining customer service. For example, the UK grocery retailer Tesco integrated its Tesco.com online customer order-processing systems using EDI. A customer order received from the website is sent to the computer server at the store nearest to the customer's home, and is then assigned to the van that will deliver the goods. It is then sent to a 'picking trolley', a shopping cart with a screen and 'shelf identifier' software that takes a picker to where each item is found. Pickers scan the items they select and the system compares bar code details with the item ordered on the customer's shopping list; sounding an alert if the wrong item is selected. Pickers also inspect expiry dates and check for damage on every item. Once the trolley is loaded, it is sent straight to the van for delivery.

Average picking time is 30 seconds per item and a typical order of 64 items could be fulfilled in about 30 minutes. Pickers work during normal store trading hours but usually go around stores when they are not crowded between 6.00 am and 10.00 am, and 11.00 am and 3.00 pm. Two to three waves of orders are filled per day so an order placed as late as noon could be delivered by 10.00 pm the same day. Tesco's picking out-of-store approach enabled the rollout of this online service, which meets the needs of 'cash rich' and 'time poor' customers.[8]

Management of Quality

Integrated information systems directly support an organization's quality management efforts by providing the customer with more accurate order fill. This occurs because the more automated the system, the less chance there is for human error. Such systems improve the quality of customer service by reducing order cycle time and improving order cycle consistency.

In addition, they create the ability to provide the customer with real-time information regarding inventory availability, order status and shipment status. Thus advanced logistics information systems support quality management.

While the types of system described focus on day-to-day operations, logistics information systems can be used to support strategic decision-making. Other systems, such as decision support systems (DSSs) and artificial intelligence (AI) provide a great deal of flexibility and support for logistics decisions based on logistics information. DSSs and AI are presented briefly in the next sections.

Decision Support Systems

Decision support systems (DSSs) encompass a wide variety of models, simulations and applications that are designed to ease and improve decision-making. These systems incorporate information from the organization's database into an analytical framework that represents relationships among data, simulates different operating environments (e.g. vehicle routing and scheduling), may incorporate uncertainty and 'what-if' analysis, and uses algorithms or heuristics. DSSs actually present an analysis and, based upon the analysis, recommend a decision.

The artificial intelligence tools can be incorporated into DSSs, which may contain decision analysis frameworks, forecasting models, simulation models and linear programming models. They can be used to assist in a wide variety of logistics decisions, such as evaluating alternative transportation options, determining warehouse location and setting levels of inventory.

While the use of DSSs is not currently widespread, it appears to be growing as the potential contribution becomes more understood and computing costs continue to decline. Figure 3.6 shows the components of a DSS.

A DSS is applications orientated. More specifically, a DSS has the following objectives:

■ to assist logistics executives in their decision processes
■ to support, but not replace, managerial judgement
■ to improve the effectiveness of logistics decisions.[9]

Perhaps the most critical element of a DSS is the quality of the data used as input into the system. DSSs require information about the environment that is both internal and external to the organization. Thus, an important first step in DSS planning, implementation and control is to

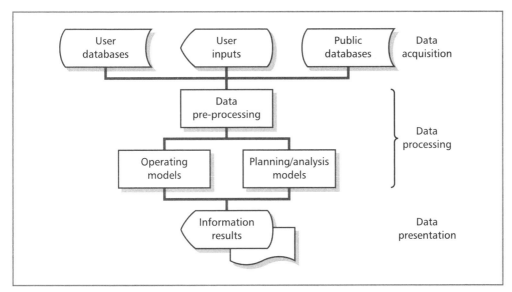

FIGURE 3.6 Decision Support System

Source: Allan F. Ayers, *Decision Support Systems: A Useful Tool for Manufacturing Management*. King of Prussia, PA: K.W. Tunnell Company, 1985, p. 2.

have good external information. This is discussed in Chapter 13 in conjunction with strategic planning. Models are also needed to provide data analysis.

Modelling can be defined as the process of developing a symbolic representation of a total system. A model must accurately represent the 'real world' and be managerially useful. The purpose of models has been described as:

> Essentially . . . to replicate reality and to assess the behavior of that reality if changes are introduced. A model supports, rather than replaces, the managerial decision-making process. By using a model, we are able to establish a current situation and then play 'what-if' games. This 'what-if' ability is significant. It allows us to quickly consider many different alternatives and test the outcome.[10]

Artificial Intelligence and Expert Systems

Developed out of the field of computer science, **artificial intelligence (AI)** is

> concerned with the concepts and methods of inference by a computer and the symbolic representation of the knowledge used in making inferences. The term *intelligence* covers many cognitive skills, including the ability to solve problems, to learn, to understand language, and in general, to behave in a way that would be considered intelligent if observed in a human.[11]

AI is a comprehensive term encompassing a number of areas, including computer-aided instruction, voice synthesis and recognition, game-playing systems, natural language translators, robotics and expert systems (ES).[12] While the number of AI applications is limited, the potential in logistics is staggering. AI has been used to model response time requirements for customer delivery; to model transportation costs and times for various transportation modes, locations and routings; to determine which warehouses should serve which plants, with which products and what inventory levels; to model customer service response with various levels of reliability; and to perform sensitivity analysis to determine how much inputs can vary without affecting the structure of the optimal solution.[13]

Of specific interest to logistics executives are the subareas of AI known as **expert systems (ES)**, natural language recognition and neural networks. An ES is defined as:

> a computer program that uses knowledge and reasoning techniques to solve problems normally requiring the abilities of human experts. An expert system is an artificial intelligence (AI) program that achieves competence in performing a specialized task by reasoning with a body of knowledge about the task and the task domain.[14]

Expert systems are capable of being applied to a variety of problems in marketing and logistics, including interpretation, monitoring, debugging, repair, instruction and control. Examples of ES applications can be found in many industries.

Five criteria aid decision-makers in determining whether expert systems should be used to solve a particular logistics problem. If any of the criteria are met, an ES may be appropriate:

1 the task or problem solution requires the use of human knowledge, judgement and experience
2 the task requires the use of heuristics (e.g. set rules), or decisions based on incomplete or uncertain information
3 the task primarily requires symbolic reasoning instead of numerical computation
4 the task is neither too easy (taking a human expert less than a few minutes) nor too difficult (requiring more than a few hours for an expert to perform)
5 substantial variability exists in people's ability to perform the task; novices gain competence with experience; experts are better than novices at performing the task.[15]

If an ES is appropriate, the next decision facing the logistics executive is whether the system can be economically justified and if EDI can be combined with other systems such as artificial intelligence (AI). At Benetton, the Italian clothing manufacturer, computers not only determine what will be included in upcoming production runs, but also designate optimum routing for all finished goods.

At each Benetton store, the point-of-sale cash registers maintain a running record of item sales, which they transmit by way of EDI to computers in branch offices. The branch offices, in turn, transmit the data to the central office computer, which uses AI and a modelling program to make decisions on production runs. If red sweaters are selling fast, the computer tells the manufacturing system to design and produce more red sweaters. The system then determines how the shipments will be routed to the stores, freeing the traffic department (which consists of only six people) to spend its time researching new routings and handling problems.[16]

Database Management

As previously mentioned, computers are excellent at managing data. A database management system allows application programs to retrieve required data stored in the computer system. The types of data stored were shown in Figure 3.5. A database management system must store data in some logical way, showing how different pieces of data are related, in order for retrieval to be efficient. This is a critical issue in logistics because of the large volume of data generated that may require analysis at a later date. For example, a buyer may want to see a history of transportation carriers with which it has placed orders for a particular item in the past six months.

The database management system must be able to use the item number to reference the order and 'pull up' the pertinent data. If the buyer sees that two suppliers have been used, the buyer may want the system to provide a transaction history with those suppliers over a given time period for all purchased items. The database management system must have the flexibility to sort data in a variety of ways that are meaningful to the user.

Relational database structures are popular today because they allow access to and sorting of data by relating the data to other data in many ways. This allows a great deal of flexibility.

Increasingly, companies are using what is known as a **local area network (LAN)**. This consists of a minicomputer linked to a number of microcomputers or terminals that allow access to a common database, software and other systems features.[17] LANs give microcomputers the power of mainframe systems.

Regardless of the sophistication of the software and hardware, a system cannot provide good results if the data in the system are not accurate and timely. Thus systems integrity is vital. If people do not use the system consistently (i.e. do not scan each bar-coded item individually) the system will quickly be inaccurate. Once a system has data-accuracy problems, it is very difficult, costly and time consuming to correct.

SUMMARY

In this chapter, we saw how the order-processing system can directly influence the performance of the logistics function. We also examined how order-processing systems can be used to improve customer communications and total order cycle time, or lead to substantial inventory reductions and transportation efficiencies. Information is vital for the planning and control of logistics systems, and we saw how the order-processing system can form the basis of a logistics information system.

Today's computer technology and communication systems make it possible for management to have the information required for strategic and operational planning of the logistics function. The order-processing system can significantly improve the quality and quantity of information for decision-making.

Computers have become an invaluable aid to the logistics executive in making various operational and strategic decisions. Decision support systems, which are computer based, provide information for the decision-making process. The DSS has three components: data acquisition, data processing and data presentation.

Computers are widely employed in many areas of logistics, including transportation, inventory control, warehousing, order processing, material handling, and so forth. Some of the most exciting areas of computerization are modelling, artificial intelligence (AI) and expert systems (ES). Improved database management contributes to the support of logistics decision-making.

KEY TERMS

QUESTIONS AND PROBLEMS

1 How do wholesalers and retailers measure the order cycle provided to them by a manufacturer?
2 Explain the impact of order cycle variability on the inventory levels of wholesalers and retailers.
3 Discuss specifically how logistics performance is affected by the order-processing system used.
4 What are the primary advantages associated with the implementation of an integrated, automated order-processing system?
5 Electronic data interchange applications have experienced significant growth in recent years. Why do you believe this growth has occurred? What are the primary benefits of EDI? Do you think that the growth rate in EDI applications will be sustained? Why or why not?
6 How does the order-processing system form the foundation of the logistics management information system?
7 How is the logistics management information system used to support planning of logistics operations?
8 Briefly describe the role of decision support systems in logistics decision-making.
9 Identify areas of logistics that use expert systems (ES) and artificial intelligence (AI) to improve efficiency and effectiveness.

THE LOGISTICS CHALLENGE!

PROBLEM: EVE OF DESTRUCTION

Adobe Systems, Inc. is a major producer of software for graphics and electronic publishing applications. It ships upwards of 400,000 boxes a year (valued at more than $100 million) into the Pacific Rim, including Japan, Southeast Asia, India, Australia and Latin America.

Like many software makers, Adobe does most of its overseas business through distributors, according to Larry C. Clopp, manager of international sales for the Pacific Rim. And it introduces regular updates of its most popular titles, rendering all previous versions obsolete.

But Adobe's method for taking back old software left much to be desired. It was getting requests from distributors to take back between 10,000 and 20,000 boxes a year, which were then shipped to California for destruction.

To discourage piracy, the company had to confirm the serial number of each returned title, then give it to a software recycler for erasure. However, Adobe was taking up to two months to credit distributors' accounts, so its distributors were hamstrung by severe reductions in their already thin credit lines.

Even worse, distributors were paying the cost (more than $2500 per shipment) of sending the product by air back to the United States. Some became fed up with the process and sold the old software locally.

How can Adobe cut its processing time on returns in half, while sharply reducing administrative costs and ensuring product security?

What Is Your Solution?

Source: 'Distribution: The Challenge', *Distribution* 95, no. 5 (May 1996), p. 72.

SUGGESTED READING

BOOKS

Allen, Mary K. and Omar Keith Helferich, *Putting Expert Systems to Work in Logistics*. Oak Brook, IL: Council of Logistics Management, 1990.

Bender, Paul S., 'Using Expert Systems and Optimization Techniques to Design Logistics Strategies', *Proceedings of the Annual Conference of the Council of Logistics Management*. Cincinnati, OH: 16–19 October, 1994.

Emmelhainz, Margaret A., *Electronic Data Interchange: A Total Management Guide*, 2nd edn. New York: Van Nostrand Reinhold, 1995.

Leenders, Michiel, Harold E. Fearon, Anna E. Flynn and P. Fraser Johnson, *Purchasing and Supply Management*, 12th edn. Burr Ridge, IL: McGraw-Hill Irwin, 2002.

McKinnon, Alan C. and Mike Forster, *European Logistical and Supply Chain Trends: 1999–2005*. Edinburgh, UK: Heriot-Watt University, 2000.

JOURNALS

Carbone, James, 'Make Way for EDI', *Electronics Purchasing*, September 1992, pp. 20–4.

Carter, Joseph R. and Gary L. Ragatz, 'Supplier Bar Codes: Closing the EDI Loop', *International Journal of Purchasing and Materials Management* 37, no. 3 (Summer 1991), pp. 19–23.

Cooke, James Aaron, 'Computers Lead the Way to Total Inbound Control', *Traffic Management* 29, no. 1 (January 1990), pp. 50–3.

Curry, Bruce and Luiz Moutinho, 'Expert Systems for Site Location Decisions', *Logistics Information Management* 4, no. 4 (1991), pp. 19–27.

Lim, Don and Prashant C. Palvia, 'EDI in Strategic Supply Chain: Impact on Customer Service', *International Journal of Information Management* 21 (2001), pp. 193–211.

Ludwick, David P. and David B. Grant, 'Simulation Models and Quality of Service in IP Networks – An Exploratory Investigation', *Operational Research: An International Journal* 2, no. 1 (2002), pp. 71–84.

Powers, Richard F. 'Optimization Models for Logistics Decisions', *Journal of Business Logistics* 10, no. 1 (1989), pp. 106–21.

Schary, Philip B. and James Coakley, 'Logistics Organization and the Information System', *The International Journal of Logistics Management* 2, no. 2 (1991), pp. 22–9.

Sheffi, Yosef, 'The Shipment Information Center', *The International Journal of Logistics Management* 2, no. 2 (1991), pp. 1–12.

Sriram, Ven and S. Banerjee, 'Electronic Data Interchange: Does its Adoption Change Purchasing Policies and Procedures?', *International Journal of Purchasing and Materials Management* 30, no. 1 (1994), pp. 31–40.

REFERENCES

[1] Helmut Baumgarten and Frank Straube, *Towards the 21st Century – Trends and Strategies in European Logistics*. Brussels: European Logistics Association, 1997; and Alan C. McKinnon and Mike Forster, *European Logistical and Supply Chain Trends: 1999–2005*. Edinburgh: Heriot-Watt University, 2000.

[2] This section is adapted from Lisa M. Ellram and Laura M. Birou, *Purchasing for Bottom-line Impact*. Burr Ridge, IL: Irwin Business One, 1995; and Don Lim and Prashant C. Palvia, 'EDI in Strategic Supply Chain: Impact on Customer Service', *International Journal of Information Management* 21, (2001), pp. 193–211.

[3] David P. Ludwick and David B. Grant, 'Simulation Models and Quality of Service in IP Networks – An Exploratory Investigation', *Operational Research: An International Journal* 2, no. 1 (2002), pp. 71–84.

[4] Lim and Palvia, 'EDI in Strategic Supply Chain: Impact on Customer Service', p. 193.

[5] James Carbone, 'Make Way for EDI', *Electronics Purchasing*, September 1992, pp. 20–4.

[6] Michael Hammer and James Champy, *Reengineering the Corporation*. New York: HarperCollins, 1993, p. 48.

[7] 'Bar Code is Here to Stay', *Logistics Business Magazine* (November/December 2004), p. 56.

[8] Padmini Maddali, 'Tesco in 2003', www.icfai.org, 2005.

[9] Eframin Turban, *Decision Support and Expert Systems*. New York: Macmillan, 1988, p. 8.

[10] John H. Campbell, 'The Manager's Guide to Computer Modeling', *Business* 32, no. 4 (October–December 1982), p. 11.

[11] Mary Kay Allen and Omar Keith Helferich, *Putting Expert Systems to Work in Logistics*. Oak Brook, IL: Council of Logistics Management, 1990, p. A6.

[12] Omar K. Helferich, Stephen J. Schon, Mary Kay Allen, Raymond L. Rowland and Robert L. Cook, 'Applications of Artificial Intelligence – Expert System to Logistics', *Proceedings of the Annual Conference of the Council of Logistics Management*, vol. 1. Oak Brook, IL: Council of Logistics Management, 1986, pp. 45–86.

[13] Paul S. Bender, 'Using Expert Systems and Optimization Techniques to Design Logistics Strategies', *Proceedings of the Annual Conference of the Council of Logistics Management*, 16–19 October, 1994, pp. 231–9.

[14] Allen and Helferich, *Putting Expert Systems to Work in Logistics*, p. A10.

[15] Ibid., p. 115.

[16] Marsha Johnston, 'Electronic Commerce Speeds Benetton Business Dealings', *Software Magazine*, January 1994, pp. 93–5.

[17] Ellram and Birou, *Purchasing for Bottom-line Impact*, p. 149.

CHAPTER 4
PURCHASING AND PROCUREMENT

CHAPTER OUTLINE

CHAPTER OBJECTIVES

■ To show how better management of purchasing activities can lead to increased profitability

■ To identify the activities that must be performed by the purchasing function

■ To describe the impact of just-in-time production on purchasing

■ To present issues in purchasing cost management

■ To illustrate the role of partnering in supplier relationship management

OBJECTIVES

INTRODUCTION

Purchasing agents for manufacturing firms in the European Union's 25 member states buy almost €4.1 trillion worth of goods each year.[1] How well this money is spent is an issue of considerable concern to companies. When one reflects that purchases consistently represent the largest single expense of doing business, it becomes evident that there is a pressing need for reliable measures of purchasing efficiency and effectiveness.

The purchasing area within many organizations is undergoing many changes, including broadening its responsibilities, which is often reflected in departmental name changes. The group that used to be called purchasing may now be called, to name a few, procurement, sourcing, strategic sourcing, supply management, strategic supply management, supplier management or materials management.

Along with the name changes has come a growing recognition of the importance of purchasing activities to the success of an organization, and a growth and shift in the types of activity performed by the sourcing area. The shifts in purchasing's activity and recognition mirror shifts in the status and responsibility of logistics activities in general.

While the sheer euro volume of purchases is impressive, it is just as significant to note that the average value of purchases in manufacturing industries has increased from approximately 40 per cent of sales in the late 1970s to over 72 per cent in 2000, reflecting a trend in outsourcing goods and subassemblies.[2] Purchase value is also significant in the service sector. For example, purchases average about 15 per cent of revenues at investment firms like Salomon Brothers and Merrill Lynch.[3]

The Role of Purchasing in the Supply Chain

As presented more fully in Chapter 1, supply chain management is an integration of business processes from end user through original suppliers that provide products, services and information that add value to customers. Figure 4.1 illustrates how purchasing, or procurement in its expanded role, supports supply chain management.

Purchasing is responsible primarily for inbound, or upstream, channel activities, whereas logistics spans both inbound and outbound relationships and material flows. The specific activities for which purchasing is frequently responsible are presented next.

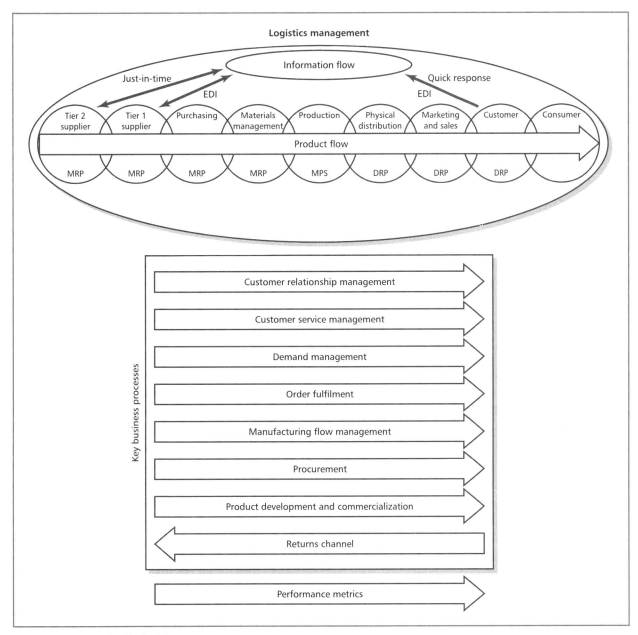

FIGURE 4.1 Supply Chain Management
Source: International Center for Competitive Excellence, University of North Florida.

Purchasing Activities

Purchasing was once looked upon primarily as a service function. As such, its responsibility was to meet the needs of the manufacturing function or other internal functions for which it was buying. It was not the responsibility of purchasing to question those needs, forge long-term relationships with suppliers or to understand the needs of the end customer.

This perspective severely limited the contribution that purchasing could make to the firm. In this scenario, purchasers had to focus primarily on a narrow set of activities to serve the needs of the internal interfaces, such as production, marketing, operations and others who needed to procure something from outside the organization. The scope of purchasing activities was defined and limited by those inside the organization.

Purchasing focused on getting the right product or service to the right place at the right time – in the right quantity, in the right condition or quality, and from the right supplier at the right price. While this may sound like a broad range of responsibilities, it really was not because the internal client, or other functional areas within the company with whom purchasing interfaces, was defining what was 'right' at each step.

While purchasing played a key role in keeping the operation running smoothly by ensuring a reliable source of supply, this was not always accomplished at the lowest total cost. In many cases, purchasing may have contributed directly to the bottom line of the organization by reducing prices paid to suppliers.

Many organizations still see this as the focal point for purchasing. For example, a survey of over 100 UK firms by A.T. Kearney and the Manchester School of Management revealed that 39 per cent of them admitted they threatened suppliers with withdrawal of business to obtain lower prices while 64 per cent said customers had used that threat against them.[4] This operational perspective focuses on the use of power in short-term, day-to-day purchasing details, rather than the big picture. Clearly, the proper purchasing perspective must be the systems approach: to look at how purchasing can support broader organizational goals for both suppliers and customers.

Typically, purchasing was not seen as an activity of strategic importance. It involved following a series of prescribed steps, which included writing up a purchase order, contacting suppliers for pricing, and sometimes following up on a supplier who failed to deliver.

Development of the Purchasing and Procurement Functions

The purchasing function has gradually evolved from simply the sourcing and buying of materials and those activities related to the buying process. As organizations increasingly automate and outsource many activities, the funds spent on external purchases increase compared to those spent on labour. Thus purchasing activities have been receiving more attention.

For the past 30 years or so, organizations have given purchasing more leeway in performing its activities. In some cases, purchasers have taken the initiative in broadening their roles in order to contribute more fully to the organization as a whole. The term procurement extends purchasing activities to a more strategic and process-orientated level and includes selection of supply source locations, determination of the form in which the material is to be acquired, timing of purchase, price determination and quality control.

In many ways, purchasing today stands at a crossroads in its development. Many of its activities that were once mainstream are being eliminated and automated in a rationalization role.[5] Activities such as purchase order placement, expediting, matching documents and calling to check stock have either been eliminated or are now possible online with electronic data interchange (EDI) or electronic point-of-sale (EPOS) exchanges.

While elimination of routine clerical activities frees up time, enabling purchasing to play a more proactive role in the firm, purchasers must recognize and seize the opportunity or they may face job elimination in an environment where downsizing and reengineering are common occurrences.

An important part of recognizing opportunities comes from understanding the organization's strategic goals and direction, so that purchasing and procurement can support those goals in a development role. It also comes from understanding the important role that purchasing plays in helping the organization achieve total customer satisfaction.

The Role of Purchasing in Total Customer Satisfaction

The major objective of any business is to create value for the owners. Many managers recognize that this can be accomplished more successfully by focusing on 'serving the customer' or 'providing a service to the customer'. This reflects the realization that if they do not serve the customer effectively by meeting some otherwise unfulfilled need, the firm will cease to exist. This is not a change in the marketing concept, but a better implementation of the concept.

Traditionally, purchasing has been separated from the firm's final customers, or end users. However, the receipt of high-quality, reliable goods and services on a timely basis at a reasonable cost often directly affects customer satisfaction. The relationship is illustrated in Figure 4.2.

An organization cannot provide its ultimate customers with better-quality goods and services than it receives from its suppliers. If a supplier is late with a delivery or has quality problems, the quality and availability of the product or service to the customer will be affected unless the firm carries higher inventory. In such cases, the supplier increases the total cost of the product or service.

It is important that purchasers understand the needs of their organization's customers. This understanding will allow purchasing to make the 'right' decisions to meet the organization's needs. The skills required and tasks performed by purchasing are very similar for buyers in the retail, manufacturing, government and service sectors.

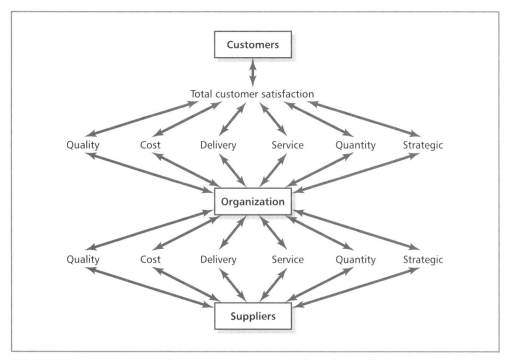

FIGURE 4.2 Total Customer Satisfaction Depends on Supplier Performance

Source: Michiel Leenders, and Anna Flynn, *Value Driven Purchasing: Managing the Key Steps in the Acquisition Process*. Burr Ridge, IL: Irwin Professional Publishing, 1994, p. 3.

The Strategic Role of Purchasing and Procurement

The strategic role of purchasing and procurement is to perform sourcing-related activities in a way that supports the overall objectives of the organization. Procurement can make many contributions to the strategic success of the organization through its key role as one of the organization's primary boundary-spanning activities of Porter's value chain, as discussed in Chapter 1.

Access to External Markets

Through external contacts with the supply market, purchasing can gain important information about new technologies, potential new materials or services, new sources of supply, and changes in market conditions.

By communicating this competitive intelligence, purchasing can help reshape the organization's strategy to take advantage of market opportunities. The Creative Solutions box illustrates the important strategic role that purchasing can play in the aviation sector.

CREATIVE SOLUTIONS

Network Innovations at Lufthansa

Global commercial aviation markets have seen an unprecedented dynamic in growth during the last 20 years after deregulation and liberalization activities. Dynamic markets have led to strong competition among the established carriers and numerous newcomers, resulting in tremendous cost pressure for all airlines and consequently for technical aftermarkets. These technical aftermarkets embrace the maintenance, repair and overhaul (MRO) business, providing services such as scheduled checks of airframes, engines, landing gears, components and cabin interiors, and repair and modification programmes including engineering services. In the MRO industry, the market structure is highly competitive. Lufthansa Technik AG (LTAG) is the global market leader with a market share of approximately 10 per cent. Other strong players are Air France Industries, ST Aerospace, FLS Aerospace and General Electric Engine Services, United Technologies/Pratt & Whitney and Rolls-Royce in the engine overhaul market.

The aircraft fleet is by far the most important asset for an airline, and punctuality of flights is of highest importance for the customers. As a result, the top priority in the MRO industry is to provide both safe and reliable aircraft in order to fulfil the airlines' preconditions. Therefore, all spare parts have to be available immediately whenever and wherever they are needed to make the aircraft fly. In the past, this has often led to excessive safety stocks no matter what costs were implied. High stock values are not only a result of limited cost awareness, but also a consequence of a significant portion of non-routine work included within major MRO tasks, with only limited predictability of parts needed to be replaced during a specific event. In combination with partly excessive lead-times for aircraft parts of up to one year, MRO shops have a wide range of parts available, many of them being slow movers. For example, LTAG keeps detailed information on 775,000 parts within its enterprise resource planning system.

In addition, for aircraft safety reasons, each of these aircraft-related parts needs to be certified by the aviation authorities and requires full traceability back to origin. These high standards for production and approval of parts, as well as other quality regulations for suppliers, in combination with high investment costs, generate an overall highly oligopolistic – and, for key parts, even monopolistic – market structure for the supply of aircraft parts and services.

However, the aviation industry is suffering from an ongoing decline of average yield per passenger for the airlines, due to the economic downturn after the events of 11 September 2001, a recession in the business cycle of the world economy and increasing competition by new market entrants such as low-fare carriers Ryanair and easyJet in Europe. The combination of these factors has massively increased economic pressure in the new millennium. LTAG has used a four-phase model to overcome this pressure, optimize processes, and to find innovative ways to increase efficiency and therefore improve airline profitability under strong competition in the long term.

Phase 1: cost-cutting programme

After market deregulation in the late 1980s and early 1990s, newly privatized inefficient flag carriers were forced to significantly cut costs and reduce inefficiencies such as heavy overstaffing, excessive inventory levels and bureaucratic processes in order to survive as profit-driven private airlines. In 1996, Deutsche Lufthansa AG set up a programme to decrease its total cost base. This included massive negotiations on big supplier contracts for aircraft and aircraft-related parts and services. The goal was unit costs per seat kilometre offered of 15 Deutsche pfennigs (approximately 7.7 euro cents) by cutting all costs by a minimum of 4 per cent per year. This goal was reached successfully in 1999. In the mid-1990s, Lufthansa started its concept of an aviation group by separating its passenger transport airline and spinning off not only LTAG as an independent company in 1994 to win new MRO markets outside Lufthansa, but also by founding other units in different fields of aviation such as cargo services, leisure travel and IT solutions.

Phase 2: cost–benefit approaches to purchasing processes

A second step was to increase overall efficiency and work on essential processes. This was based on a net present value analysis for each project process and a total cost of ownership approach for all long-term investments such as contracting the parts supply and the customer support for new aircraft for more than a decade. This cost–benefit approach broadened the view from simple cost-cutting to a quality-based, long-term assessment of business cases. During this phase, major original equipment manufacturers (OEMs) started offering parts pools and on-site support services for the aftersales market of engine parts, components, expendables and consumables in order to contribute to common cost-saving initiatives, and also to secure their aftermarkets. LTAG launched programmes such as 'Programm 150' within the strategic purchasing department, which not only focuses on price reductions, but also systematically tackles inventory levels, logistics and transaction costs. Innovative new tools for purchasing enabled LTAG to head in this direction, especially the invention of a state-of-the-art vendor monitoring/purchasing controlling system, as well as the use of Internet-based platforms for reverse auctions such as the international aviation e-marketplace Aeroxchange.

Phase 3: supply chain development

New tools – together with the ongoing economic pressure that could not be absorbed by MRO providers themselves but needed the combined efforts of both MRO providers and suppliers – have led to strategic projects that considered not only purchasing, but included the supplier base as well as selected airline customers. This required a new level of frankness with key suppliers concerning information exchange and openness in mind, resulting in long-term reciprocal agreements, improving predictability of demand and reduced 'bullwhip effects'. During this phase, LTAG began to exchange forecasting information with selected key suppliers to allow production planning according to the expected demand in the long run, and to place orders corresponding to the actual just-in-time demand from production. Consequently, making use of online, on-time availability of information has lowered stock levels drastically. The customized lead-times programme of Airbus Spares Support and Services in Hamburg allows airlines and their MRO providers to order Airbus proprietary parts without long catalogue lead-times according to their actual demand. Currently, LTAG and Airbus are working on a joint project to connect their parts management systems in such a way that, for a wide range of parts, the actual demand of LTAG's overhaul shops directly leads to a supply by Airbus without manual interference or additional processes.

To support ongoing supply chain initiatives, LTAG also promotes the extension of already existing, and the invention of new, electronic communication and purchasing tools. Currently, supplier communication via the Internet using the XML-SPEC2000 standard is introduced in cooperation with Aeroxchange. As a result, more, and especially smaller, suppliers that have not run a SPEC2000 converter for cost reasons will be in a position to use electronic communication tools when dealing with LTAG. All participating parties will save costs in ordering and accounting processes as well as finding intelligent supply chain solutions like just-in-time ordering/delivery.

Phase 4: integrated supply chain solutions

The next step in evolving purchasing and SCM will be a systematic cross-functional collaboration within the complete MRO organization including all affiliates; cross-border to the supplier base including their sub-suppliers; and with airline/alliance customers worldwide. LTAG is working on establishing a network with its more than 30 different affiliates and also with selected customers. Airlines and MRO providers are also looking for global cooperation – for example, within the Star Alliance Network or by using cooperative e-auctions.

These purchasing networks inherently face growing complexity of different organizational structures, non-standardized IT systems and global production sites. The necessity of stringent supply chain coordination and standardization is one of the most urgent challenges for the near future in the aviation industry. The MRO sector in the global aviation industry will keep changing due to numerous reasons that will affect the market structure and challenge purchasing and supply chain strategies. For MRO providers in the aviation business, a key success factor will be to build up and orchestrate these supply chain networks between major OEM groups on the one hand and their heterogeneous global airline customer base on the other. Finally, future challenges to survive changing conditions and install efficiently working supply chain networks are a matter of active change management within the organization.

Question: **What potential problems do you foresee with these types of network?**

Source: adapted from Dr Jörg Rissiek and Joachim Kressel, 'New Developments in Purchasing and Supply Chain Strategies for the Aviation Industry', *Business Briefing: Global Purchasing & Supply Chain Strategies*. London: Touch Briefings, 2004, pp. 52–5.

Supplier Development and Relationship Management

Purchasing can help support the organization's strategic success by identifying and developing new and existing suppliers. Getting suppliers involved early in the development of new products and services or modifications to existing offerings can reduce development times. The idea of time compression – getting to market quickly with new ideas – can be very important to the success of those ideas and perhaps to the organization's position as a market leader or innovator.

Among the primary purchasing activities that influence the ability of the firm to achieve its objectives are supplier selection, evaluation and ongoing management (sourcing), management of quality, and purchasing planning and research.

Relationship to Other Functions

Virtually every department within an organization relies on the purchasing function for some type of information or support. Purchasing's role ranges from a support role to a strategic function. To the extent that purchasing provides value to other functional areas, it will be included in important decisions and become involved early in decisions that affect purchasing. Being well informed allows the purchasing function to better anticipate and support the needs of other functional areas. This support in turn leads to greater recognition and participation.

Purchasing often has the same functional reporting relationship as logistics, which is helpful for coordinating materials management. Purchasing and logistics need to work closely in coordinating inbound logistics and associated material flows. The following sections apply to purchases of goods and services; they apply equally to purchasing of logistics services and managing relationships with logistics service providers.

Supplier Selection and Evaluation

In the acquisition process, perhaps the most important activity is selecting the best supplier from among a number of suppliers that can provide the needed materials. The buying process is complex because of the variety of factors that must be considered when making a purchase. The process includes both decision-makers and decision-influencers, who combine to form the **decision-making unit**. Increasingly, organizations are using cross-functional teams to make important decisions. The use of teams is described more fully in Chapter 10, which deals with organizational structures.

Figure 4.3 shows some of the many information flows between purchasing and other internal functions that may affect the supplier selection and evaluation system. These flows exist at many levels, from dealing with users on order commitments, to verifying contractual terms with the legal department, to ensuring adequate materials availability, to supporting marketing's sales promotions.

Figure 4.4 shows a basic, five-step purchasing process for managing supplier relationships from the identification of a need to make a purchase through ongoing evaluation and follow-up. Purchasing managers may consider a broad range of factors when making the purchasing decision. These may include issues such as lead-time, on-time delivery performance, ability to expedite, price competitiveness and post-purchase sales support.

Purchase Categories

There are six major purchase categories in most companies: (1) component parts, (2) raw materials, (3) operating supplies, (4) support equipment, (5) process equipment, and (6) services. These may be routine, ongoing purchases or non-routine purchases that may require special attention because they represent a new buy, an infrequent purchase, a major acquisition, or if there are problems or major opportunities (strategic, cost savings) associated with the buy.

In the 1980s and 1990s, increased concern for productivity improvements and cost reduction caused management attention to focus on the purchasing function and on the development of closer ties with a reduced number of suppliers. To determine the impact of supplier performance on productivity, performance must be measured and evaluated (see phase 5 in Figure 4.4). Next, the data can be used to identify those suppliers with whom the firm wishes to develop long-term relationships, to identify problems so that corrective action can be taken and to realize productivity improvements.[6]

Evaluating Suppliers

A variety of evaluation procedures are possible; there is no best method or approach for all firms. Most important, always use consistent procedures to increase the objectivity of the process. Table 4.1 presents an example of an evaluation procedure.

The manager must identify all potential suppliers for the items being purchased. The next step is to develop a list of factors to evaluate each supplier. These should complement the factors used

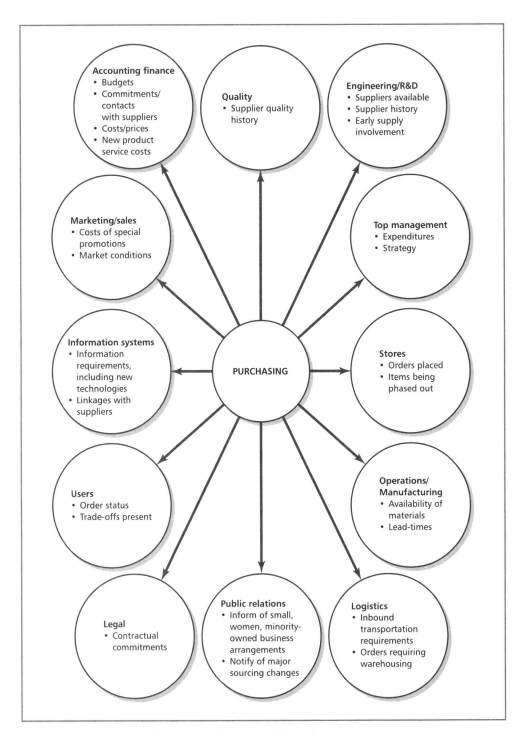

FIGURE 4.3 Overview of Internal Information Flows from Purchasing

Source: adapted from Lisa M. Ellram and Laura M. Birou, *Purchasing for Bottom Line Impact*. Burr Ridge, IL: Irwin Professional Publishing, 1995, p. 74.

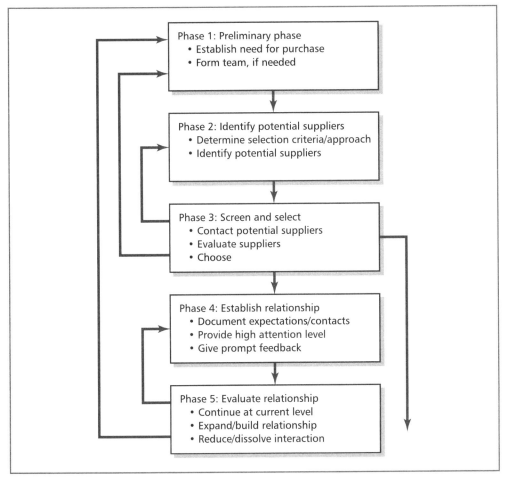

FIGURE 4.4 Five Phases in the Selection, Development and Management of Purchasing Relationships

Source: Lisa M. Ellram, 'A Managerial Guideline for the Development and Implementation of Purchasing Partnerships', *International Journal of Purchasing and Materials Management* 31, no. 2 (1995), p. 12.

earlier in supplier selection. Once the factors have been determined, the performance of individual suppliers should be evaluated on each factor (e.g. product reliability, price, ordering convenience). Table 4.1 uses a five-point scale (1 = worst rating; 5 = highest rating), but other scales may be used.

Prior to evaluating suppliers, management must determine the relative importance of the factors to its particular situation, and assign each a weight. For example, if product reliability were of paramount importance to the firm, it would be given the highest importance rating. If price were not as important as product reliability, management would assign price a lower importance rating. A factor of no importance to the firm would be assigned a zero.

The next step is to develop a weighted composite measure for each factor by multiplying the supplier's evaluation by the factor's importance. The addition of the composite scores for each supplier provides an overall rating that can be compared to the ratings of other suppliers. The

Factor	Rating of supplier (1 = worst rating; 5 = highest rating) 1 2 3 4 5	×	Importance of factor to your firm (0 = no importance; 5 = highest importance) 0 1 2 3 4 5	=	Weighted composite rating (0 = minimum; 25 = maximum)

Supplier A
  Product reliability
  Price
  Ordering convenience
  ■
  ■
  ■
 	
After-sale service
  Total for supplier A
		×		=	Total: _____

Supplier B
  Product reliability
  Price
  Ordering convenience
  ■
  ■
  ■
 	
After-sale service
  Total for supplier B
		×		=	Total: _____

Supplier C
  Product reliability
  Price
  Ordering convenience
  ■
  ■
  ■
 	
After-sale service
  Total for supplier C
		×		=	Total: _____

TABLE 4.1 Evaluating Suppliers in a Typical Manufacturing Firm
Decision rule: select the supplier with highest composite rating.

higher the composite score, the more closely the supplier meets the needs and specifications of the procuring company. Going through the process itself is one of the major benefits of this approach. This forces management to formalize the important elements of the purchasing decision and to question existing methods, assumptions and procedures.

Implementation of a supplier performance evaluation methodology in a company that assembled kits for the health care industry resulted in a reduction in the number of suppliers, closer relationships with remaining suppliers, and a 34 per cent reduction in component inventories within the first few months.[7] After two full years of using the quarterly performance reports, buyers had reduced component inventories by more than 60 per cent.

Selecting Suppliers

Selecting the right suppliers has an immediate and long-term impact on the firm's ability to serve its customers. A formal selection process, similar to the formal evaluation process presented in the previous section, is advisable.

The supplier selection process is more difficult when materials are being purchased in international markets or for international operations. Firms buy raw materials, components and subassemblies from foreign sources because of cost and availability issues.

Some of the complexities of doing business internationally are presented in Chapter 12, which looks at global logistics. The Global box demonstrates how Diageo Scotland uses a framework of 12 key enablers to maximize its procurement efforts for the benefit of the entire Diageo corporation.

The rewards associated with the proper selection and evaluation of suppliers can be significant. As we saw in Chapter 1, logistics cost savings can be leveraged into substantial improvements in profits. Similarly, purchasing activities can have positive effects on the firm's profits. Not only will a reduction in the cost of materials increase the profit margin on every unit manufactured and sold, but the lower cost logistics associated with the materials purchased will reduce the investment in inventories by decreasing the cost per unit and the number of units in inventory.

In addition, customer service improvements are possible because the manufacturing process can operate smoothly, with no slowdowns or shutdowns. Since effective purchasing management results in the acquisition of high-quality materials, there is less likelihood that customers will return finished goods due to product failure.

GLOBAL

Diageo Scotland Uses 12 Enablers for World-class Procurement

Diageo is the world's leading premium drinks business. With its global vision and local marketing focus, Diageo brings to consumers an outstanding collection of beverage alcohol brands across the spirits, wine and beer categories, including Smirnoff, Guinness, Johnnie Walker, Baileys, J&B, Cuervo, Captain Morgan and Tanqueray, and Beaulieu Vineyard and Sterling Vineyards wines. Diageo trades in some 180 markets around the world and is listed on both the New York Stock Exchange (DEO) and the London Stock Exchange (DGE).

As a global company Diageo recognizes the importance of the world-class procurement required to support the company's entire value chain and supply costs, and has invested heavily in improving its total procurement effectiveness accordingly.

One of the many examples of this investment can be seen in the company's operations in Scotland. A procurement team of 25 people have the responsibility for ensuring the seamless supply of a £350 million per annum spend on the goods and services required in order to support 27 malt and two grain distilleries, and three major bottling plants, which provide 120 Diageo brands to 180 markets globally.

To ensure the procurement team continually improves the contribution it makes to the business through cost savings, supply chain risk management, materials quality, service improvement and support to brand growth through supplier innovation, the team uses a framework of 12 enablers of procurement capability to agree priority areas for improvement.

Philip Boyd, procurement director, Diageo Scotland, says, 'The 12 enablers are not exhaustive but they are relevant to diagnosing the capability of the procurement team as well as identifying key areas for future focus.'

The 12 enablers are as follows.

1 *Sponsorship and support*: where does the real drive and support for improving procurement's capability, contribution, standing and rewards come from?
2 *Total expenditures impacted*: what is the real organization reach of procurement within the company?
3 *Consistent processes*: what is the level of routine application of codified processes to drive source selection, supplier management and capability development?
4 *Aggregation of expenditures*: to what extent are purchases being aggregated across business unit and geographic boundaries wherever justified?
5 *Supplier selection criteria*: what key criteria determine sourcing and which functions actively input to decision-making?
6 *Supplier relationship*: what is the nature of prevailing supplier relationships; how and by whom are they managed?
7 *Size of the supply base*: what concerted effort has gone into consolidation of the total supply base?
8 *Business and strategic planning processes*: what direct linkages are there between procurement and overall company budgeting, annual and strategic plans?
9 *People capability*: what is the principal source of talent and what is the depth and breadth of functional, organizational and leadership capability contained within procurement?
10 *Procurement information systems and e-technology systems*: how sophisticated are information systems and e-enablement platforms supporting sourcing strategy, supplier selection decisions, supplier relationship management and capability development?
11 *Measurement and reporting*: what key performance indicators measure procurement's contribution against imperative, and when and by whom are they monitored and reviewed?
12 *Communications*: how effective are procurement's communications to key internal and external stakeholders?

The procurement leadership team uses these enabling statements in a framework as a diagnostics tool to measure progress. Current priority areas of focus for Diageo Scotland are the organizational impact of expenditures, the leverage of expenditures, e-technology and people capability.

Question: **How could these 12 steps be successfully implemented in small and medium-sized enterprises (SMEs)?**

Sources: interview with Philip Boyd, procurement director Scotland, Diageo, April 2005, and Diageo website at www.diageo.com (2005). Reprinted with kind permission.

Quality Management

Although cost is an important consideration in materials acquisition, so is quality management. The initial purchase price of an item is only one element of the total cost. For example, some items are easier to work with than others and can save production costs. Materials of higher quality may require fewer fabrication processes or have a longer life span, resulting in lower overall product costs or higher prices for finished products. Companies must achieve some balance between the components of the acquisition process – namely, price, cost and value received.

After the required quality level has been determined and specifications developed, usually by manufacturing, it becomes purchasing's responsibility to secure the proper materials. The correct quality specification must be given to suppliers. The supplier that offers the best cost–quality combination that meets the specifications should be selected.

The firm should never pay higher prices to obtain materials with quality levels greater than those specified by manufacturing unless justifiable marketing or logistics reasons exist for doing so. Purchasing materials that needlessly exceed quality specifications adds unnecessary costs to products and increased inventory costs.

One way that firms might ensure quality is through inspection of incoming materials parts. However, this is costly and time consuming. Inspection requires human resources, space and perhaps test equipment. In addition, incoming inventory is tied up or delayed awaiting inspection. For these reasons, purchasing managers have turned to **supplier certification**. In the certification process, the supplier's quality levels and processes are closely evaluated by members of the buying firm. If they 'pass', the buying organization no longer inspects that supplier's incoming material.

Quality is even more critical for firms pursuing a JIT philosophy, where little or no inventory is held. Improper quality in a JIT environment can shut down processes immediately, creating excessive costs and delays. This topic is discussed more fully in Chapter 6, on materials management, but is also presented here as it relates to purchasing and procurement activities concerned with quality of inbound materials.

Just-in-time

Just-in-time (JIT) manufacturing is more a philosophy of doing business than a specific technique. The JIT philosophy focuses on the identification and elimination of waste wherever it is found in the manufacturing system. The concept of continuous improvement becomes the central managerial focus.

Typically, JIT implementation involves the initiation of a 'pull' system of manufacturing (matching production to known demand) and the benefits include: significant reductions of raw material, work-in-process and finished goods inventories; significant reductions in throughput time; and large decreases in the amount of space required for the manufacturing process.

A company implementing JIT can usually make the greatest improvement in the area of quality. The JIT focus on the elimination of waste includes the supplier, with the aim of reducing waste and cost throughout the entire supply chain. If a manufacturer decides it will no longer carry a raw materials inventory and that henceforth its suppliers must carry this inventory, the supply chain cost is reduced because inventory with lower value added is being held.

One example of this is Nissan, which requires its suppliers to hold a certain amount of inventory at factories or warehouses near its northeast England production facilities in Sunderland, so suppliers can respond quickly in case of problems. It is preferable that this inventory be eliminated altogether because while those additional inventory carrying costs may be borne in the short term by the seller, eventually they may be passed on to the buyer in the form of higher prices. The supplier needs to reduce its own manufacturing and supplier lead-times. The major differences between 'traditional' and JIT purchasing are summarized in Table 4.2.

Difficulties in Implementing JIT

One of the reasons most frequently cited to do with difficulty in the implementation of JIT is a lack of cooperation from suppliers, due to the changes required in the supplier's system. In addition to changing from traditional quality-control inspection practices to the implementation of statistical process control, the supplier is asked to manufacture in quantities that may differ from the usual lot sizes and to make frequent deliveries of small lots with precise timing. The supplier and buyer are normally required to provide each other with access to their master production planning system, shop floor schedule and material requirements planning system.

Purchasing activity	Traditional approach	JIT approach
Supplier selection	Minimum of two suppliers; price is central	Often one local supplier; frequent deliveries
Placing the order	Order specifies delivery time and quality	Annual order; deliveries made as needed
Change of orders	Delivery time and quality often changed at the last moment	Delivery time and quality fixed, quantities are adjusted within predetermined margins if necessary
Follow-up of orders	Many phone calls to solve delivery problems	Few delivery problems thanks to sound agreements; quality and delivery problems are not tolerated
Incoming inspection	Inspection of quality and quantities of nearly every delivered order	Initial sample inspections; later, no inspections necessary
Supplier assessment	Qualitative assessment; delivery deviations of up to 10% are tolerated	Deviations are not accepted; price is fixed based on open calculation
Invoicing	Payment per order	Invoices are collected and settled on a monthly basis

TABLE 4.2 Differences Between the Traditional Approach and the JIT Approach in Purchasing

Source: A.J. van Weele, *Purchasing Management: Analysis, Planning and Practice*. New York: Chapman and Hall, 1994, p. 132.

Importance of Buyer–supplier Communication

Under JIT, close and frequent buyer–supplier communication is essential. Suppliers are given long-range insight into the buyer's production schedule. Often, this look ahead spans months, but the schedule for the nearest several weeks is frozen. This allows the supplier to acquire raw materials in a stockless production mode and to supply the buyer without inventory build-ups. Suppliers provide daily updates of progress, production schedules and problems. Purchasers and suppliers must cooperate and have a trusting relationship in order to convert supply chains to JIT operations.

Supplier selection, single sourcing, supply management and supplier communication become critical issues for purchasing and materials managers in implementing JIT. Issues relating to supplier selection include quality-control methods, supplier proximity, manufacturing flexibility and lead-time reliability.

JIT manufacturers and their suppliers generally develop close collaborative relationships supported by long-term, single-source contracts. The concept of partnering, described in further detail later in the chapter, is often applied to JIT buyer–supplier relationships.

Following supplier selection, careful supplier performance measurement and management often lead to supplier certification – a designation reserved for those suppliers whose quality, on-time delivery and reliability have proven acceptable over long periods of time.

Under JIT, the purchasing department has significantly changed from the processing of orders to a focus on supplier selection and long-term contract negotiation. Many times these close communications are supported by electronic data interchange (EDI) capabilities to facilitate the timely and accurate transmittal of information. The remaining sections apply to purchasing in general, and are critical to support the success of JIT purchasing.

Purchase Agreements

JIT purchasing is facilitated by an even, repetitive master production schedule. Repetitive manufacture of products evens out the demand for individual parts. The steady demand for parts has an impact on shipping quantities, containers and purchasing paperwork. In Japan, JIT purchase agreements usually involve little paperwork. The purchase order may specify price and overall quantity, but the supplier will deliver in accordance with a schedule or with daily production needs, which are telephoned from the buying plant. The JIT purchase agreement does not permit variability. In most cases, the buyer expects and receives the exact quantity. Having a purchase agreement in place saves much time in negotiating and pricing each order.

Value Analysis

Value analysis is a respected purchasing practice that may receive more attention as a result of the interest in JIT purchasing. When negotiating a purchase agreement, the supplier receives the buyer's specifications and provides a bid price. If the price is too high, the buyer may visit the supplier's plant to review its processes. The objective is to identify areas where the supplier's costs exceed the value added and, if possible, to modify the minimal specifications in order to reduce the supplier's cost and the bid price.

'Loose' Engineering Specifications/Early Supplier Involvement

Manufacturing engineers in the USA and Europe tend to specify tolerances for almost every design feature for which parts are purchased. The Japanese place more importance on how the item actually performs than on conformance to tight design specifications. The supplier is permitted to innovate on the premise that the supplier is the expert.

The concept of getting the supplier involved in the design process is often called **early supplier involvement** (ESI). **Concurrent engineering** is a type of early supplier involvement where the engineers in the buying and selling firms work together on product development or product improvement.

The benefits of closer coordination of engineering and quality matters are significant. Engineers and quality control people may pay frequent visits to a supplier's plant to answer engineering questions and identify potential quality problems before they surface. Xerox Corporation utilizes these practices with its key suppliers, resulting in better supplier quality, responsiveness and competitiveness.

Control of Inbound Transportation

Inbound freight decisions such as delivery and routing are frequently left to the supplier's traffic department. This is often the case when materials are purchased 'Free On Board (FOB) shipping point' and the buyer owns the goods and absorbs the inventory carrying costs from the date of shipment.

JIT purchasing requires steady, reliable incoming deliveries. The objective is to avoid excessive inventory carrying costs for materials that arrive early and to avoid disruptions in manufacturing operations when goods arrive late. Therefore, the buying firms must become involved in selecting both the transportation mode and the specific carrier.

Supplier Development

Supplier development is a when buyer and supplier jointly:

> . . . make dedicated investments in the relationship and create technical bonds . . . in order to create new product and service offerings [and where] the buyer takes the lead in setting 'stretch' [improvement] targets on functionality and costs.[8]

Sometimes organizations find that their current suppliers are unable to support stringent JIT quality and delivery requirements. Such organizations may search for other suppliers or work with suppliers to develop the skills needed to support JIT. Supplier development efforts are increasing as organizations form longer-term relationships with suppliers.

JIT II and JIT III

JIT II is an innovative type of purchasing relationship that aims JIT principles at the purchasing function. Like JIT, JIT II attempts to eliminate waste, redundancy and excess paperwork, and to improve quality, responsiveness and innovation in the purchasing arena. It represents a type of alliance relationship between a buying and selling organization. The term *JIT II* was coined by Bose Corporation to describe this type of relationship.[9] The steps in developing JIT II are shown in Box 4.1.

Box 4.1 Steps in JIT II Information Flow

Step 1 & 2: Supplier reassigns its sales representative to new duties, and customer reassigns its purchaser.

Step 3: In full JIT II implementation, the customer reassigns its material planner to new duties.

Step 4: Supplier replaces purchaser, planner and salesperson with a full-time professional at the customer's location. At Bose Corporation, supplier professionals are called 'in-plants'. Although supplier replaces purchaser with an in-plant rep, this step actually assists existing purchasing personnel as more people address the overall department workload.

Step 5: The in-plant representative works 40 hours a week at the customer's location, usually in its purchasing department.

Step 6: Customer empowers the in-plant within its planning and purchasing systems. The in-plant works directly from the customer's MRP (or similar) system, and uses the customer's purchase order to place material orders on his or her own company. Note: the customer typically prohibits the in-plant from placing purchasing orders with other companies.

Step 7: Customer provides the in-plant with an employee badge (or equivalent), providing free access to customer engineering and manufacturing personnel. When not planning and ordering material, the in-plant practises concurrent engineering by working with the customer's design engineering staff.

Step 8: Customer and supplier understand that many more steps lie ahead. JIT II will cause change in both organizations.

Source: *Purchasing*, 6 May, 1993, p. 17.

In JIT II, the supplier places one of its employees, called an 'in-plant', in the buying company's office, replacing a purchaser, planner and salesperson. In addition to co-location, the concurrent engineering and continuous improvement aspects of JIT II distinguish it from other supplier relationships. One of the companies with which Bose has established this in-plant relationship is G&F Industries, an injection moulder. The in-plant representative places orders, practises concurrent engineering, and has full access to all of Bose's facilities, information and employees. The supplier benefits include greater integration with the customer, improved communications, more efficient administrative processes and savings on 'sales effort'.[10]

JIT III is a further extension of JIT II, where the supplier locates its own factory close to its customer to facilitate sequential and frequent deliveries.[11] Suppliers to Nissan's factory in Sunderland, northeast England, are located close to the factory. Sunderland also has good road and rail links that enable Nissan to readily receive supplies from over 100 separate UK component and subassembly suppliers and also to ship completed vehicles.

Purchasing Research and Planning

Uncertainty in the business environment is making the purchasing decisions for key items more complex and the effects of these decisions more long lasting. Important environmental considerations include uncertainty of supply and dependence on foreign sources for key commodities, price increases on key commodities, extended and variable lead-times, energy shortages or price increases, government regulation such as environmental laws, and increasing worldwide competition.

The changing environment makes it necessary for purchasing management to do a more effective job of researching the supply market. Purchasing needs to provide information about supply conditions (e.g. availability, lead-times, prices, technology) to different groups within the firm, including top management, logistics, engineering, design, manufacturing and marketing. This information is important when formulating long-term strategy and making short-term decisions. Key materials for which availability, pricing and quality problems may occur should be identified, so that action plans can be developed before problems become critical and costly.

Strategic planning for purchasing involves the identification of critical purchases, supply market analysis, risk assessment, and strategy development and implementation. It is important to determine whether materials problems or shortages might jeopardize current or future production of new or existing products, whether materials quality can be expected to change, whether prices are likely to increase or decrease, and the appropriateness of forward buying. Management should develop specific plans to ensure an uninterrupted flow of materials.

Typical criteria to use in identifying critical purchases are percentage of product cost, percentage of total purchase expenditure and use on high-margin end items. Criteria used for determining the risk in the supply market include number of suppliers, availability of raw materials to suppliers, supplier cost and profitability needs, supply capacity and technological trends. The more critical the purchase and the riskier the supply market, the greater attention the purchase requires.

Risk assessment requires that the purchaser determine the probability of best or worst conditions occurring. Supply strategies like those shown in Table 4.3 should be developed for the predicted events. Asking these questions for any given strategy or situation can help purchasing ensure that it has considered the important issues. Implementation of a particular strategy requires the involvement of top management and integration with the firm's overall business plan.

Purchasing Cost Management

Given the large percentage of the firm's dollars that purchasing spends, purchasing departments must manage and reduce costs. Purchasing can use a number of methods to reduce administrative costs, purchase prices and inventory carrying costs, but the most prevalent are purchase cost reduction programmes, price change management programmes, volume leverage (time or quantity) contracts, systems contracts and stockless purchasing, and establishing long-term relationships with suppliers.

Purchasing savings have the same sort of profit leverage effect as logistics cost savings (see Chapter 1). If top management calls for a set percentage of cost reduction in all areas of spending, the potential impact of purchasing is large. Because purchasing spends such a large percentage of a firm's revenue, a 10 per cent cost reduction in purchase expenditures has a much greater impact than a 10 per cent reduction in labour or overhead expenses.

| 1. What?
   Make or buy
   Standard versus special

2. Quality?
   Quality versus cost
   Supplier involvement

3. How much?
   Large versus small quantities (inventory)

4. Who?
   Centralize or decentralize
   Quality of staff
   Top management involvement

5. When?
   Now versus later
   Forward buy

6. What price?
   Premium
   Standard
   Lower
   Cost-based
   Market-based
   Lease/make/buy

7. Where?
   Local, regional
   Domestic, international
   Large versus small | Single versus multiple source
High versus low supplier turnover
Supplier relations
Supplier certification
Supplier ownership

8. How?
   Systems and procedures
   Computerization
   Negotiations
   Competitive bids
   Fixed bids
   Blanket orders/open orders
   Systems contracting
   Blank check system
   Group buying
   Materials requirements planning
   Long-term contracts
   Ethics
   Aggressive or passive
   Purchasing research
   Value analysis

9. Why?
   Objectives congruent
   Market reasons
   Internal reasons
   1. Outside supply
   2. Inside supply |

TABLE 4.3 Supply Strategy Questions

Source: Michiel E. Leenders, Harold E. Fearon, Anna E. Flynn and P. Fraser Johnson, *Purchasing and Supply Management*, 12th edn. Burr Ridge, IL: McGraw-Hill Irwin, 2002, p. 698.

Cost-reduction Programmes

An effective cost-reduction programme by purchasing requires top management support, clear definition of goals, visibility of savings to top management, measurement of savings, reporting on the process and its results, and incorporation of cost-reduction goals in the individual performance appraisal process.

For a cost-reduction programme to succeed, top management must communicate the need for cost-saving accomplishments in both good and bad economic times. The programme must define cost-reduction objectives adequately, so that accomplishments can be measured and performances evaluated.

In many firms, for example, a 'cost reduction' is defined as a decrease in prior purchase price. This means a cost reduction occurs only when the firm is paying a lower price. Cost avoidance is the amount that *would have* been paid less the amount actually paid. The distinction between cost savings and cost avoidance is shown in Table 4.4.

Cost-reduction and avoidance programmes may include any of the following:

- supplier development
- development of competition
- requirement of supplier cost reduction
- early supplier involvement in new product design and design changes
- substitution of materials
- standardization
- make-or-buy analysis
- value analysis, including supplier involvement
- the reduction of scrap
- a change in tolerances
- improvement of payment terms and conditions
- volume buying
- process changes.

The appropriateness of each technique will vary with the purchase situation and type of supplier relationship.

	Per unit cost
Scenario 1 – Cost savings:	
Current price paid	€20.00
New price	19.00
Cost savings	€1.00
Scenario 2 – Cost avoidance:	
Current price paid	€20.00
New price quoted by supplier	25.00
Price obtained from alternate supplier	22.00
Cost savings	
Current price paid	€20.00
New price actually paid	22.00
Cost savings [Actually a €2.00 price increase]	– €2.00
Cost avoidance	
New price quoted	€25.00
New price actually paid	22.00
Cost avoidance	€3.00

TABLE 4.4 Cost Savings Versus Cost Avoidance

Price Change Management

Purchasing managers must challenge supplier price increases and not treat them as pass-through costs. It is important to work with suppliers to restrict price increases to a reasonable and equitable level. Furthermore, purchasing should establish a systematic method of handling all price increase requests from suppliers. At a minimum, the system should require:

- determination of the reason for the price change request
- specification of the total euro value impact on the firm
- justification of the price change by suppliers
- review of the price change by management
- strategies to deal with price increases
- alternatives for reducing other price elements or improving processes to offset the price increase.

Purchasing should work with the supplier to offset price increases through other improvements, such as reduced delivery lead-times, better service or other opportunities. To restrict price increases, management should require price-protection clauses and advance notification of 30, 60 or 90 days for price increases. As part of a programme of price change management, purchasing should determine the impact of engineering changes on product costs before it recommends making these changes.

Forward Buying Versus Speculative Buying

Frequently, conditions such as potential supply constrictions or inflationary markets cause purchasing managers to buy more of a product than is required for current consumption. This practice, called **forward buying**, serves to protect the organization from anticipated shortages or to delay the impact of rising prices. The trade-off, of course, is increased inventory carrying costs. When using this strategy, the purchasing manager must evaluate the trade-off between inventory carrying cost increases and the risk of supply constriction or increased prices.

Speculative buying refers to purchases made not for internal consumption, but to resell at a later date for profit. Speculative goods may be the same as goods purchased for consumption, but the quantities purchased will be in excess of current or future needs. An example occurs in the diverting of retail goods.

Companies may offer special discounts to retailers only in certain areas of the country. Retailers will buy substantial quantities of goods to ship to other locations or even to sell to other retailers in different parts of the country where the discount is unavailable. The retailer makes a profit by selling for more money or saving enough money to offset the increased freight and handling costs. The fundamental intent is to take advantage of expected increases in price to profit from the resale of the goods.

Volume Contracts

Volume contracts are a way to leverage purchase requirements over time, between various business units or locations in the company, or on different line-item requirements. As a result of combining purchases, the buyer's leverage with suppliers can lead to reductions in purchase prices and administrative costs. Cumulative volume discounts allow a buyer to combine purchase volume over time, getting lower prices with successive buys as it places additional orders throughout the year. More companies are using this approach to support smaller, more frequent buys in JIT purchasing.

In non-cumulative discounts, the price is based on the amount of each order. A review of purchase prices for a particular item often identifies the opportunity for suppliers to provide quotes on a semi-annual or contract basis. An increase in the purchase quantity can enable suppliers to reduce their costs and prices as a result of production or purchasing economies. In addition, the supplier may be willing to accept lower per unit margins on a higher volume of business.

Past purchase patterns should be available from computer-generated requirement plans and from suppliers. Management needs to review the firm's purchase history systematically and regularly to find new opportunities for volume contracting.

Stockless Purchasing

Stockless purchasing, or **blanket orders**, are a means of reducing materials-related costs such as unit purchase price, transportation, inventory and administration. Contracts are arranged for a given volume of purchases over a specified period of time. The supplier provides products to individual plant locations as ordered, and payment is arranged through purchasing. Logistics is involved in the actual order release, which simplifies repetitive ordering considerably and lowers transaction costs.

Stockless purchasing implies that the firm does not carry inventory of purchased materials. While it may or may not result in 'zero' inventory, the underlying principles of stockless purchasing support improved inventory management. The objectives of stockless purchasing are to:

- lower inventory levels
- reduce the supplier base
- reduce administrative cost and paperwork
- reduce the number of purchases of small euro value and requisitions that purchasers have to handle, freeing up time for more important activities
- achieve volume leverage with suppliers, lowering costs and improving service
- provide for timely delivery of material directly to the user
- standardize purchase items where possible
- have suppliers manage inventory and, in some cases, place orders.[12]

Stockless purchasing systems are best suited to frequently purchased items of low euro value where administrative processing costs are relatively high compared with unit prices. In many cases, the combined administrative, processing and inventory carrying costs may exceed the item's cost. Stockless purchasing may lead to larger supplier discounts, reduced processing costs and increased product availability.

Going beyond systems contracts is the concept of **integrated supply**. Under this concept, a purchaser will combine all buys of like items with one supplier, further reducing administrative costs and increasing leverage. Examples of items well suited to integrated supply are office supplies, lab supplies, small tools, screws, nuts and bolts, and standard electrical components.

Usually, the length of the contract for purchasing agreements varies from one to five years and includes price-protection clauses. The purchaser should have the right to research the market to ensure that suppliers' unit prices are reasonable.

In the European Union the Utilities Directive and the proposed consolidated public-sector Directive both define a **framework agreement**, which is an agreement with public-sector suppliers, the purpose of which is to establish the terms governing contracts to be awarded during a given period, in particular with regard to price and quantity. In other words, a framework agreement is a general term for agreements with suppliers that set out terms and conditions under which specific purchases (call-offs) can be made throughout the term of the agreement. Framework agreements can be concluded with a single supplier or with several suppliers, for the same goods, works or services. The length of call-offs is not specifically limited by the directives – for example, call-offs might be for three, six or 12 months, or longer.

Vendor Managed Inventory (VMI) and Supplier Managed Inventory

VMI and SMI are two initiatives within a broad class of automatic replenishment programs (ARPs).[13] Other ARPs of continuous replenishment, quick response and efficient consumer response were discussed in Chapter 2. VMI involves the coordinated management of finished goods inventories outbound from a manufacturer, distributor or reseller to a retailer or other merchandiser, while SMI involves the flow of raw materials and component parts inbound to a manufacturing process.

VMI is the more common of the two replenishment programs. VMI vendors generate purchase orders on an as-needed basis by closely monitoring customer inventory levels and replenishing supplies based on an established inventory plan. The inventory plan accounts for dynamic changes in demand associated with the product forecast, life cycle and related promotional activity.

SMI employs a similar logic to VMI but on the inbound side to the manufacturing operation. The key difference between SMI and VMI is that rather than replenishing finished goods on a reorder point basis, the manufacturer's production schedule triggers the replenishment of materials in the SMI program. Suppliers are provided with production schedules in advance with regular updates on 'takt' time (desired time between output units of production that are synchronized to customer demand), production assortments, and total volume adjusted to changes in demand.

E-procurement

The widespread use of the Internet has created numerous opportunities for improving supply chain performance, particularly in purchasing and procurement. E-procurement is the use of the Internet in purchasing and procurement. There are six forms of e-procurement applications: e-sourcing, e-tendering, e-informing, e-reverse auctions, e-MRO and web-based enterprise resource planning (ERP).[14]

E-sourcing consists of identifing new suppliers using Internet technology. E-tendering is the process of sending requests for information and proposals to suppliers, and receiving responses via the Internet. E-informing is the process of gathering and distributing purchasing information among internal and external parties, using Internet technology.

E-reverse auctioning is the Internet-based equivalent of a reverse auction, which enables a supplier to sell surplus goods and services to a number of known or unknown buying organizations. Lastly, e-MRO and web-based ERP are processes for creating and approving purchasing requisitions, placing purchase orders, and receiving goods and services ordered using Internet-based software systems.

The Technology box gives an example of how suppliers to the German government can take advantage of an initiative to tender bids and supply on an e-procurement basis.

TECHNOLOGY

E-procurement in Public Purchasing

Germany's e-procurement flagship project, which has been supported since 2000 with around €4.5 million of funding from the Federal Ministry of Economics and Technology, has led the country's public administration into new territory. For the first time, the pilot project shows that even the complexities of public procurement procedures can be handled via the Internet, while adherence to all the procurement rules and a high level of security remain assured. All the types of procurement, be they the purchasing of paper-clips or the construction of town halls, can be dealt with fully online.

The e-procurement system covers all the necessary steps, from the contract notice to the contract award, and does so in conformity with the law. The economic impact of the new electronic technologies is immense. The federal government, the Länder and the municipalities award public contracts worth a total of €250 billion a year and are therefore Germany's largest purchaser of goods and services. These contracts account for around 13 per cent of gross national product and 25 per cent of public-sector spending.

The e-procurement project pursues several objectives. First, it aims to use modern information and communications technologies and multimedia to enhance quality and efficiency, and thus to contribute towards an efficient and economic public administration. A reference model for the electronic award of public contracts is being developed, which can be used by contracting authorities at federal, Länder and municipal level and by other public agencies.

E-procurement also aims to turn the public sector into a driving force for the spread of more electronic business processes in Germany. E-procurement is one of the most important examples of applications in the government-to-business part of e-business. This project makes an important contribution to ensuring that e-business and e-government fit together. The digital structures of public administration and business are designed in the project in such a way that they can be linked up throughout the process and that interactions do not require the use of different media. Ultimately, the aim is to integrate the entire process chain in a holistic e-business and e-procurement concept.

Both the contracting authorities and the companies submitting bids benefit from e-procurement. The companies obtain simpler access to the invitations to tender via the Internet. This shortens their response times and facilitates direct contact with further partners when drawing up a bid. Also, the processing times in the authorities are shortened substantially since the move from the submission of the bid to the evaluation and award process occurs seamlessly and, in many cases, in parallel and can also be documented without any difficulty.

Finally, e-procurement has the aim of encouraging the increased use of multimedia and the Internet in business. After all, the opportunity to participate in the contracts issued by the federal government, Länder and municipalities is a major incentive for the continued expansion of modern information and communications technologies in business. The achievement of greater transparency in the field of contract notices also suggests that more small and medium-sized enterprises (SMEs) will participate in public procurement procedures. Greater competition creates expectations of lower procurement prices and thus savings.

The German government will do all it can to make the electronic award of public contracts via e-procurement the prevailing form of modern procurement. There will be no alternative in future to e-procurement as a key element of e-business. The experience gained must be deepened by further electronic invitations to tender so that this method can replace the paper form in the foreseeable future. It is already apparent that the platform created by the e-procurement flagship project can also become a model for public procurement in Europe.

Question: **Discuss the use and potential applications of electronic purchasing in other, non-public sectors.**

Source: adapted from Dr Andreas Goerdeler, 'Electronic Public Procurement in Germany', *Business Briefing: Global Purchasing & Supply Chain Strategies*. London: Touch Briefings, January 2003, pp. 51–4.

Managing Supplier Relationships

Supplier partnerships have become one of the hottest topics in inter-firm relationships. Business pressures such as shortened product life cycles and global competition are making business too complex and expensive for one firm to go it alone. Despite all the interest in partnerships, a great deal of confusion still exists about what constitutes a partnership and when it makes the most sense to have one. This section will present a model that can be used to identify when a partnership is appropriate as well as the type of partnership that should be implemented.

While there are countless definitions of partnerships in use today, we prefer this one:

A partnership is a tailored business relationship based on mutual trust, openness, shared risk and shared rewards that yields a competitive advantage, resulting in business performance greater than would be achieved by the firms individually.[15]

Types of Partnership[16]

Relationships between organizations can range from arm's length relationships (consisting of either one-time exchanges or multiple transactions) to vertical integration of the two organizations, as shown in Figure 4.5. Most relationships between organizations have been at arm's length where the two organizations conduct business with each other, often over a long period of time and involving multiple exchanges. However, there is no sense of joint commitment or joint operations between the two companies. In arm's length relationships, a seller typically offers standard products/services to a wide range of customers who receive standard terms and conditions. When the exchanges end, the relationship ends. While arm's length represents an appropriate option in many situations, there are times when a closer, more integrated relationship, called a partnership, would provide significant benefits to both firms.

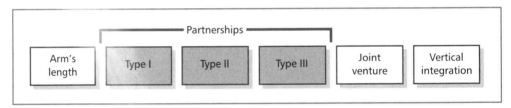

FIGURE 4.5 Types of Relationship

Source: Douglas M. Lambert, Margaret A. Emmelhainz and John T. Gardner, 'Developing and Implementing Supply Chain Partnerships', *The International Journal of Logistics Management* 7, no. 2 (1996), p. 2.

A partnership is not the same as a joint venture or strategic alliance, which normally entails some degree of shared ownership across the two parties. Nor is it the same as vertical integration. Yet a well-managed partnership can provide benefits similar to those found in joint ventures or vertical integration.

While most partnerships share some common elements and characteristics, there is no one ideal or 'benchmark' relationship that is appropriate in all situations. Because each relationship has its own set of motivating factors as well as its own unique operating environment, the duration, breadth, strength and closeness of the partnership will vary from case to case and over time. Research has indicated that three types of partnerships exist.

Type I

The organizations involved recognize each other as partners and, on a limited basis, coordinate activities and planning. The partnership usually has a short-term focus and involves only one division or functional area within each organization.

Type II

The organizations involved progress beyond coordination of activities to integration of activities. Although not expected to last 'for ever', the partnership has a long-term horizon. Multiple divisions and functions within the firm are involved in the partnership.

Type III

The organizations share a significant level of integration. Each party views the other as an extension of their own firm. Typically no 'end date' for the partnership exists.

Normally, a firm will have a wide range of relationships spanning the entire spectrum, the majority of which will not be partnerships but arm's length associations. Of the relationships that are partnerships, the largest percentage will be of Type I, and only a limited number will be of Type III. Type III partnerships should be reserved for those suppliers or customers who are critical to an organization's long-term success. The relationship between Coca-Cola and McDonald's restaurants, for instance, has been evaluated as a Type III partnership.

The Partnership Model

The partnership model shown in Figure 4.6 has three major elements that lead to outcomes: drivers, facilitators and components. **Drivers** are compelling reasons to partner. **Facilitators** are supportive corporate environmental factors that enhance partnership growth and development. **Components** are joint activities and processes used to build and sustain the partnership. Outcomes reflect the performance of the partnership.

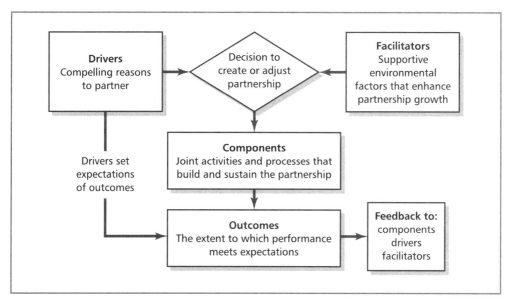

FIGURE 4.6 The Partnering Process
Source: Douglas M. Lambert, Margaret A. Emmelhainz and John T. Gardner, 'Developing and Implementing Supply Chain Partnerships', *The International Journal of Logistics Management* 7, no. 2 (1996), p. 4.

Drivers

Both parties must believe that they will receive significant benefits in one or more areas and that these benefits would not be possible without a partnership. The primary potential benefits that drive the desire to partner include: (1) asset/cost efficiencies, (2) customer service improvements, (3) marketing advantage, and (4) profit stability/growth (see Box 4.2 for examples).

While the presence of strong drivers is necessary for successful partnerships, the drivers by themselves do not ensure success. The benefits derived from the drivers must be sustainable over the long term. If, for instance, the marketing advantage or cost efficiencies resulting from the relationship can easily be matched by a competitor, the probability of long-term partnership success is reduced.

Box 4.2 Partnership Drivers, Facilitators, and Components

Partnership drivers

- Asset/cost efficiency: What is the probability that this relationship will substantially reduce channel costs or improve asset utilization – for example, product costs, distribution costs savings, handling cost savings, packing costs savings, information-handling costs savings, managerial efficiencies and assets devoted to the relationship?
- Customer service: What is the probability that this relationship will substantially improve the customer service level as measured by the customer – for example, improved on-time delivery, better tracking of movement, paperless order processing, accurate order deliveries, improved cycle times, improved fill rates, customer survey results and process improvements?
- Marketing advantage: What is the probability that this relationship will lead to substantial marketing advantages – for example, new market entry, promotion (joint advertising, sales promotion), price (reduced price advantage), product (jointly developed product innovation, branding opportunities), place (expanded geographic coverage, market saturation), access to technology and innovation potential?
- Profit stability/growth: What is the probability that this relationship will result in profit growth or reduced variability in profit – for example, growth, cyclical levelling, seasonal levelling, market share stability, sales volume and assurance of supply?

Partnership facilitators

- Corporate compatibility: What is the probability that the two organizations will mesh smoothly in terms of: (1) culture – for example, both firms place a value on keeping commitments, constancy of purpose, employees viewed as long-term assets and external stakeholders considered important, and (2) business – for example, strategic plans and objectives consistent, commitment to partnership ideas and willingness to change?
- Management philosophy and techniques: What is the probability that the management philosophy and tech-niques of the two companies will match smoothly – for example, organizational structure, use of TQM, degree of top management support, types of motivation used, importance of teamwork, attitudes toward 'personnel churning' and degree of employee empowerment?
- Mutuality: What is the probability both parties have the skills and predisposition needed for mutual relationship building? Is management skilled at two-sided thinking and action, taking the perspective of the other company, expressing goals and sharing expectations, and taking a longer-term view, for example, or is management willing to share financial information and integrate systems?
- Symmetry: What is the probability that the parties are similar on the following important factors that will affect the success of the relationship: relative size in terms of sales, relative market share in their respective industries, financial strength, productivity, brand image/reputation and technological sophistication?

Partnership components

- Planning (style, level and content)
- Joint operating controls (measurement and ability to make changes)
- Communications (non-routine and day-to-day: organization, balanced flow and electronic)
- Risk/reward sharing (loss tolerance, gain commitment and commitment to fairness)
- Trust and commitment to each other's success
- Contract style (time frame and coverage)
- Scope (share of partner's business, value added and critical activities)
- Investment (financial, technology and people)

Partnership outcomes

- Global performance outcomes (enhancement of profits, levelling of profits over time)
- Process outcomes (improved service, reduced costs)
- Competitive advantage (market positioning, market share, access to knowledge)

In evaluating a relationship, how does a manager know if there are enough drivers to pursue a partnership? First, drivers must exist for each party. It is unlikely that the drivers will be the same for both parties, but they need to be strong for both. Second, the drivers must be strong enough to provide each party with a realistic expectation of significant benefits through a strengthening of the relationship. Each party should independently assess the strength of its specific drivers.

Facilitators

Drivers provide the motivation to partner. But even with a strong desire for building a partnership, the probability of success is reduced if the corporate environments are not supportive of a close relationship. Just as the relationship of a young couple with a strong desire to marry can be derailed by unsupportive in-laws, different communication styles and dissimilar values, so can a corporate relationship be side-tracked by a hostile environment. On the other hand, a supportive environment that enhances integration of the two parties will improve the success of the partnership.

Facilitators are elements of a corporate environment that allow a partnership to grow and strengthen. They serve as a foundation for a good relationship. In the short run, facilitators cannot be developed: they either exist or they don't. And the degree to which they exist often determines whether a partnership succeeds or fails. Facilitators include: (1) corporate compatibility, (2) similar managerial philosophy and techniques, (3) mutuality and (4) symmetry (see Box 4.2 for details).

Facilitators apply to the combined environment of the two potential partners. Therefore, unlike drivers, which are assessed by managers in each firm independently, facilitators should be assessed jointly. The discussion of corporate values, philosophies and objectives often leads to an improved relationship even if no further steps towards building a partnership are taken. The more positive the facilitators, the better the chance of partnership success.

If both parties realistically expect benefits from a partnership and if the corporate environments appear supportive, then a partnership is warranted. The appropriateness of any one type of partnership is a function of the combined strength of the drivers and facilitators. A combination of strong drivers and strong facilitators would suggest a Type III partnership, while low drivers and low facilitators suggest an arm's length relationship.

While it might seem, from all of the literature on the importance of integrated relationships and alliances, that managers should attempt to turn all of their corporate relationships into Type III partnerships, this is not the case. In partnering, more is not always better. The objective in establishing a partnership should not be to have a Type III partnership; rather it should be to have the *most appropriate* type of partnership given the specific drivers and facilitators. In fact, in situations with low drivers and/or facilitators, trying to achieve a Type III partnership is likely to be counter-productive. The necessary foundation is just not there. Having determined that a partnership of a specific type is warranted and should be pursued, the next step is to actually put the partnership into place. This is done through the components.

An assessment of drivers and facilitators is used to determine the potential for a partnership, but the components describe the type of relationship that has actually been implemented.

Components

Components are the activities and processes that management establishes and controls throughout the life of the partnership. Components make the relationship operational and help managers create the benefits of partnering. Every partnership has the same basic components, but the way in which the components are implemented and managed varies. Components include: planning, joint operating controls, communications, risk/reward sharing, trust and commitment, contract style, scope and financial investment. Box 4.2 summarizes the drivers, facilitators and components of partnership.

Outcomes and Feedback

Whatever type of supplier partnership is implemented, the effectiveness of the relationship must be evaluated and possibly adjusted. The key to effective measurement and feedback is how well the drivers of partnership were developed at the outset. At this beginning point, the measurement and metrics of relating to each driver should have been made explicit. These explicit measures then become the standard in evaluation of the partnership outcomes. Feedback can loop back to any step in the model. Feedback can take the form of periodic updating of the status of the drivers, facilitators and components.

SUMMARY

In this chapter, we saw how better management of purchasing activities can lead to increased profitability. We described the activities, such as supplier selection, evaluation and management, that must be performed by the purchasing function, and explored the implications of just-in-time and JIT II in purchasing. We examined various types of supplier relationship and the role of purchasing in supply chain management, with an emphasis on various types of partnership. Because the costs of purchased materials represent a significant cost of doing business, we devoted a considerable amount of attention to the management of purchasing cost. Effective logistics plays an important role in effective purchasing management. In addition, third-party logistics service providers are 'suppliers' to the firm. Much can be learned from purchasing about better managing those relationships.

KEY TERMS

A full Glossary can be found at the back of the book.

QUESTIONS AND PROBLEMS

1 Explain why supplier selection and evaluation is frequently considered the most important activity in the purchasing function.

2 What are some of the reasons that purchasing is taking on a more strategic role in organizations?

3 Explain the concept of forward buying and its relationship to total cost trade-off analysis.

4 What are the major advantages of just-in-time purchasing? What are the possible difficulties in implementing a JIT system?

5 Why is cost measurement an important activity for purchasing management?

6 Why is it necessary for two firms each to have strong drivers if they are considering forming a partnership?

7 The chapter stated that the majority of a firm's relationships would be arm's length. Why do you think this would be the case?

8 Why are quality suppliers important to a firm? Why is this even more true in a JIT environment?

THE LOGISTICS CHALLENGE!

PROBLEM: PAPERLESS PURCHASING

The not-so-humble invoice is at the heart of the entire economic system. It might only be a bit of paper but it says 'Pay Me' in the clearest possible way. A supplier provides a customer with a service and then sends them an invoice. The two are as inextricably linked as bees and honey, and as sequential as night and day.

The basic rule of business, probably the only one everybody understands, is: no invoice, no payment. Without the invoice, money would stop moving, companies would stop working and business would grind to a halt.

Yet Björn Algkvist, former chief executive officer at the Swedish software company Intentia International AB, has posed the question, 'Why can't we do away with invoices?'

But how can you bill someone without an invoice? Of course, this is a reactionary view and one that Mr Algkvist simply does not accept. He argues that business has become unreasonably wedded to the concept of the invoice and is simply fearful of abandoning it.

The technology is readily available to handle transactions without bits of paper littering people's desks and occasionally falling down the back of the photocopier. Then there is the effort involved in printing out invoices, stuffing them into envelopes and posting them. And, at the other end, the envelopes have to be opened and the invoices processed.

The interesting part of all this activity is that the invoice will likely have been generated by computer and its information likely re-keyed into a computer at the customer end. That is no doubt good for employment but it will hardly win a business efficiency award.

There are serious questions to be asked: 'Are we failing to make the most of the purchasing technologies available to us through basic conservatism or even prejudice? Are we simply too unwilling to give up these comforting bits of paper?'

Perhaps Mr Algkvist is right. Perhaps we should simply accept that the invoice is rapidly going the way of videotape technology, and leave it to the technology to handle the transaction.

Do you think invoiceless purchasing is the way of the future and do you trust it to work properly?

What Is Your Solution?

Source: adapted from Malory Davies, 'No Invoice, No Payment', *Distribution Business* 15, no. 8 (October 2002), p. 3.

SUGGESTED READING

BOOKS

Axelsson, Björn and Finn Wynstra, *Buying Business Services*. Chichester, UK: John Wiley & Sons Ltd, 2002.

Day, Marc (ed.), *Gower Handbook of Purchasing Management*, 3rd edn. Aldershot, UK: Gower Publishing Ltd, 2002.

Dixon, Lance and Anne Millen Porter, *JIT II: Revolution in Buying and Selling*. Newton, MA: Cahners, 1994.

Ellram, Lisa M. and Laura M. Birou, *Purchasing for Bottom Line Impact*. Burr Ridge, IL: Business One Irwin, 1995.

Gadde, Lars-Erik and Håkan Håkansson, *Supply Network Strategies*. Chichester, UK: John Wiley & Sons Ltd, 2001.

Leenders, Michiel, Harold E. Fearon, Anna E. Flynn and P. Fraser Johnson, *Purchasing and Supply Management*, 12th edn. Burr Ridge, IL: McGraw-Hill Irwin, 2002.

Leenders, Michiel R. and Anna E. Flynn, *Value Driven Purchasing*. Burr Ridge, IL: Irwin Professional Publishers, 1995.

Lysons, Kenneth, *Purchasing and Supply Chain Management*, 5th edn. Harlow, UK: Financial Times Prentice Hall, 2000.

Raedels, Alan R., *Value Focused Supply Management*. Burr Ridge, IL: Irwin Professional Publishers, 1995.

JOURNALS

Harrington, Thomas C., Douglas M. Lambert and Martin Christopher, 'A Methodology for Measuring Vendor Performance', *Journal of Business Logistics* 12, no. 1 (1991), pp. 83–104.

Lambert, Douglas M., Margaret A. Emmelhainz and John T. Gardner, 'Developing and Implementing Supply Chain Partnerships', *The International Journal of Logistics Management* 7, no. 2 (1996), pp. 1–17.

Newbourne, Paul T., 'The Role of Partnerships in Strategic Account Management', *The International Journal of Logistics Management* 8, no. 1 (1997), pp. 67–74.

Tersine, Richard J. and Albert B. Schwarzkopf, 'Optimal Transition Ordering Strategies with Announced Price Increases', *The International Journal of Logistics Management* 2, no. 1 (1991), pp. 26–34.

Tully, Shawn, 'Purchasing's New Muscle', *Fortune*, 20 February 1995, pp. 75–83.

REFERENCES

[1] Eurostat, 'Industry, Trade and Services Annual Enterprise Statistics for Manufacturing in 2000', http://europa.eu.int (2005).

[2] Kenneth Lysons, *Purchasing and Supply Chain Management*, 5th edn. Harlow, UK: Financial Times Prentice Hall, 2000), pp. 3–4; and Eurostat, 'Industry, Trade and Services Annual Enterprise Statistics for Manufacturing in 2000'.

[3] Shawn Tully, 'Purchasing's New Muscle', *Fortune*, 20 February 1995, p. 76.

[4] A.T. Kearney and Manchester School of Management, *Partnership or Power Play?* Manchester, UK: Manchester School of Management, 1994, p. 14.

[5] Lars-Erik Gadde and Håkan Håkansson, *Supply Network Strategies*. Chichester, UK: John Wiley & Sons Ltd, 2001), p. 8. Copyright John Wiley & Sons Ltd. Reproduced with permission.

[6] Thomas C. Harrington, Douglas M. Lambert and Martin Christopher, 'A Methodology for Measuring Vendor Performance', *Journal of Business Logistics* 12, no. 1 (1991), p. 83.

[7] Ibid., pp. 97–8.

[8] Andrew Cox, 'The Art of the Possible: Relationship Management in Power Regimes and Supply Chains', *Supply Chain Management: An International Journal* 9, no. 5 (2004), p. 349.

[9] Lance Dixon and Anne Porter Millen, *JIT II: Revolution in Buying and Selling*. Newton, MA: Cahners, 1994, pp. 9–15.

[10] Ibid., pp. 143–50, 159–60.

[11] Gadde and Håkansson, *Supply Network Strategies*, p. 70.

[12] Lisa M. Ellram, *Fundamentals of Purchasing*, seminar workbook provided to the National Association of Purchasing Management, 1995.

[13] Terrance L. Pohlen and Thomas J. Goldsby, 'VMI and SMI Programs: How Economic Value Added can Help Sell the Change', *The International Journal of Physical Distribution & Logistics Management* 33, no. 7 (2003), pp. 565–81.

[14] Luitzen de Boer, Jeroen Harink and Govert Heijboer, 'A Conceptual Model for Assessing the Impact of Electronic Procurement', *European Journal of Purchasing and Supply Management* 8, no. 1 (2002), pp. 25–33.

[15] Douglas M. Lambert, Margaret A. Emmelhainz and John T. Gardner, 'Developing and Implementing Supply Chain Partnerships', *The International Journal of Logistics Management* 7, no. 2 (1996), p. 2.

[16] This and the following section are taken from Lambert, Emmelhainz and Gardner, 'Developing and Implementing Supply Chain Partnerships', pp. 1–17.

CHAPTER 5
INVENTORY CONCEPTS AND MANAGEMENT

CHAPTER OBJECTIVES

OBJECTIVES

INTRODUCTION

Inventory is a large and costly investment. Better management of corporate inventories can improve cash flow and return on investment. Nevertheless, most companies (retailers, wholesalers and manufacturers) suffer through periodic inventory rituals – that is, crash inventory-reduction programmes are instituted every year or so. However, the lack of comprehensive understanding of inventory management techniques and trade-offs often causes customer service levels to drop, so the programmes are abandoned.

Inventories represent the largest single investment in assets for many manufacturers, wholesalers and retailers. Inventory investment can represent over 20 per cent of the total assets of manufacturers, and more than 50 per cent of the total assets of wholesalers and retailers. Competitive markets of the past 30 years have led to a proliferation of products as companies have attempted to satisfy the needs of diverse market segments. Customers have come to expect high levels of product availability. For many firms, the result has been higher inventory levels.

With the growing popularity of just-in-time (JIT) manufacturing, the reduction of product life cycles and an increased emphasis on time-based competition, firms that hold large amounts of inventory have been much criticized. Inventory does serve some very important purposes, but carrying excessive levels of inventory is costly. Organizations frequently do not identify or capture all of the many costs associated with holding inventory.

Since capital invested in inventories must compete with other investment opportunities available to the firm, and because of the out-of-pocket costs associated with holding inventory, the activity of inventory management is extremely important. Management must have a thorough knowledge of inventory carrying costs to make informed decisions about logistics system design, customer service levels, the number and location of distribution centres, inventory levels, where to hold inventory and in what form, transportation modes, production schedules, and minimum production runs. For example, ordering in smaller quantities on a more frequent basis will reduce inventory investment, but may result in higher ordering costs and increased transportation costs.

It is necessary to compare the savings in inventory carrying costs to the increased costs of ordering and transportation to determine how the decision to order in smaller quantities will affect profitability. A determination of inventory carrying costs also is necessary for new product evaluation, the evaluation of price deals/discounts, make-or-buy decisions and profitability reports. It is thus imperative to accurately measure a firm's inventory carrying costs.

Obviously, a better approach to inventory management is necessary. This chapter will provide the reader with the knowledge required to understand inventory concepts and improve the practice of inventory management.

Basic Inventory Concepts

In this section, we will consider basic inventory concepts such as the reasons for holding inventory and various types of inventory.

Why Hold Inventory?

Formulation of an inventory management policy requires an understanding of the role of inventory in production and marketing. Inventory serves five purposes within the firm: (1) it enables the firm to achieve economies of scale, (2) it balances supply and demand, (3) it enables specialization in manufacturing, (4) it provides protection from uncertainties in demand and order cycle, and (5) it acts as a buffer between critical interfaces within the channel of distribution.

Economies of Scale

Inventory is required if an organization is to realize economies of scale in purchasing, transportation or manufacturing. For example, ordering large quantities of raw materials or finished goods inventory allows the manufacturer to take advantage of the per unit price reductions associated with volume purchases. Purchased materials have a lower transportation cost per unit if ordered in large volumes because less handling is required.

Finished goods inventory makes it possible to realize manufacturing economies. Plant utilization is greater and per unit manufacturing costs are lower if a firm schedules long production runs with few line changes. Manufacturing in small quantities leads to short production runs and high changeover costs.

The production of large quantities, however, may require that some of the items be carried in inventory for a significant period of time before they can be sold. The production of large quantities may also prevent an organization from responding quickly to stockouts, since large production runs mean that items are produced less frequently. The cost of maintaining this inventory must be 'traded off' against the production savings realized.

Although frequent production changeovers reduce the quantity of inventory that must be carried and shorten the lead-time that is required in the event of a stockout, they require time that could be used for manufacturing a product. In addition, at the beginning of a production run, the line often operates less efficiently due to fine-tuning the process and equipment settings.

When a plant is operating at or near capacity, frequent line changes that create machine downtime may mean that contribution to profit is lost because there is not enough product to meet demand. In such situations, the costs of lost sales and changeovers must be compared to the increase in inventory carrying costs that would result from longer production runs. To respond to this, many companies have made a major effort toward reducing changeover times. This allows production of small lots, eliminating the penalty of higher set-up costs.

Balancing Supply and Demand

Seasonal supply or demand may make it necessary for a firm to hold inventory. For example, a producer of plastic garden furniture experiences significant sales volume increases during the months of May and June. The cost of establishing production capacity to handle the volume at these peak periods would be substantial. In addition, substantial idle capacity and wide fluctuations in the workforce would result if the company were to produce to meet demand when it occurs. The decision to maintain a relatively stable workforce and produce at a somewhat constant level throughout the year creates significant inventory build-up at various times during the year, but at a lower total cost to the firm. The seasonal inventories are usually stored in a warehouse, which may be adjacent to the plant.

On the other hand, demand for a product may be relatively stable throughout the year, but raw materials may be available only at certain times during the year (e.g. producers of canned fruits and vegetables). This makes it necessary to manufacture finished products in excess of current demand and hold them in inventory.

Specialization

Inventory makes it possible for each of a firm's plants to specialize in the products it manufactures. The finished products can be shipped to field warehouses where they are mixed to fill customer orders. The economies that result from the longer production runs and from savings in transportation costs more than offset the costs of additional handling. Companies have found significant cost savings in the operation of consolidation warehouses that allow the firm to specialize manufacturing by plant location. Specialization by facility is known as **focused factories**.

Protection from Uncertainties

Inventory is held as protection from uncertainties – that is, to prevent a stockout in the case of variability in demand or variability in the replenishment cycle. Raw materials inventories in excess of those required to support production can result from speculative purchases made because management expects a price increase or supply shortage, perhaps due to a potential strike. Another reason to hold raw materials inventory is to maintain a source of supply. Regardless of the reason for maintaining inventory, the costs of holding the inventory should be compared to the savings realized or costs avoided by holding it.

Work-in-process inventory is often maintained between manufacturing operations within a plant to avoid a shutdown if a critical piece of equipment were to break down and to equalize flow, since not all manufacturing operations produce at the same rate. The stockpiling of work-in-process within the manufacturing complex permits maximum economies of production without work stoppage. Increasingly, organizations are focusing on rebalancing production processes to minimize or eliminate the need for work-in-process inventory. This is supportive of JIT manufacturing initiatives, as presented in Chapters 4 and 6.

Inventory planning is critical to successful manufacturing operations because a shortage of raw materials can shut down the production line or lead to a modification of the production schedule; these events may increase expenses or result in a shortage of finished product. While shortages of raw materials can disrupt normal manufacturing operations, excessive inventories can increase inventory carrying costs and reduce profitability. Organizations are working closely with suppliers and carriers to improve supply reliability, allowing a reduction in the amount of raw materials held to cover delivery uncertainty.

Finally, finished goods inventory can be used as a means of improving customer service levels by reducing the likelihood of a stockout due to unanticipated demand or variability in lead-time. If the inventory is balanced, increased inventory investment will enable the manufacturer to offer higher levels of product availability and less chance of a stockout. A balanced inventory is one that contains items in proportion to expected demand.

Inventory as a Buffer

Inventory is held throughout the supply chain to act as a buffer for the following critical interfaces:

- supplier–procurement (purchasing)
- procurement–production
- production–marketing
- marketing–distribution
- distribution–intermediary
- intermediary–consumer/user.

Because channel participants are separated geographically, it is necessary for inventory to be held throughout the supply chain to successfully achieve time and place utility (see Chapter 1). Figure 5.1 shows the typical inventory positions in a supplier–manufacturer–intermediary–consumer supply chain. Raw materials must be moved from a source of supply to the manufacturing location, where they will be input into the manufacturing process. In many cases, work-in-process inventory will be necessary at the plant.

Once the manufacturing process has been completed, product must be moved into finished goods inventory at plant locations. The next step is the strategic deployment of finished goods

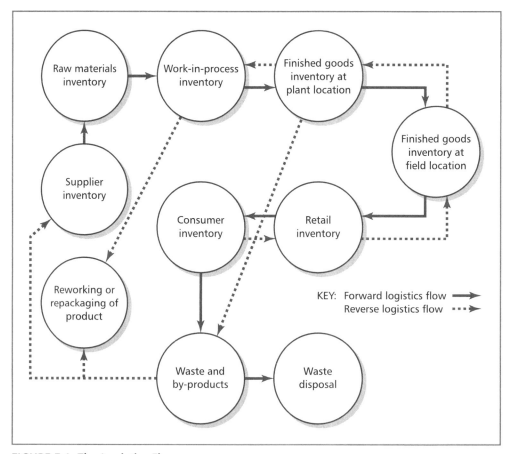

FIGURE 5.1 The Logistics Flow

inventory to field locations, which may include corporate-owned or leased distribution centres, public warehouses, wholesalers' warehouses, retail chain distribution centres or delivery directly to the retail location. Inventory is then positioned to enable customer purchase. Similarly, the customer maintains an inventory to support individual or institutional consumption.

All of these product flows are the result of a decision by the ultimate consumer or user to purchase the product. The entire process depends on the information flow from the customer to the firm and to the firm's suppliers. Communication is an integral part of a logistics system because no product flows until information flows.

It is often necessary to move a product backwards through the channel for a number of reasons. For example, a customer may return a product because it is damaged or a manufacturer may need to recall a product because of defects. Alternatively, wastes and by-products may be returned for disposition in accordance with various European Union directives. The process for dealing with returned or waste goods is referred to as 'reverse logistics' and is discussed more fully in Chapter 9.

Types of Inventory

Inventories can be classified based on the reasons for which they are accumulated. The categories of inventories include cycle stock, in-transit inventories, safety or buffer stock, speculative stock seasonal stock and dead stock.

Cycle Stock

Cycle stock is inventory that results from replenishment of inventory sold or used in production. It is required in order to meet demand under conditions of certainty – that is, when the firm can predict demand and replenishment times (lead-times). For example, if the rate of sales for a product is a constant 20 units per day and the lead-time is always 10 days, no inventory beyond the cycle stock would be required. While assumptions of constant demand and lead-time remove the complexities involved in inventory management, let's look at Figure 5.2 for an example to clarify the basic inventory principles. The example shows three alternative reorder strategies.

Since demand and lead-time are constant and known, orders are scheduled to arrive just as the last unit is sold. Thus no inventory beyond the cycle stock is required. The average cycle stock in all three examples is equal to half the order quantity. However, the average cycle stock will be 200, 100 or 300 units depending on whether management orders in quantities of 400 (part A), 200 (part B) or 600 (part C), respectively.

In-transit Inventories

In-transit inventories are items that are en route from one location to another. They may be considered part of cycle stock even though they are not available for sale or shipment until after they arrive at the destination. For the calculation of inventory carrying costs, in-transit inventories should be considered as inventory at the place of shipment origin since the items are not available for use, sale or subsequent reshipment.

Safety or Buffer Stock

Safety or buffer stock is held in excess of cycle stock because of uncertainty in demand or lead-time. Average inventory at a stockkeeping location that experiences demand or lead-time variability is equal to half the order quantity plus the safety stock.

In Figure 5.3, for example, the average inventory would be 100 units if demand and lead-time were constant. But if demand was actually 25 units per day instead of the predicted 20 units per

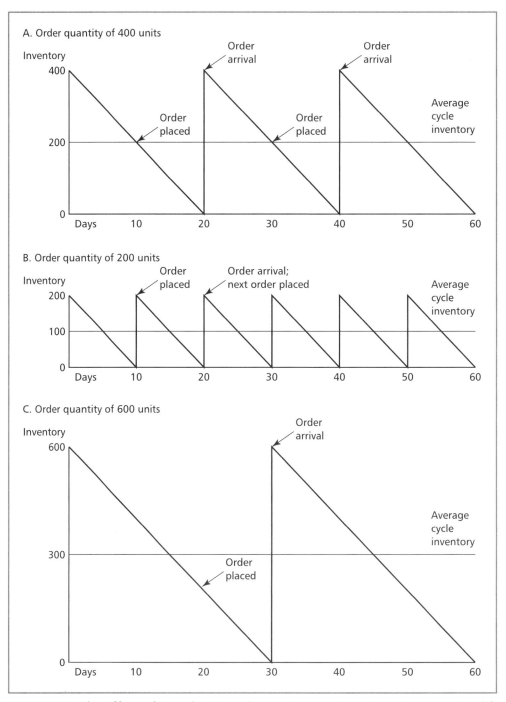

FIGURE 5.2 The Effect of Reorder Quantity on Average Inventory Investment with Constant Demand and Lead-time

day with a 10-day lead time, inventory would be depleted by day 8 (200/25). Since the next order would not arrive until day 10 (order was placed on day zero), the company would be out of stock for two days. At 25 units of demand per day, this would be a stockout of 50 units in total. If management believed that the maximum variation in demand would be plus or minus 5 units, a safety stock of 50 units would prevent a stockout due to variation in demand. This would require holding an average inventory of 150 units (100 units average inventory + 50 units safety stock).

Now consider the case in which demand is constant but lead-time can vary by plus or minus two days (part B of Figure 5.3). If the order arrives two days early, the inventory on hand would be equal to a 12-day supply, or 240 units, since sales are at a rate of 20 units per day and 40 units would remain in inventory when the new order arrived. However, if the order arrived two days late, on day 12 – which is a more likely occurrence – the firm would experience stockouts for a period of two days (40 units). If management believed that shipments would never arrive more than two days late, a safety stock of 40 units would ensure that a stockout due to variation in lead-time would not occur if demand remained constant. This would require holding an average inventory of 140 units.

In most business situations, management must be able to deal with variability in demand and lead-time. Forecasting is rarely accurate enough to predict demand, and demand is seldom, if ever, constant. In addition, transportation delays, and supplier and production problems, make lead-time variability a fact of life. Consider part C of Figure 5.3, in which demand uncertainty (part A) and lead-time uncertainty (part B) are combined.

Combined uncertainty is the worst of all possible worlds. In this case, demand is above forecast by the maximum, 25 units instead of 20 units per day, and the incoming order arrives two days late. The result is a stockout period of four days at 25 units per day. If management wanted to protect against the maximum variability in both demand and lead-time, the firm would need a safety stock of 100 units. This policy (no stockouts) would result in an average inventory of 200 units.

In sum, variability in the order cycle requires safety stock. Since holding safety stock costs firms money, managers will try to reduce or eliminate variability. Forecasting can be used to better predict demand, resulting in less safety stock. Utilizing transportation carriers that provide consistent on-time deliveries will reduce lead-time variability. The goal is not necessarily to have the fastest delivery, but the most dependable, allowing safety stock reduction and the ability to plan more accurately.[1]

Speculative Stock

Speculative stock is inventory held for reasons other than satisfying current demand. For example, materials may be purchased in volumes larger than necessary in order to receive quantity discounts, because of a forecasted price increase or materials shortage, or to protect against the possibility of a strike. Production economies may also lead to the manufacture of products at times other than when they are in demand.

Seasonal Stock

Seasonal stock is a form of speculative stock that involves the accumulation of inventory before a seasonal period begins. This often occurs with agricultural products and seasonal items. The fashion industry is also subject to seasonality with new fashions coming out many times a year. The back-to-school season is a particularly important time. See the Global box for a presentation of how Benetton tries to take some of the guesswork out of demand for fashions.

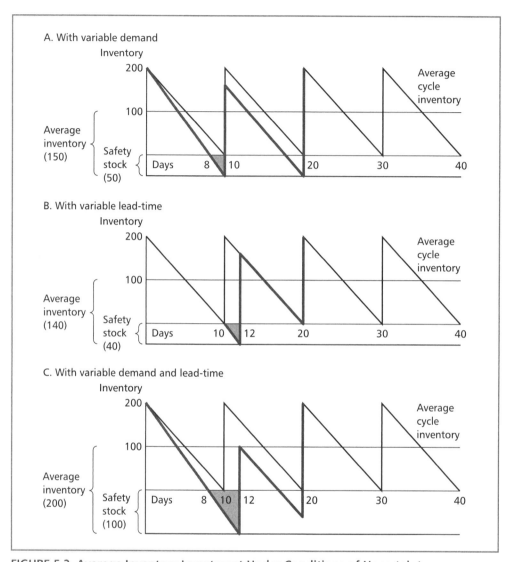

FIGURE 5.3 Average Inventory Investment Under Conditions of Uncertainty

GLOBAL

Benetton Uses Postponement to Improve Global Inventory Management

The concept of postponement – delaying commitment of the product to its final form until the last possible moment – can be an outstanding technique for responding to variability in customer demand. Benetton has only one distribution centre – in Castrette, Italy – to serve stores in 120 countries. Benetton produces its most popular styles as 'grey' goods, or undyed garments. When it sees what the actual demand pattern is for a certain sweater or pair of leggings by colour, it can dye the grey goods quickly and speed them off to market. That way, Benetton doesn't end up with too many aqua sweaters and not enough black sweaters. This allows Benetton to minimize its inventory while meeting customer demand and reducing end-of-season markdowns. Since it has only one distribution centre and ships reorders based on actual demand, inventory is less likely to be located where it is not needed.

Question: **How would this strategy affect Benetton's customer service?**

Source: Carla Rapoport and Justin Martin, 'Retailers Go Global', *Fortune*, 20 February 1995, pp. 102–8.

Dead Stock

Dead stock refers to items for which no demand has been registered for some specified period of time. Dead stock might be obsolete throughout a company or only at one stockkeeping location. If it is the latter, the items may be transshipped to another location to avoid the obsolescence penalty or markdown at their current location.

Basic Inventory Management

Inventory is a major use of working capital. Accordingly, the objectives of inventory management are to increase corporate profitability through improved inventory management, to predict the impact of corporate policies on inventory levels, and to minimize the total cost of logistics activities while meeting customer service requirements.

Measures of Inventory Management Effectiveness

The key measure of effective inventory management is the impact that inventory has on corporate profitability. Effective inventory management can improve profitability by lowering costs or supporting increased sales.

Measures to decrease inventory-related costs include reducing the number of back-orders or expedited shipments, purging obsolete or dead stock from the system, and improving the accuracy of forecasts. Transshipment of inventory between field warehouses and small-lot transfers can be reduced or eliminated by better inventory planning. Better inventory management can increase the ability to control and predict how inventory investment will change in response to management policy.

Inventory turnover is another measure of inventory performance. It is measured as:

$$\frac{\text{Annual euro sales volume at cost}}{\text{Average euro inventory investment}}$$

All else being equal, a higher number is preferred, indicating that inventory moves through the firm's operations quickly, rather than being held for an extensive period. Turnover should not be used as the only measure of inventory effectiveness, but should be combined with other measures that reflect customer service issues.

Increased sales are often possible if high levels of inventory lead to better in-stock availability and more consistent service levels. **Fill rate** is a common measure of the customer service performance of inventory. As presented in Chapter 2 a fill rate is often presented as the percentage of units available when requested by the customer. A 96 per cent fill rate means that 4 per cent of requested units were unavailable when ordered by the customer. Low inventory levels can reduce fill rates, hurting customer service and creating lost sales.

Finally, total cost integration should be the goal of inventory planning – that is, management must determine the inventory level required to achieve least total cost logistics, given the required customer service objectives.

Impact of Demand Patterns on Inventory Management

Whether inventory is 'pulled' or 'pushed' through a system, and whether the demand is 'dependent' or 'independent' has implications for inventory management methods.

Pull versus push systems are distinguished by the way the company's production is driven. If a company waits to produce products until customers demand it, that is a pull system – customer demand 'pulls' the inventory. If a firm produces to forecast or anticipated sales to customers, that is a 'push' system – the firm is 'pushing' its inventory into the market in anticipation of sales.

Independent versus dependent demand inventory focuses on whether the demand for an item depends on demand for something else. An independent demand item is a finished good, while dependent demand items are the raw materials and components that go into the production of that finished good. The demand for raw materials or components is 'derived' based on the demand for the finished good. The need for dependent demand items doesn't have to be forecast; it can be calculated based on the production schedule of the finished good. The need for production of the finished good may be forecast or based on customer demand/orders.

Inventory managers must determine how much inventory to order and when to place the order. To illustrate the basic principles of reorder policy, let's consider inventory management under conditions of certainty. In reality, the more common situation is inventory management under uncertainty, but the management process will be similar in both instances.

Inventory Management under Conditions of Certainty

Replenishment policy under conditions of certainty requires the balancing of ordering costs against inventory carrying costs when the supplier pays the freight cost. For example, a policy of ordering large quantities infrequently may result in inventory carrying costs in excess of the savings in ordering costs. Ordering costs for products purchased from an outside supplier typically include (1) the cost of transmitting the order, (2) the cost of receiving the product, (3) the cost of placing it in storage, and (4) the cost of processing the invoice for payment.

In restocking its own field warehouses, a company's ordering costs typically include (1) the cost of transmitting and processing the inventory transfer, (2) the cost of handling the product if it is in stock, or the cost of setting up production to produce it, and the handling cost if the product is not in stock, (3) the cost of receiving at the field location, and (4) the cost of documentation. Remember that only direct out-of-pocket expenses should be included in ordering costs. Inventory carrying costs will be explained in detail later in this chapter.

Economic Order Quantity

The best ordering policy can be determined by minimizing the total of inventory carrying costs and ordering costs using the **economic order quantity (EOQ)** model. The EOQ is a

> concept which determines the optimal order quantity on the basis of ordering and carrying costs. When incremental ordering costs equal incremental carrying costs, the most economic order quantity exists. It does not optimize order quantity and thus the shipment quantity, on the basis of total logistics costs, but only ordering and carrying costs.[2]

Two questions seem appropriate in reference to the example in Figure 5.2.
1 Should we place orders for 200, 400 or 600 units, or some other quantity?
2 What is the impact on inventory if orders are placed at 10-, 20- or 30-day intervals, or some other time period? Assuming constant demand and lead-time, sales of 20 units per day and 240 working days per year, annual sales will be 4800 units, assuming the plant is closed for four weeks each year. If orders are placed every 10 days, 24 orders of 200 units will be placed. With a 20-day order interval, 12 orders of 400 units are required. If the 30-day order interval is selected, eight orders of 600 units are necessary. The average inventory is 100, 200 and 300 units, respectively. Which of these policies would be best?

The cost trade-offs required to determine the most economical order quantity are shown graphically in Figure 5.4. By determining the EOQ and dividing the annual demand by it, the frequency and size of the order that will minimize the two costs are identified.

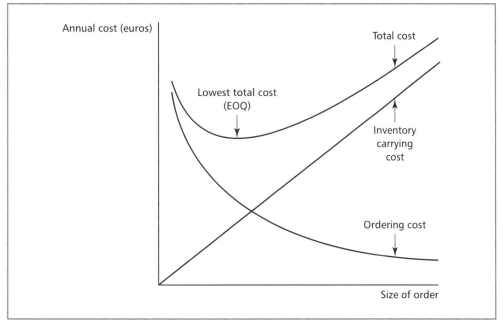

FIGURE 5.4 Cost Trade-offs Required to Determine the Most Economical Order Quantity

The EOQ in units can be calculated using the following formula:

$$EOQ = \sqrt{\frac{2\,PD}{CV}}$$

where

 P = the ordering cost per order
 D = annual demand or usage of product (number of units)
 C = annual inventory carrying cost (as a percentage of product cost or value)
 V = average cost or value of one unit of inventory.

Now, using the EOQ formula, we will determine the best ordering policy for the situation described in Figure 5.2:

 P = €40
 D = 4800 units
 C = 25 per cent
 V = €100 per unit

$$EOQ = \sqrt{\frac{2(€40)(4800)}{(25\%)(€100)}}$$

$$= \sqrt{\frac{384{,}000}{25}}$$

$$= 124 \text{ units}$$

If 20 units fit on a pallet, then the reorder quantity of 120 units would be established. This analysis is shown in Table 5.1.

Order quantity	Number of orders (D/Q)	Ordering cost P × (D/Q) (€)	Inventory carrying cost $\frac{1}{2}$ Q × C × V (€)	Total cost (€)
40	120	4800	500	5300
60	80	3200	750	3950
80	60	2400	1000	3400
100	48	1920	1250	3170
120	40	1600	1500	3100
140	35	1400	1750	3150
160	30	1200	2000	3200
200	24	960	2500	3460
300	16	640	3750	4390
400	12	480	5000	5480

TABLE 5.1 Cost Trade-offs Required to Determine the Most Economical Order Quantity

The EOQ model has received significant attention and use in industry, but it is not without its limitations. The simple EOQ model is based on the following assumptions:
1 a continuous, constant and known rate of demand
2 constant and known replenishment or lead-time
3 constant purchase price independent of the order quantity or time
4 constant transportation cost independent of the order quantity or time
5 the satisfaction of all demand (no stockouts are permitted)
6 no inventory in transit
7 only one product in inventory or at least no interaction between products (independent demand items)
8 an infinite planning horizon
9 no limit on capital availability.

It would be very unusual to find a situation where demand and lead-time are constant, both are known with certainty and costs are known precisely. However, the simplifying assumptions are of great concern only if policy decisions will change as a result of the assumptions made. The EOQ solution is relatively insensitive to small changes in the input data.

Referring to Figure 5.4, one can see that the EOQ curve is relatively flat around the solution point. This is often referred to as the 'bathtub effect'. Although the calculated EOQ was 124 units (rounded to 120), an EOQ variation of 20 or even 40 units does not significantly change the total cost (see Table 5.1).

Adjustments to the EOQ
Typical refinements made to the EOQ model include adjustments for volume transportation rates and for quantity discounts. The simple EOQ model does not consider the impact of these two factors. A firm's uncertainty concerning its production set-up and other order costs, customer demand, supplier lead-times and orders for discrete numbers of items also affect the simple EOQ model and must be taken into account.[3]

Fixed Order Point Versus Fixed Order Interval Policy

The EOQ represents a **fixed order point** policy. Once the EOQ has been determined, we order a fixed quantity each time, based on the EOQ. Actual demand may cause the time between orders to vary. An order is placed when inventory on hand reaches the predetermined minimum level necessary to satisfy demand during the order cycle. The automated inventory control system normally generates an order, or at least a management report, when the reorder point is reached.

Another reorder policy is the **fixed order interval** approach. Under this approach, inventory levels are reviewed at a certain, set time interval, perhaps every week. An order is placed for a variable amount of inventory, whatever is required to get the company back to its desired inventory level. This approach is common where many items are purchased from the same supplier. A weekly order may be placed to reduce ordering costs and take advantage of purchase volume discounts and freight consolidation.

Inventories and Customer Service

The establishment of a service level, and thus a safety stock policy, is a matter of managerial judgement. Management should consider factors such as customer relations, customer wants and needs, competitive service levels and the ability of the firm to support continuous production processes.

In many companies, management improves customer service levels simply by adding safety stock because the cost of carrying inventory has often not been calculated for the firm or has been set arbitrarily at an artificially low level. Figure 5.5 illustrates the relationship between customer service levels and inventory investment. The calculations are shown in Table 5B.5 in Appendix B (ignore for the moment the broken line and arrows in Figure 5.5).

Although inventory investment figures will vary from situation to situation, relationships similar to those in the example will hold. As customer service levels move towards 100 per cent, inventory levels increase disproportionately. It becomes obvious that customer service levels should not be improved solely by the addition of inventory. The need to develop an accurate inventory carrying cost for the purpose of planning should be clear.

One way of resolving this problem is to substitute information for inventory carrying costs by using customers' point-of-sale scanner data to plan short-term production and restocking of customer inventory locations. Another possibility is to recognize the wide differences in demand levels and demand variation associated with each product, and manage their inventories differently.

Managers often make the mistake of treating all products the same. Generally, a more economical policy is to stock the highest volume items at retail locations, high- and moderate-volume items at field warehouse locations, and slow-moving items at centralized locations. The centralized location may be a distribution centre or a plant warehouse. This type of multi-echelon stocking procedure is referred to as **ABC analysis**, and will be discussed later in this chapter. The broken line in Figure 5.5 shows how the relationship between inventory investment and customer service levels can be shifted, using some of these strategies.

Production Scheduling

Earlier in this chapter, we discussed how inventory levels can be influenced by production policies. The reverse also is true. In many cases, logistics policy changes – especially those that decrease inventory levels – can create significant increases in total production costs that are beyond the control of manufacturing management.

An inventory policy decision that reduces logistics costs by less than the increase in production set-up cost results in lower overall profit performance for the company. Logistics managers must be aware of the impact of their decisions on the efficiency of manufacturing operations, and consider associated changes in manufacturing costs when establishing logistics policies.

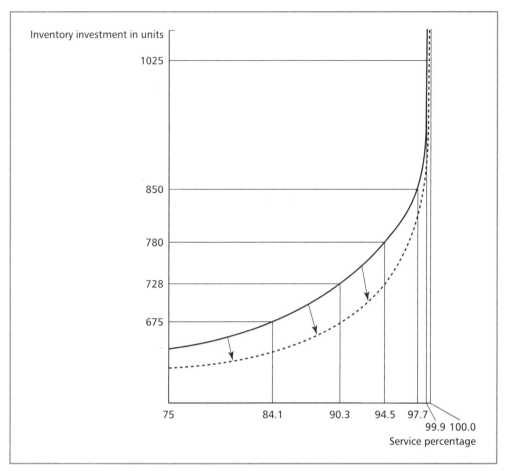

FIGURE 5.5 Relationship Between Inventory Investment and Customer Service Levels

Financial Aspects of Inventory Strategy

The quality of inventory management and the inventory policies a firm sets have a significant impact on corporate profitability and the ability of management to implement its customer service strategies at least total cost logistics.

Inventory and Corporate Profitability

Inventory represents a significant portion of a firm's assets. Consequently, excessive inventory levels can lower corporate profitability in two ways: (1) net profit is reduced by out-of-pocket costs associated with holding inventory, such as insurance, taxes, storage, obsolescence, damage and interest expense, if the firm borrows money specifically to finance inventories; and (2) total assets are increased by the amount of the inventory investment, which decreases asset turnover, or the opportunity to invest in other more productive assets is foregone. In any case, the result is a reduction in return on net worth.

Inventory and Least Total Cost Logistics

Least total cost logistics is achieved by minimizing the total of the logistics costs illustrated in Figure 5.6 for a specified level of customer service. However, successful implementation of cost trade-off analysis requires that adequate cost data be available to management. Management should not set inventory levels and inventory turnover policies arbitrarily, but should do so with full knowledge of inventory carrying costs, total logistics system costs and necessary customer service policies.

The cost of carrying inventory has a direct impact not only on the number of warehouses that a company maintains, but on all of the firm's logistics policies, including stockouts and associated

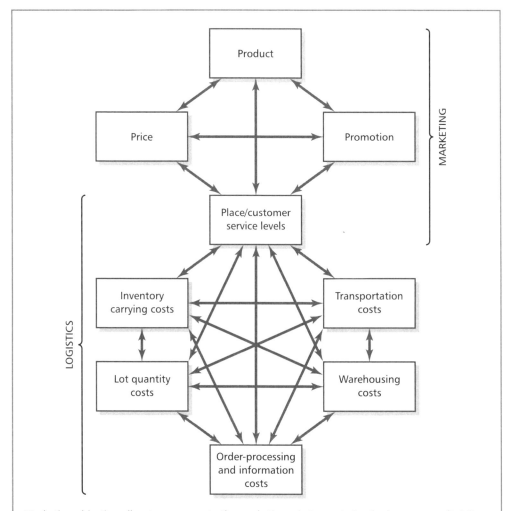

Marketing objective: allocate resources to the marketing mix to maximize the long-run profitability of the firm.

Logistics objectives: minimize total costs given the customer service objective where: Total costs = Transportation costs + Warehousing costs + Order-processing and information costs + Lot quantity costs + Inventory carrying costs.

FIGURE 5.6 Costs Trade-offs Required in a Logistics System

Source: adapted from Douglas M. Lambert, *The Development of an Inventory Costing Methodology: A Study of the Costs Associated with Holding Inventory*. Chicago: National Council of Physical Distribution Management, 1976, p. 7.

customer service costs. Inventory carrying costs are being traded off with other logistics costs, such as transportation and customer service. For example, given the same customer service level, firms with low inventory carrying costs will likely hold more inventory and use slower modes of transportation, such as rail, because this provides the least total cost logistics.

High inventory carrying costs likewise result in a reduction in inventory investment and require faster means of transportation, such as motor or air carriers, to minimize total costs while achieving the desired customer service level. Without an accurate assessment of the costs of carrying inventory, then, it is difficult for a company to implement logistics policies that mini-mize costs.

In addition, knowledge of the cost of carrying inventory is required to accurately determine economic manufacturing quantities, economic order quantities and sales discounts, all of which are usually calculated on the basis of estimated costs in the majority of companies that use these formulas.

Inventory Carrying Costs

Inventory carrying costs are those costs associated with the amount of inventory stored. They are made up of a number of different cost components and generally represent one of the high-est costs of logistics.[4] The magnitude of these costs and the fact that inventory levels are directly affected by the configuration of the logistics system shows the need for accurate inventory car-rying cost data. Without such data, appropriate trade-offs cannot be made within the organization or the supply chain. Nevertheless, most managers who consider the cost of holding inventory use estimates or traditional industry benchmarks.

We have seen how inventory levels can affect corporate profit performance and have dis-cussed the need for assessment of inventory carrying costs in logistics system design. Unfortunately, many companies have never calculated inventory carrying costs, even though these costs are both real and substantial. When inventory carrying costs are calculated, they often include only the current interest rate plus expenditures such as insurance and taxes. Many managers use traditional textbook percentages or industry averages. All of these approaches have problems.

First, there are only a few special circumstances in which the current interest rate is the rele-vant cost of money (we will explore these shortly). Traditional textbook percentages also have serious drawbacks.

Most of the carrying cost percentages presented in published sources between 1951 and 1997 were about 25 per cent. If 25 per cent was an accurate number in 1951, how could it be accurate in 1997, when the prime interest rate had fluctuated between 3 and 20 per cent during that period?

There is also the method of using inventory carrying costs that are based on 'benchmarking' with industry averages. For the most part, businesspeople seem to find comfort in such num-bers, but many problems are inherent with this practice. For example, would the logistics executive of a cosmetics manufacturer want to compare his or her firm to: Avon, a company that sells its products door to door; Revlon, a company that sells its products through major depart-ment stores; or – even worse – use an average of the two companies? The last approach would compare the executive's firm to a nonentity – no company at all.

Even if two companies are very similar in terms of the manufacture and distribution of their products, the availability of capital may lead to two different inventory strategies – that is, one firm may experience shortages of capital (capital rationing) while the other may have an abun-dance of cash. If capital is short, the cost of money for inventory decisions may be 35 per cent pre-tax, which is the rate of return the company is earning on new investments. If capital is plentiful, the cost of money may be 8 per cent pre-tax, which is the interest rate the company is earning on its cash. If both of these companies are well managed, the company whose cost of money is 8 per cent will have more inventory.

As presented above, the lower the cost of money, the more attractive it is to increase inventory levels. The company with the 35 per cent cost of money will have lower inventories by making different trade-offs such as incurring production set-up costs more frequently, choosing more expensive transportation modes or reducing customer service levels. Each company may have what represents least total cost logistics, and yet one may turn its inventories six times a year and the other 12 times. However, if either company were to change any component of its logistics system in order to match the other's performance, total costs could increase and return on net worth could decrease.

Calculating Inventory Carrying Costs

Because each company faces a unique operating environment, each should determine its own logistics costs and strive to minimize the total of these costs, given its customer service objectives. Inventory carrying costs should include only those costs that vary with the quantity of inventory and that can be categorized into the following groups: (1) capital costs, (2) inventory service costs, (3) storage space costs, and (4) inventory risk costs. The elements to be considered in each of these categories are identified in Figure 5.7.

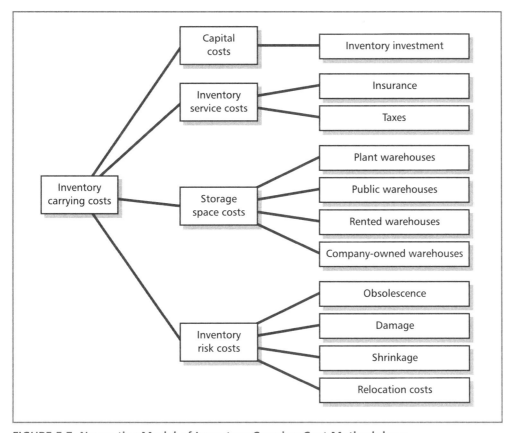

FIGURE 5.7 Normative Model of Inventory Carrying Cost Methodology

Source: Douglas M. Lambert, *The Development of an Inventory Costing Methodology: A Study of the Costs Associated with Holding Inventory*. Chicago: National Council of Physical Distribution Management, 1976, p. 68.

Capital Costs on Inventory Investment

Holding inventory ties up money that could be used for other types of investment. This holds true for both internally generated funds and capital obtained from external sources, such as debt from banks and insurance companies, or from the sale of common stock. Consequently, the company's **opportunity cost of capital**, the rate of return that could be realized from some other use of the money, should be used to accurately reflect the true cost involved. Virtually all companies seek to reduce inventory because management recognizes that holding excessive inventory provides no value added to the firm. The company must consider what rate of return it is sacrificing on the cash invested in inventory.

Some companies differentiate among projects by categorizing them according to risk and looking for rates of return that reflect the perceived level of risk. For example, management could group projects into high-, medium- and low-risk categories. High-risk projects, such as investments in new products or technology, may have a desired rate of return of 25 per cent after tax. Investment in new product inventory should reflect that higher perceived risk level.

Medium-risk projects may be required to obtain an 18 per cent after-tax return while low-risk projects, which may include such investments as warehouses, private trucking and inventory of established, stable product lines, might be expected to achieve an after-tax return of 10 per cent. (Keep in mind that all inventory carrying cost components must be stated in pre-tax numbers because all of the other costs in the trade-off analysis, such as transportation and warehousing, are reported in pre-tax dollars.)

In some very special circumstances, such as the fruit-canning industry, short-term financing may be used to finance the seasonal build-up of inventories. Fruit must be packaged as it is harvested to meet all customer demand through to the end of the next growing season. In this situation, the inventory build-up is short term and the actual cost of borrowing is the acceptable cost of money.

Once management has established the cost of money, it must determine the out-of-pocket (cash) value of the inventory for which the inventory carrying cost is being calculated. For wholesalers or retailers, the out-of-pocket value of the inventory is the current replacement cost of the inventory, including any freight costs paid, or the current market price if the product is being phased out. For manufacturers, the relevant cost is only the cost directly associated with producing the inventory and making it available for sale. Thus, it is necessary to know whether the company is using direct costs to determine the inventory value or using some form of absorption costing.

Direct costing is a method of cost accounting based on separating costs into fixed and variable components. For management planning and control purposes, the fixed–variable cost breakdown provides more information than that obtained from current financial statements designed for external reporting. Under direct costing, the fixed costs of production are excluded from inventory values. Therefore, inventory values more closely reflect the out-of-pocket cost of their replacement. With **absorption costing** (otherwise known as full costing or full absorption costing), the traditional approach used by most manufacturers, fixed manufacturing overhead is included in the inventory value.

In addition to the distinction between direct costing and absorption costing, companies may value inventories based on actual costs or standard costs. There are, then, four distinct costing alternatives.

1 *Actual absorption costing* includes actual costs for direct material and direct labour, plus predetermined variable and fixed manufacturing overhead.

2 *Standard absorption costing* includes predetermined direct material and direct labour costs, plus predetermined variable and fixed manufacturing overhead.

3 *Actual direct costing* includes actual costs for direct material and direct labour, plus predetermined variable manufacturing overhead; it excludes fixed manufacturing overhead.

4 *Standard direct costing* includes predetermined costs for direct material and direct labour, plus predetermined variable manufacturing overhead; it excludes fixed manufacturing overhead.

The preceding material on methods of inventory valuation supports the conclusion that using industry averages for inventory carrying costs is not a good policy. This is so because the various component percentages may not be calculated using comparable inventory valuation systems.

In summary, inventory requires capital that could be used for other corporate investments; by having funds invested in inventory, a company forgoes the rate of return that it could obtain in such investments. Therefore, the company's opportunity cost of capital should be applied to the investment in inventory. The cost of capital should be applied to the out-of-pocket cost investment in inventory.

Inventory Service Costs

Inventory service costs are comprised of value-added or personal property taxes and fire and theft insurance paid as a result of holding the inventory. Taxes vary depending on the country in which inventories are held. In general, taxes vary directly with inventory levels.

Insurance rates are not strictly proportional to inventory levels because insurance is usually purchased to cover a certain value of product for a specified time period. Nevertheless, an insurance policy will be revised periodically based on expected inventory level changes. In some instances, an insurance company will issue policies in which premiums are based on the monthly amounts insured. Insurance rates depend on the materials used in the construction of the storage building, its age and considerations such as the type of fire-prevention equipment installed.

The actual amount spent on insurance and taxes during the past year can be calculated as a percentage of that year's inventory value and added to the cost-of-money component of the carrying cost. If budgeted figures are available for the coming year, they can be used as a percentage of the inventory value based on the inventory plan (the forecasted inventory level) in order to provide a future-orientated carrying cost. In most cases, there will be few, if any, significant changes from year to year in the tax and insurance components of the inventory carrying cost.

Storage Space Costs

Storage space costs relate to four general types of facility: (1) plant warehouses, (2) public warehouses, (3) rented or leased (contract) warehouses, and (4) company-owned (private) warehouses.

Plant warehouse costs are primarily fixed. If any costs are variable, they are usually variable with the amount of product that moves through the facility, **throughput**, and not with the quantity of inventory stored. If some variable costs, such as the cost of taking inventory or any other expenses, change with the level of inventory, management should include them in inventory carrying costs. Fixed charges and allocated costs are not relevant for inventory policy decisions. If the firm can rent out the warehouse space or use it for some other productive purpose instead of using it for storing inventory, an estimate of the relevant opportunity costs would be appropriate.

Public warehouse costs are usually based on the amount of product moved into and out of the warehouse (handling charges) and the amount of inventory held in storage (storage charges). In most cases, handling charges are assessed when the products are moved into the warehouse, and storage charges are assessed on a periodic basis (e.g. monthly). Usually, the first month's storage must be paid when the products are moved into the facility. In effect, this makes the first month's storage a handling charge since it must be paid on every case of product regardless of how long it is held in the warehouse.

The use of public warehouses is a management policy decision because it may be the most economical way to provide the desired level of customer service without incurring excessive transportation costs. For this reason, handling charges, which represent the majority of costs related to the use of public warehouses, should be considered as throughput costs – that is, they should be thought of as part of the warehousing cost category of the cost trade-off analysis, and not as part of inventory carrying costs. Only charges for *warehouse storage* should be included in inventory carrying costs because these are the public warehouse charges that will vary with the level of inventory.

Where a throughput rate (handling charge) is given based on the number of inventory turns, it is necessary to estimate the storage cost component by considering how the throughput costs per case will change if the number of inventory turns changes. Of course, the public warehouse fees that a company pays when its inventory is placed into field storage should be included in the value of its inventory investment.

Rented or leased warehouse space is normally contracted for a specified period of time. The amount of space rented is based on the maximum storage requirements during the period covered by the contract. Thus, warehouse rental charges do not fluctuate from day to day with changes in the inventory level, although rental rates can vary from month to month or year to year when a new contract is negotiated. Most costs, such as rent payment, the manager's salary, security costs and maintenance expenses, are fixed in the short run. But some expenses, such as warehouse labour and equipment operating costs, vary with throughput. During the term of the contract, few, if any, costs vary with the amount of inventory stored.

All of the costs of leased warehouses could be eliminated by not renewing the contract and are therefore a relevant input for logistics decision-making. However, operating costs that do not vary with the quantity of inventory stored, such as those outlined in the preceding paragraph, should not be included in the carrying costs. Rather, these costs belong in the warehousing cost category of the cost trade-off analysis. The inclusion of fixed costs, and those that are variable with throughput in inventory carrying costs, has no conceptual basis. Such a practice is simply incorrect and will result in erroneous decisions.

The costs associated with *company-owned* or **private warehouses** are primarily fixed, although some may vary with throughput. All operating costs that can be eliminated by closing a company-owned warehouse or the net savings resulting from a change to public warehouses should be included in warehousing costs, not inventory carrying costs. Only those costs that vary with the quantity of inventory belong in inventory carrying costs. Typically, these costs are negligible in company-owned warehouses.

Inventory Risk Costs

Inventory risk costs vary from company to company, but typically include charges for (1) obsolescence, (2) damage, (3) shrinkage, and (4) relocation of inventory.

Obsolescence cost is the cost of each unit that must be disposed of at a loss because it can no longer be sold at a regular price. In essence, it is the cost of holding products in inventory beyond their useful life. Obsolescence cost is the difference between the original cost of the unit and its salvage value, or the original selling price and the reduced selling price if the price is lowered (marked down) to move the product. Generally, obsolescence costs are buried in the 'cost of goods manufactured' account or the 'cost of goods sold' account instead of being shown as a separate item on profit-and-loss statements. Consequently, managers may have some difficulty arriving at this figure. However, it is a relevant cost of holding inventory, especially as product life cycles decrease.

Damage costs incurred during shipping should be considered a throughput cost, since they will continue regardless of inventory levels. Damage attributed to a public warehouse operation is usually charged to the warehouse operator if it is above a specified maximum amount. Damage is often identified as the net amount after claims.

Shrinkage costs have become an increasingly important problem for American businesses. Many authorities think inventory theft is a more serious problem than cash embezzlement. Theft is far more common, involves far more employees and is hard to control. Shrinkage can also result from poor record-keeping, or shipping the incorrect products or quantities to customers. In the case of agricultural products, natural ores or similar items that are shipped in bulk, shrinkage may result from loss in weight or spillage that occurs during transportation and handling. However, shrinkage costs may be more closely related to company security measures than inventory levels, even though they definitely will vary with the number of warehouse locations. For this reason, management may find it more appropriate to assign some or all of these costs to the warehouse locations than to the amount of inventory.

Relocation costs are incurred when inventory is transshipped from one warehouse location to another to avoid obsolescence. By shipping the products to the location where they will sell, the company avoids the obsolescence cost but incurs additional transportation costs. Transshipments to avoid obsolescence or markdowns are the result of having too much inventory, and the cost should be included in inventory carrying costs. Often, transshipment costs are not reported separately, but are simply included in transportation costs. In such cases, a managerial estimate or a statistical audit of freight bills can isolate the transshipment costs.

The frequency of these types of shipment will determine which approach is more practical in any given situation. That is, if such shipments are rare, the percentage component of the carrying cost will be very small and a managerial estimate should suffice.

In some cases, firms may incur transshipment costs as a result of inventory stocking policies. For example, if inventories are set too low in field locations, stockouts may occur and may be rectified by shipping product from the nearest warehouse location that has the items in stock. The transportation costs associated with transshipment to avoid stockouts are a result of decisions that involve trade-offs among transportation costs, warehousing costs, inventory carrying costs and/or stockout costs. They are transportation costs and should not be classified as inventory carrying costs.

Because managers do not always know just how much of the costs of damage, shrinkage and relocation are related to the amount of inventory held, they may have to determine mathematically if a relationship exists. For example, a cost for damage may be available, but the amount of this cost attributed to the volume of inventory may be unknown.

Damage can be a function of such factors as throughput, general housekeeping, the quality and training of management and labour, the type of product, the protective packaging used, the material handling system, the number of times the product is handled, how it is handled, and the amount of inventory (which may lead to damage as a result of overcrowding in the warehouse). To say which of these factors is most important and how much damage each one accounts for is extremely difficult.

Even an elaborate reporting system may not yield the desired results, as employees may try to shift the blame for the damaged product. The quality of damage screening during the receiving function, and the possible hiding of damaged product in higher inventories until inventories are reduced, may contribute to the level of damage reported, regardless of the cause.

An Example from the Consumer Packaged Goods Industry

Table 5.2 summarizes the methodology that should be used to calculate inventory carrying costs, as well as the values for an example that follows. The model is normative, because using it will lead to a carrying cost that accurately reflects a firm's costs. Now let's examine an actual application of the methodology for a manufacturer of packaged goods.

Step no.	Cost of category	Source	Explanation	Amount (current study)
1.	Cost of money	Comptroller	This represents the cost of money invested in inventory. The return should be comparable to other investment opportunities.	30% pre-tax
2.	Average monthly inventory valued at variable costs delivered to the distribution centre	1. Standard cost data – comptroller's department 2. Freight rates and product specs are from distribution reports 3. Average monthly inventory in cases from printout received from sales forecasting	Only want variable costs since fixed costs go on regardless of the amount of product manufactured and stored – follow steps outlined in body of report.	€7,800,000 valued at variable cost delivered to the DC (variable manufactured cost equalled 70% of full manufactured cost). Variable cost FOB the DC averaged 78% of full manufactured cost
3.	Taxes	The comptroller's department	Personal property taxes paid on inventory.	€90,948, which equals 1.17%
4.	Insurance	The comptroller's department	Insurance rate €100 of inventory (at variable costs).	€4,524, which equals 0.06%
5.	Recurring storage (public warehouse)	Distribution operations	This represents the portion of warehousing costs that are related to the volume of the inventory stored.	€225,654 annually, which equals 2.89%
6.	Variable storage (plant warehouses)	Transportation services	Only include those costs that are variable with the amount of inventory stored.	Nil
7.	Obsolescence	Distribution department reports	Cost of holding product inventory beyond its useful life.	0.800% of inventory
8.	Shrinkage	Distribution department reports	Only include the portion attributable to inventory storage.	€100,308 which equals 1.29%
9.	Damage	Distribution department reports	Only include the portion attributable to inventory storage.	
10.	Relocation costs	Not available	Only include relocation costs incurred to avoid absolescence.	Not available
11.	Total inventory carrying costs	Calculate the numbers generated in steps 3, 4, 5, 6, 7, 8, 9 and 10 as a percentage of average inventory valued at variable cost delivered to the distribution centre, and add them to the cost of money (step 1)		36.21%

TABLE 5.2 Summary of Inventory Carrying Cost Data-collection Procedure

Source: Douglas M. Lambert and Robert H. Quinn, 'Profit Orientated Inventory Policies Require a Documented Inventory Carrying Cost', *Business Quarterly* 46, no. 3 (Autumn 1981), p. 71.

Using the methodology summarized in Table 5.2, we calculate costs for the following four basic categories: (1) capital costs, (2) inventory service costs, (3) storage space costs, and (4) inventory risk costs.

Capital Costs

The opportunity cost of capital should be applied only to the out-of-pocket investment in inventory. The *out-of-pocket investment* is the direct variable expense incurred up to the point at which the inventory is held in storage. Using the company's inventory plan and standard variable product cost for the coming year, it is possible to calculate the average inventory for each product. This can be determined for each storage location and for the total system.

Next, the average transportation and warehousing cost per case of product is added to the variable manufactured cost. This is necessary because the transportation and warehousing costs are out-of-pocket costs. If any warehousing costs are incurred moving product into storage in field locations, these costs should be added on a per-case basis to the standard variable manufactured cost. When public warehousing is used, any charges paid at the time products are moved into the facility should be added on a per-case basis to all products held in inventory. In corporate facilities, only the variable out-of-pocket costs of moving the products into storage should be included.

The company had €10 million in average system inventory valued at full manufactured cost. Annual sales were €175 million, or approximately €125 million at manufactured cost. The inventory value based on variable manufacturing costs and the forecasted product mix was €7 million – the average annual inventory held at plants and field locations in order to achieve least cost distribution. The variable costs associated with transporting the strategically deployed field inventory and moving it into public warehouses totalled €800,000. Therefore, the average system inventory was €7.8 million when valued at variable cost delivered to the storage location. All of the remaining inventory carrying cost components should be calculated as a percentage of the variable delivered cost (€7.8 million) and added to the capital cost percentage.

To establish the opportunity cost of capital – the minimum acceptable rate of return on new investments – an interview was conducted with the company's comptroller. Due to capital rationing, the current hurdle rate on new investments was 15 per cent after taxes, 30 per cent before taxes (Table 5.2, step 2). A pre-tax cost of money is required because all of the other components of inventory carrying cost and the other categories in logistics cost trade-off analysis, such as transportation and warehousing, are pre-tax numbers.

Inventory Service Costs

Return to the example of the manufacturer of packaged goods, with system inventories of €7.8 million valued at variable cost delivered to the storage location. Taxes for the year were €90,948, which is 1.17 per cent of the €7.8 million inventory value; this figure was added to the 30 per cent capital cost (see Table 5.2, step 3). Insurance costs covering inventory for the year were €4,524, which is 0.06 per cent of the inventory value (Table 5.2, step 4).

Storage Space Costs

The storage component of the public warehousing cost was €225,654 for the year, which equals 2.89 per cent of the inventory value (Table 5.2, step 5). Variable storage costs in plant warehouses should include only those costs that are variable with the amount of inventory stored. The vast majority of plant warehousing expenses were fixed in nature. Those costs that were variable fluctuated with the amount of product moved into and out of the facilities (throughput) and were not variable with inventory levels. Consequently, variable storage costs were negligible in plant warehouses.

Inventory Risk Costs

Obsolescence cost was being tracked in this company and represented 0.80 per cent of inventory for the past 12 months (Table 5.2, step 7).

Shrinkage and damage costs were not recorded separately. Regression analysis would have been a possible means of isolating the portion of these costs that were variable with inventories. However, management was confident that no more than 10 per cent of the total shrinkage and damage, €100,308, was related to inventory levels; therefore, a managerial estimate of 10 per cent was used. This was equal to 1.29 per cent of the inventory value of €7.8 million (Table 5.2, steps 8 and 9).

Relocation costs, incurred transporting products from one location to another to avoid obsolescence, were not available. Management said that such costs were incurred so infrequently that they were not recorded separately from ordinary transportation costs.

Total Inventory Carrying Costs

When totalled, the individual percentages gave an inventory carrying cost of 36.21 per cent. Thus, management would use a 36 per cent inventory carrying cost when calculating cost trade-offs in the logistics system.

Up to this point, we have assumed that the company has a relatively homogeneous product line – that is, similar products are manufactured at each plant location, shipped in mixed quantities and stored in the same facilities. Consequently, if the company has a 12-month inventory plan and standard costs are available, a weighted-average inventory carrying cost can be used for all products and locations. This figure would require updating on an annual basis when the new inventory plan, updated standard costs, and the budgeted expenditures for insurance, taxes, storage and inventory risk costs become available.

In companies with heterogeneous product lines, however, inventory carrying costs should be calculated for each individual product. For example, bulk chemical products cannot be shipped in mixed quantities or stored in the same tanks. For this reason, transportation and storage costs should be included on a specific product/location basis, instead of on the basis of average transportation and storage costs as one would use for homogeneous products.

The previous example should clarify the methodology and how it can be applied. At this point, you should be able to calculate an inventory carrying cost percentage for a company.

The Impact of Inventory Turnover on Inventory Carrying Costs

In Chapter 1 we saw that, in many firms, management tries to improve profitability by emphasizing the need to improve inventory turnover. But pushing for increased inventory turnover without considering the impact on total logistics system costs may actually lead to decreased profitability. Often, management expects inventory turns to increase each year. If the company is inefficient and has too much inventory, increasing inventory turns will lead to increased profitability. In the absence of a systems change, however, continued improvements in inventory turns may eventually result in the firm cutting inventories below the optimal level.

As shown in Table 5.3, if a logistics system is currently efficient and the goal is to increase turns from 11 to 12, the annual savings in carrying costs would be €2,273. Care must be taken that the costs of transportation, lot quantity, warehouse picking, and order processing and information do not increase by more than this amount; that lower customer service levels do not result in lost profit contribution in excess of the carrying cost savings; or that some combination of the above does not occur.

Figure 5.8 illustrates the relationship between inventory carrying costs and the number of inventory turnovers. The example shows that improvement in the number of inventory turns has the greatest impact if inventory is turned fewer than six times a year. Indeed, beyond eight turns,

Inventory turns	Average inventory (€)	Carrying cost at 40% (€)	Carrying cost savings (€)
1	750,000	300,000	–
2	375,000	150,000	150,000
3	250,000	100,000	50,000
4	187,500	75,000	25,000
5	150,000	60,000	15,000
6	125,000	50,000	10,000
7	107,143	42,857	7,143
8	93,750	37,500	5,357
9	83,333	33,333	4,167
10	75,000	30,000	3,333
11	68,182	27,273	2,727
12	62,500	25,000	2,273
13	57,692	23,077	1,923
14	53,571	21,428	1,649
15	50,000	20,000	1,428

TABLE 5.3 The Impact of Inventory Turns on Inventory Carrying Costs
Source: Douglas M. Lambert and Robert H. Quinn, 'Profit Oriented Inventory Policies Require a Documented Inventory Carrying Cost', *Business Quarterly* 46, no. 3 (Autumn 1981), p. 65.

the curve becomes relatively flat. Increasing inventory turns from five to six times generates the same savings in inventory carrying costs as improving them from 10 to 15 times. When establishing inventory turnover objectives, it is necessary to fully document how each alternative strategy will increase the other logistics costs and to compare this with the savings in inventory carrying costs.

For a number of management decisions, it may be useful to calculate inventory carrying costs on a per unit basis. The inventory carrying costs associated with each item sold can vary dramatically between high-turnover items and low-volume, low-turn items. The previous analysis of inventory carrying costs and inventory turns is repeated, using a carrying cost per unit example to illustrate this point.

For example, if the manufacturer's selling price of an item to a retailer is €100, the retailer sells the item for €150, giving it a potential contribution of €50 before costs. If the annual inventory carrying cost is 30 per cent, the monthly cost to carry the item in inventory is €2.50 (€30 ÷ 12). Annual turns of one (12 months in inventory) would consume €30 in carrying costs, whereas two turns a year would cost €15, four turns €7.50, and eight turns €3.75. It is not uncommon for some speciality items in a retail store to turn less than once a year. This can easily place an otherwise profitable retailer into a loss position. This example should make it clear how inventory turns affect the profitability of each item sold. Next, we will review methods for improving the profitability of products by better managing their inventories.

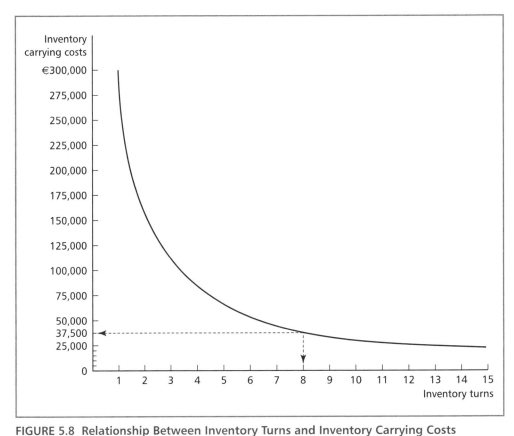

FIGURE 5.8 Relationship Between Inventory Turns and Inventory Carrying Costs

Source: Douglas M. Lambert and Robert H. Quinn, 'Profit Oriented Inventory Policies Require a Documented Inventory Carrying Cost', *Business Quarterly* 46, no. 3 (Autumn, 1981), p. 65.

Symptoms of Poor Inventory Management

This section deals with how to recognize improper management of inventories. Recognition of problem areas is the first step in determining where opportunities exist for improving logistics performance. If a firm is experiencing continuing problems associated with inventory management, a change in processes or systems may be in order.

The following symptoms may be associated with poor inventory management:

- increasing numbers of back-orders
- increasing euro investment in inventory with back-orders remaining constant
- high customer turnover rate
- increasing number of orders cancelled
- periodic lack of sufficient storage space
- wide variance in turnover of major inventory items between distribution centres.
- deteriorating relationships with intermediaries, as typified by dealer cancellations and declining orders
- large quantities of obsolete items.

In many instances inventory levels can be reduced by one or more of the following steps:
- multi-echelon inventory planning (ABC analysis is an example of such planning)
- lead-time analysis
- delivery time analysis; this may lead to a change in carriers or negotiation with existing carriers
- elimination of low-turnover and/or obsolete items
- analysis of pack size and discount structure
- examination of procedures for returned goods
- encouragement/automation of product substitution
- installation of formal reorder review systems
- measurement of fill rates by stockkeeping units (SKUs)
- analysis of customer demand characteristics
- development of a formal sales plan and demand forecast by predetermined logic
- expand view of inventory to include inventory management and information sharing at various levels in the supply chain
- reengineering inventory management practices (include warehousing and transportation) to realize improvements in product flow.

In many companies, the best method of reducing inventory investment is to reduce order cycle time by using advanced order-processing systems (see Chapter 3). If the order cycle currently offered to customers is satisfactory, the time saved in the transmittal, entry and processing of orders can be used for inventory planning. The result will be a significant reduction in inventory.

Improving Inventory Management

Inventory management can be improved by using one or more of the following techniques: ABC analysis, forecasting, inventory models and advanced order-processing systems.

ABC Analysis

In his study of the distribution of wealth in Milan, Villefredo Pareto (1848–1923) found that 20 per cent of the people controlled 80 per cent of the wealth. The concept that critical issues, wealth, importance, and so on, are concentrated among a few is termed Pareto's law. This applies in our daily lives – most of the issues we face have little importance, but a few are critical, long-term issues – and it certainly applies to inventory systems.[5]

This type of ABC analysis should not be confused with activity-based costing, also abbreviated as ABC (see Chapter 11). The logic behind ABC analysis is that 20 per cent of the firm's customers or products account for 80 per cent of the sales, and perhaps an even larger percentage of profits. The first step in ABC analysis is to rank products by sales or, preferably, by contribution to corporate profitability if such data are available. The next step is to check for differences between high-volume and low-volume items that may suggest how certain items should be managed.

Inventory levels increase with the number of stockkeeping locations. By stocking low-volume items at a number of logistics centres, the national demand for these products is divided by the number of locations. Each of these locations must maintain safety stock. If one centralized location had been used for these items, the total safety stock would be much lower. For example, if only one centralized warehouse is used and sales are forecast on a national basis, a sales increase in Paris may offset a sales decrease in Marseilles. However, safety stock is required to protect against variability in demand, and there is greater variability in demand when we forecast demand by region. The total system inventory will increase with the number of field warehouse locations because the variability in demand must be covered at each location.

When a firm consolidates slow-moving items at a centralized location, transportation costs often increase. However, these costs may be offset by lower inventory carrying costs and fewer stockout penalties. Customer service can be improved through consolidation of low-volume items by decreasing the probability of experiencing a stockout. ABC analysis is a method for deciding which items should be considered for centralized warehousing. See the Creative Solutions box for an example of ABC analysis in use at a major Scottish whisky producer.

CREATIVE SOLUTIONS

Forecasting and ABC Inventory Analysis for Whisky

A Scottish whisky producer, whom we shall call 'WhiskCo', specializes in the production of single-malt whiskies and employs about 300 people. It has a market share of around 17 per cent in the United Kingdom with annual turnover of about £65 million. The brand portfolio of WhiskCo primarily consists of three single-malt whisky products. All bottling and warehouse activities are performed at one site in central Scotland. The company also sells its products around the globe, however its UK volume is two-thirds of company output.

WhiskCo uses primarily qualitative methods for forecasting taking previous year's sales, stock on hand and incoming orders as inputs to produce outputs of projected shipments of each item to all customers for the next six months. A member of the forecasting team forecasts shipment quantities by product, customer and month going out six months, thus a rolling horizon is used. WhiskCo's forecasts may be categorized as medium term, and while forecasts go out for six months only forecasts for one or two months affect resource and capacity planning. Consequently, demand prediction is used for tactical decisions.

WhiskCo has a bottom–up approach as forecasts are first made at each product level and then aggregated into an overall forecast. Forecasters predict on the basis of the above inputs and their own intuition and experience. Although WhiskCo does not generally perform market surveys to identify the opinions and buying intentions of its customers, it occasionally contacts large retail customers such as Sainsbury's, Tesco or Asda regarding their opinion on how successful various sales promotions might be. The previous year's sales figure of a respective forecasted month is used when the forecaster does not have any information on likely demand or sales promotion activity.

There is an exchange of information between WhiskCo's sales and logistics departments of forecast data. When the forecast is prepared it is also distributed to various areas inside the organization so that they can be aware of the next month's forecasted demand and control their respective activities appropriately.

Seasonal variations, general market trends, promotional activities and other economic and environmental factors affect demand for its products and an effort is made to incorporate these factors into forecasts. The pattern of WhiskCo's demand is seasonal. WhiskCo does not measure actual sales but shipments of cases during each month. Shipments have a large peak in November followed by a large decrease in December and January. The Christmas holiday period causes this peak as WhiskCo ships its products to retailers early enough for the season since various sales promotions may be undertaken.

WhiskCo uses a form of average percentage error (APE) to measure accuracy. The measure is calculated for all stockkeeping units (SKUs), and an average is calculated to obtain WhiskCo's overall forecast accuracy. Forecasts of total shipments are also desegregated for analysis of individual product SKUs.

An ABC analysis of the 37 United Kingdom SKUs that are forecasted yielded four fast-moving 'A' SKUs, six medium-moving 'B' SKUs and 27 slow-moving 'C' SKUs. The four, fast-moving 'A' SKUs represented 78 per cent of sales and 11 per cent of items shipped during the 12-month period of analysis. These items also constituted almost 20 per cent of shipments.

The forecast accuracy error for the four fast-moving products averaged about 13 per cent. WhiskCo places a greater importance on maintaining customer service than on having excess inventory resulting from over-forecasting. Its methodology appears sound, is reasonably accurate and certainly meets its customers needs.

Question: **Assess the advantages and disadvantages of qualitative versus quantitative forecasting.**

Source: **Charoula Karagianni and David B. Grant, 'Quantitative versus Qualitative Forecasting in Whisky Distilling', in Håkan Aronsson (ed.),** *Challenging Boundaries with Logistics – Proceedings of the XVIth Annual Conference for Nordic Researchers in Logistics.* **Linköping, Sweden: Linköping University, 2004, pp. 373–90.**

Forecasting

Forecasting the amount of each product that is likely to be purchased is an important aspect of inventory management. One forecasting method is to *survey buyer intentions*, using mail questionnaires, telephone interviews or personal interviews. These data can be used to develop a sales forecast. This approach is not without problems, however. It can be costly and the accuracy of the information may be questionable.

Another approach is to solicit the opinions of salespeople or known experts in the field. This method, termed **judgemental sampling**, is relatively fast and inexpensive. However, the data are subject to the personal biases of the individual salespeople or experts.

Most companies simply project future sales based on past sales data. Because most inventory systems require only a one- or two-month forecast, short-term forecasting is acceptable. A number of techniques are available to aid the manager in developing a short-term sales forecast. One method for developing the forecast is shown in Figure 5.9. Rather than trying to forecast at the stockkeeping unit (SKU) level, which would result in large forecasting errors, management can improve forecast accuracy significantly by forecasting at a much higher level of aggregation.

For example, in Figure 5.9 a forecasting model is used to develop the forecast at the total company or product-line level. The next step is to break down that forecast by product class and SKU based on past sales history. The inventory is then 'pushed out' from the central distribution centre to branch or regional distribution centres using one of the following methods:

- going rate – the rate of sales that the SKU is experiencing at each location
- weeks/months of supply – the number of weeks/months of sales based on expected future sales that management wishes to hold at each location
- available inventory – currently available inventory less back-orders.

The only certainty in developing a forecast is that the forecast will not be 100 per cent accurate. For this reason, many firms are developing time-based strategies that focus on reducing the total time from sourcing of materials to delivery of the final product. The shorter this time period can be made, the less critical forecasting becomes, because the firm can respond more quickly to changes in demand. Time-based competitive strategies will be discussed throughout the text. We will examine the most frequently used forecasting techniques in Chapter 10, providing a description of each technique and its advantages and disadvantages.

Firms are increasingly moving towards 'demand pull' production where they continuously replenish inventory based on actual customer demand/sales. The Technology box discusses two **continuous replenishment (CR)** programmes: efficient consumer response (ECR) and **collaborative planning and forecasting (CPFR)** in which daily sales data are combined with forecast data.

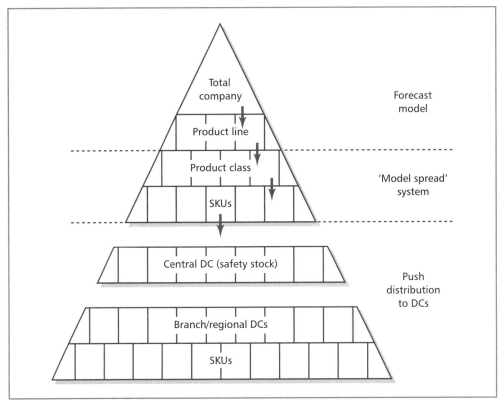

FIGURE 5.9 Building a Forecast

Source: Professor Jay U. Sterling, University of Alabama. Reproduced with permission.

TECHNOLOGY

Continuous Replenishment Programmes for Managing Inventory

Continuous replenishment programmes such as efficient consumer response (ECR) began in the early 1990s. ECR was developed by consultants Kurt Salmon Associates for a working group of grocery industry representatives concerned about losses in market share and declining productivity. Leading European retailers and manufacturers founded ECR-Europe in the mid-1990s to consider ECR for the European business situation. ECR is defined as a grocery industry strategy in which distributors and suppliers are working closely together (i.e. in partnership to bring better value to the grocery consumer through a seamless delivery of products at a total low cost). This seamless delivery is consumer-driven through a paperless information flow initiated by a retailer's electronic point-of-sale (EPOS) that also sets and manages production levels for suppliers. Benefits of ECR include lower total system inventories and costs, enhanced consumer value in terms of choice and quality of products, and more successful development of new consumer-driven products.

However, implementation of ECR, while easy in theory, has proved difficult in practice. An early ECR pilot programme at UK grocery retailer Somerfield saw inventory levels reduced by up to 25 per cent but service levels improved by only about 2.5 per cent. Despite integration difficulties, some 'soft' benefits occurred, such as improved management of seasonal events. Stockouts and availability continue to be a problem in some settings, and the product category management that is a feature of some ECR applications has been criticized as being too time and data intensive. ECR is still perceived as a technique suitable only for large manufacturers and retailers.

Collaborative planning and forecasting (CPFR) is a follow-on to ECR that was developed by the Voluntary Inter-industry Commerce Standards (VICS) group in the United States to minimize out-of-stocks by synchronizing forecasting and planning between retailers and manufacturers. This enhancement is therefore a step beyond ECR or other continuous replenishment programmes that rely on inventory restocking triggered by actual needs rather than relying on long-range forecasts and layers of safety stock just in case.

CPFR, as presently configured between only manufacturers and retailers, is currently unsuitable for every firm as firms require sufficient revenue and product volumes to be economically feasible and real-time information sharing on a common platform such as the Internet. This will require collaboration and technological sophistication throughout the entire supply chain. The reported number of active CPFR partnerships in the USA is only 20 but a survey published in 2000 indicates that 80 per cent of grocery executives intend to increase collaboration in the future. The uptake in Europe has been slow, with only five pilots reported. This lack of progress in Europe may be a result of the Y2K phenomenon and the 'dotcom' retreat slowing progress, differences in retailers' economic status, cultural issues, existing ECR implementation and supply chain structures, or sustained business consolidations and increased market competition with foreign entrants.

Question: **What benefits will initiatives such as ECR and CPFR bring to consumers?**

Sources: adapted from Herbert Kotzab, 'Improving Supply Chain Performance by Efficient Consumer Response? A Critical Comparison of Existing ECR Approaches', *Journal of Business & Industrial Marketing* 14, no. 5/6 (1999), pp. 364–77; R. Marzian and E. Garriga, *A Guide to CPFR Implementation*. Brussels: ECR Europe, 2001; Theodore P. Stank, Patricia J. Daugherty and Chad W. Autry, 'Collaborative Planning: Supporting Automatic Replenishment Programs', *Supply Chain Management: An International Journal* 4, no. 2 (1999), pp. 75–85; and R. Younger, *Logistics Trends in European Consumer Goods: Challenges for Suppliers, Retailers and Logistics Companies*. London: Financial Times Management Report, 1997.

Order-Processing/Inventory Systems

Many companies have not undertaken comprehensive and ongoing analysis and planning of inventory policy because of a lack of time and information. Many times a poor communications system is a contributing factor. A primary goal of inventory management is to achieve an optimum balance between inventory carrying costs and customer service. The essential task of determining the proper balance requires continuous and comprehensive planning. It hinges on the availability of information. Communications make information available.

An automated and integrated order-processing system that utilizes up-to-date customer demand data and is linked to forecasting and production scheduling can reduce the time needed to perform certain elements of the order cycle and to reduce information lags in order processing and inventory replenishment. Cycle-time reductions in the performance of these activities can be used for inventory planning, assuming the current order cycle time is satisfactory to the manufacturer's customers. In this way, the firm can gain substantial cost savings by reducing its levels of safety stock.

In addition, an automated and integrated logistics information system can reduce message errors and unexpected time delays. This facilitates better decision-making and improves internal coordination in the firm.

With full, up-to-the-second information on orders, raw materials inventory and production scheduling can be better managed. The distribution centre can meet customer commitments without increasing inventories. The firm can prepare more accurate invoices, invoice customers sooner, and receive payments more quickly and with fewer reconciliations. Reduced inventories and faster invoicing improve cash flow. Inventory management is improved by placing vital information into the hands of decision-makers and providing them with the necessary time to use this information in planning inventory strategies.

Impact of an Inventory Reduction on Corporate Profit Performance

Inventory reductions have far-reaching implications for organizational **return on investment (ROI)** and profitability. Suppose, for example, that through the implementation of EDI, an organization can reduce its required cycle stock by €5 million. This will reduce the order transmission time without reducing lead-time to its customers. What are some of the effects this would have? This is illustrated in Table 5.4.

	Before inventory reduction	After inventory reduction
	Income statement (€ millions)	
Sales	100.0	100.0
Cost of goods sold (COGS)	60.0	60.0
Cost margin	40.0	40.0
Operating variable	18.0	16.5
Operating fixed	18.0	18.0
Pre-tax profit	4.0	5.5
	Balance sheet (€millions)	
Current assets inventory	14.0	9.0
Other	8.0	8.0
Total	22.0	17.0
Fixed assets	18.0	23.0
Total assets	40.0	40.0
	Profit margin	
Profit/sales	€4/€100 = 4%	€5.5/€100 = 5.5%
	Return on assets	
Profit/total assets	€4/€40 = 10%	€5.5/€40 = 13.8%
	Inventory turnover	
COGS/inventory	€60/€14 = 4.3%	€60/€9 = 6.7 times

TABLE 5.4 Selected Company X Financial Data for Analysis Purposes

chapter 5 Inventory Concepts and Management

First, profitability is improved due to a reduction in inventory carrying costs. With an incremental annual inventory carrying cost of 30 per cent, company X had a €1.5 million reduction in inventory carrying costs (30 per cent × €5 million inventory reduction). This went right to the bottom line to improve pre-tax profit by €1.5 million. The pre-tax profit margin would thus go up from 4.0 per cent to 5.5 per cent as the company increases earnings on the same sales value.

Return on assets would also go up because the company could make the same return (profit) with fewer assets. It changes from 10.0 per cent to 13.8 per cent. Second, inventory turnover would go up because average inventory would be lower on the same sales euros. Analysts like to see such numbers. Here, the figure goes from 4.3 per cent to 6.7 per cent!

Many other ratios of improvements could be calculated in this simple example. If the company took its savings and invested them in assets that could improve production, marketing, or research and development to generate increased sales, even more changes could be seen.

SUMMARY

In this chapter and the accompanying appendices (see pages 163–171), we have examined the basic concepts of inventory management. The EOQ model was introduced, along with methods for adjusting it. In addition, we discussed demand and order cycle uncertainty. Appendices 5.A, 5.B and 5.C examine both types of uncertainty when calculating safety stock requirements. We saw that the traditional approach to improving customer service, increasing inventory investment, is costly. We closed with an overview of the impact of inventory investment on production scheduling. In the next chapter, we will further examine inventory concepts, focusing on inventory management techniques and the cost of holding inventory.

We also saw how to determine the impact of inventory investment on a firm's corporate profit performance. We examined the way in which inventory policy affects least total cost logistics and described a methodology that can be used to calculate inventory carrying costs. We looked at the relationship between inventory turnover and inventory carrying costs. It should now be apparent that inventory is a costly investment. The chapter concluded with an explanation of techniques to use to improve inventory management, and a method to determine the impact of an inventory reduction on corporate profit performance. In the next chapter, we will see how a knowledge of materials management can improve logistics performance.

KEY TERMS

A full Glossary can be found at the back of the book.

QUESTIONS AND PROBLEMS

1   Why is inventory so important to the efficient and effective management of a firm?

2   How does uncertainty in demand and lead-time affect inventory levels?

3   How does the economic order quantity model mathematically select the most economical order quantity?

4   One of the product lines carried by Farha Wholesale Foods was a line of canned fruit manufactured by Spanish Canners. Mr Jones, the canned goods buyer, knew that the company did not reorder from its suppliers in a systematic manner and wondered if the EOQ model might be appropriate. For example, the company ordered 200 cases of fruit cocktail each week, and the annual volume was about 10,000 cases. The purchase price was €8 per case, the ordering cost was €15 per order, and the inventory carrying cost was 35 per cent. Spanish Canners paid the transportation charges, and there were no price breaks for ordering quantities in excess of 200 cases. Does the economic order quantity model apply in this situation? If so, calculate the economic order quantity.

5   Explain the basic differences between a fixed order point, fixed order quantity model and a fixed order interval inventory model. Which is likely to lead to the largest inventory levels? Why?

6   Calculate the economic order quantity, the safety stock and the average inventory necessary to achieve a 98 per cent customer service level, given the following information.

a) The average daily demand for a 25-day period was found to be:

Day	Units demand	Day	Units demand
1	8	14	9
2	5	15	10
3	4	16	5
4	6	17	8
5	9	18	11
6	8	19	9
7	9	20	7
8	10	21	7
9	7	22	6
10	6	23	8
11	7	24	10
12	8	25	11
13	12		

b) There is no variability in order cycle.
c) The ordering cost is £20 per order.
d) The annual demand is 2000.
e) The cost is £100 per unit.
f) The inventory carrying cost is 35 per cent.

7 Explain how excessive inventories can erode corporate profitability.

8 Many businesspeople rely on industry averages or textbook percentages for the inventory carrying cost that they use when setting inventory levels. Why is this approach wrong?

9 What problems do you foresee in gathering the cost information required to calculate a company's inventory carrying costs?

10 Calculate the inventory carrying cost percentage for XYZ Company, given the following information.
- Finished goods inventory is €26 million, valued at full manufactured cost.
- Based on the inventory plan, the weighted-average variable manufactured cost per case is 70 per cent of the full manufactured cost.
- The transportation cost incurred by moving the inventory to field warehouse locations was €1.5 million.
- The variable handling cost of moving this inventory into warehouse locations was calculated to be €300,000.
- The company is currently experiencing capital rationing, and new investments are required to earn 15 per cent after taxes.
- Personal property tax paid on inventory was approximately €200,000.
- Insurance to protect against loss of finished goods inventory was €50,000.
- Storage charges at public warehouses totalled €450,000.
- Variable plant storage was negligible.
- Obsolescence was €90,000.
- Shrinkage was €100,000.

- Damage related to finished goods inventory levels was €10,000.
- Transportation costs associated with the relocation of field inventory to avoid obsolescence was €50,000.

11 What are some of the key symptoms of poor inventory management?

12 Describe ABC analysis of inventory, and how it can be used to improve inventory management.

SUGGESTED READING

BOOKS

Lambert, Douglas M., *The Development of an Inventory Costing Methodology: A Study of the Costs Associated with Holding Inventory*. Chicago: National Council of Physical Distribution Management, 1976.

Makridakis, Spyros G., Steven C. Wheelwright and Rob J. Hyndman, *Forecasting: Methods and Applications* 3rd edn. Chichester, UK: John Wiley & Sons Ltd, 1998.

Shank, John K. and Vijay Govindarajan, *Strategic Cost Management: The New Tool for Competitive Advantage*. New York: Free Press, 1993.

Slack, Nigel, Stuart Chambers and Robert Johnson, *Operations Management*, 4th edn. Harlow, UK: FT Prentice Hall, 2004.

Waters, Donald, *Inventory Control and Management*, 2nd edn. Chichester, UK: John Wiley & Sons Ltd, 2003.

JOURNALS

Chambers, John C., Satinder K. Mullick and Donald D. Smith, 'How to Choose the Right Forecasting Technique', *Harvard Business Review* 49, no. 4 (July–August 1971), pp. 45–74.

Copacino, William C., 'Moving Beyond ABC Analysis', *Traffic Management* 33, no. 3 (March 1994), pp. 35–6.

Farris, M. Theodore, II, 'Utilizing Inventory Flow Models with Suppliers', *Journal of Business Logistics* 17, no. 1 (1996), pp. 35–61.

Hau, Lee L. and Corey Billington, 'Managing Supply Chain Inventories: Pitfalls and Opportunities', *Sloan Management Review* 33, no. 3 (Spring 1992), pp. 65–73.

Hsleh, Pei Jung, 'New Developments in Inventory and Materials Management', *Logistics Information Management* 5, no. 2 (1992), pp. 32–41.

Krupp, James A., 'Measuring Inventory Management Performance', *Production and Inventory Management Journal* 35, no. 4 (1994), pp. 1–6.

Loar, Tim, 'Patterns of Inventory Management and Policy: A Study of Four Industries', *Journal of Business Logistics* 13, no. 2 (1993), pp. 69–82.

Tersine, Richard J. and Michele G. Tersine, 'Inventory Reduction: Preventive and Corrective Strategies', *The International Journal of Logistics Management* 1, no. 2 (1990), pp. 17–24.

'What Level of Inventory Should be Held?', *Logistics Information Management* 4, no. 2 (1992), pp. 50–5.

Zinn, Walter, 'Developing Heuristics to Estimate the Impact of Postponement on Safety Stock', *The International Journal of Logistics Management* 1, no. 2 (1990), pp. 11–16.

REFERENCES

[1] Helen L. Richardson, 'Trust Time-Definite, Reduce Inventory', *Transportation and Distribution* 35, no. 1 (January 1994), pp. 41–4.

[2] Kenneth B. Ackerman, *Words of Warehousing*. Columbus, OH: K.B. Ackerman Company, 1992, p. 28.

[3] For additional examples of special-purpose EOQ models and amendments, see Donald Waters, *Inventory Control and Management*, 2nd edn. Chichester, UK: John Wiley & Sons Ltd, 2003, pp. 65–191. Copyright John Wiley & Sons. Reproduced with permission.

[4] This section draws heavily on Douglas M. Lambert, *The Development of an Inventory Costing Methodology: A Study of the Costs Associated with Holding Inventory*. Chicago: National Council of Physical Distribution Management, 1976.

[5] Nigel Slack, Stuart Chambers and Robert Johnson, *Operations Management*, 4th edn. Harlow, UK: FT Prentice Hall, 2004.

APPENDICES

These appendices contain some valuable, but slightly more advanced, calculations for dealing with inventory management under conditions of uncertainty. Detailed, step-by-step examples allow the reader to develop an excellent grasp of how to calculate safety stock requirements for desired levels of customer service, and how to calculate fill rates. Question 6 illustrates these concepts.

INVENTORY MANAGEMENT UNDER UNCERTAINTY   APPENDIX 5A

As we have noted, managers rarely, if ever, know for sure what demand to expect for the firm's products. Many factors, including economic conditions, competitive actions, changes in government regulations, market shifts, and changes in consumer buying patterns, may influence forecast accuracy. Order cycle times also are not constant. Transit times vary; it may take more time to assemble an order or wait for scheduled production on one occasion than another, supplier lead-times for components and raw materials may be inconsistent, and suppliers may not have the capability of responding to changes in demand.

Consequently, management has the option of either maintaining additional inventory in the form of safety stocks (see Figure 5.3) or risking a potential loss of sales revenue due to inventory stockouts. We must thus consider an additional cost trade-off: inventory carrying costs versus stockout costs.

The uncertainties associated with demand and lead-time cause most managers to concentrate on *when* to order rather than how much to order. The order quantity is important to the extent that it influences the number of orders and consequently the number of times that the company is exposed to a potential stockout at the end of each order cycle. The point at which the order is placed is the primary determinant of the future ability to fill demand while waiting for replenishment stock. As presented in the chapter, order policy may be based on fixed order quantity or fixed interval. Figure 5A.1 illustrates these two methods.

A review of part A of Figure 5A.1 shows that replenishment orders are placed on days 15, 27 and 52, respectively, under the fixed order point, fixed order quantity model. In contrast, when the fixed order interval model is used (part B), orders are placed at 20-day intervals on days 15, 35 and 55. With the fixed order interval model, it is necessary to forecast demand for days 20 through to 40 on day 15, for days 40 through to 60 on day 35, and so on. The fixed order interval system is more adaptive because management is forced to consider changes in sales activity and make a forecast for every order interval.

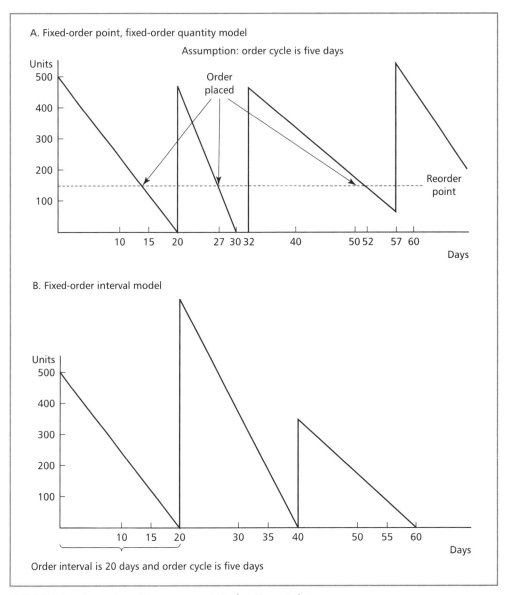

FIGURE 5A.1 Inventory Management Under Uncertainty

APPENDIX 5B   CALCULATING SAFETY STOCK REQUIREMENTS

The amount of safety stock necessary to satisfy a given level of demand can be determined by computer simulation or statistical techniques. In this illustration, we address the use of statistical techniques. In calculating safety stock levels, it is necessary to consider the joint impact of demand and replenishment cycle variability. This can be accomplished by gathering statistically valid samples of data on recent sales volumes and replenishment cycles. Once the data are gathered, it is possible to determine safety stock requirements by using this formula:[1]

$$\sigma c = \sqrt{\bar{R}(\sigma S^2) + \bar{S}^2(\sigma R^2)}$$

where

σc	=	Units of safety stock needed to satisfy 68 per cent of all probabilities (one standard deviation)
$\bar{R}$	=	Average replenishment cycle
σR	=	Standard deviation of the replenishment cycle
$\bar{S}$	=	Average daily sales
σS	=	Standard deviation of daily sales

Assume that the sales history contained in Table 5B.1 has been developed for market area 1. From this sample, we can calculate the standard deviation of sales as shown in Table 5B.2. The formula is:

$$\sigma S = \sqrt{\frac{\Sigma f d^2}{n - 1}}$$

where

σS	=	Standard deviation of daily sales
f	=	Frequency of event
d	=	Deviation of event from mean
n	=	Total observations

Applying this formula to the data yields a standard deviation of sales equal approximately to 20 units:

$$\sigma S = \sqrt{\frac{10,000}{25 - 1}}$$
$$= 20$$

This means that 68 per cent of the time, daily sales fall between 80 and 120 units (100 units ± 20 units). A protection of two standard deviations, or 40 units, would protect against 95 per cent of all events. In setting safety stock levels, however, it is important to consider only events that exceed the mean sales volume. Thus, a safety stock level of 40 units actually protects against almost 98 per cent of all possible events (see Figure 5B.1). Given a distribution of measurements that is approximately bell-shaped, the mean, plus or minus one standard deviation, will contain approximately 68 per cent of the measurements. This leaves 16 per cent in each of the tails, which means that inventory sufficient to cover sales of one standard deviation in

Day	Sales in cases	Day	Sales in cases
1	100	14	80
2	80	15	90
3	70	16	90
4	60	17	100
5	80	18	140
6	90	19	110
7	120	20	120
8	110	21	70
9	100	22	100
10	110	23	130
11	130	24	110
12	120	25	90
13	100		

TABLE 5B.1 Sales History for Market Area 1

Daily sales in cases	Frequency (f)	Deviation from mean (d)	Deviation squared (d²)	fd²
60	1	−40	1600	1600
70	2	−30	900	1800
80	3	−20	400	1200
90	4	−10	100	400
100	5	0	0	0
110	4	+10	100	400
120	3	+20	400	1200
130	2	+30	900	1800
140	1	+40	1600	1600
$\bar{S}$ = 100	n = 25			Σfd^2 = 10,000

TABLE 5B.2 Calculation of Standard Deviation of Sales

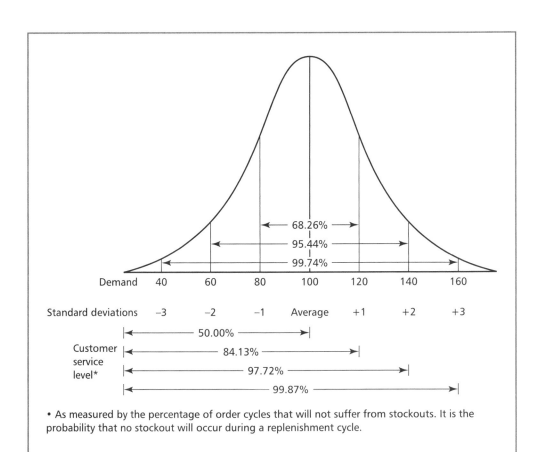

• As measured by the percentage of order cycles that will not suffer from stockouts. It is the probability that no stockout will occur during a replenishment cycle.

FIGURE 5B.1 Area Relationship for the Normal Distribution

166

excess of mean daily sales will actually provide a customer service level of 84 per cent. (If the sample does not represent a normal distribution, refer to a basic statistics book for an alternative treatment.)

The same procedure can be used to arrive at the mean and standard deviation of the replenishment cycle. Once this is accomplished, the formula shown previously can be used to determine safety stock requirements at a certain level of demand. For example, an analysis of replenishment cycles might yield the results shown in Table 5B.3. The standard deviation of the replenishment cycle is:

$$(\sigma R) = \sqrt{\frac{\Sigma fd^2}{n-1}}$$
$$= \sqrt{2.67}$$
$$= 1.634$$

The average replenishment cycle is:

$$(\bar{R}) = 10$$

The combined safety stock required to cover variability in both demand and lead-time can be found using the formula:

$$\sigma c = \sqrt{\bar{R}(\sigma S^2) + S^2(\sigma R^2)}$$
$$= \sqrt{4,000 + 26,700} = \sqrt{30,700}$$
$$= 175 \text{ cases}$$

Thus, in a situation in which daily sales vary from 60 to 140 cases and the inventory replenishment cycle varies from 7 to 13 days, a safety stock of 175 cases will allow the manufacturer to satisfy 84 per cent of all possible occurrences. To protect against 98 per cent of all possibilities, 350 cases of safety stock are required. Table 5B.4 shows alternative customer service levels and safety stock requirements.

To establish the average inventory for various levels of customer service, we must first determine the EOQ. The projected yearly demand is found by multiplying the average daily demand by 250 working days,[2] which equals 25,000 cases (250 × 100). The inventory carrying cost was calculated to be 32 per cent, the average value of a case of product was €4.37, and the ordering cost was €28.00. The average inventory required to satisfy each service level is shown in Table 5B.5.

Note that the establishment of a safety stock commitment is really a policy of customer service and inventory availability. Although we have demonstrated a quantitative method of calculating safety stock requirements to protect the firm against stockouts at various levels of probability, additional calculations are necessary to determine the specific fill rate when stockouts occur. Fill rate represents the percentage of units demanded that are on hand to fill customer orders.

Lead-times in days	Frequency (f)	Deviation from mean (d)	Deviation squared (d²)	fd²
7	1	−3	9	9
8	2	−2	4	8
9	3	−1	1	3
10	4	0	0	0
11	3	+1	1	3
12	2	+2	4	8
13	1	+3	9	9
$\bar{R} = 10$	n = 16			$\Sigma fd^2 = 40$

TABLE 5B.3 Calculation of Standard Deviation of Replenishment Cycle

Service levels (%)	Number of standard deviations (C) needed	Safety stock requirements (cases)
84.1	1.0	175
90.3	1.3	228
94.5	1.6	280
97.7	2.0	350
98.9	2.3	403
99.5	2.6	455
99.9	3.0	525

TABLE 5B.4 Summary of Alternative Service Levels and Safety Stock Requirements

Service levels (%)	Average cycle stock ($1/_2 \times EOQ$)	Safety stock (units)	Total average inventory (units)
84.1	500	175	675
90.3	500	228	728
94.5	500	280	780
97.7	500	350	850
98.9	500	403	903
99.5	500	455	955
99.9	500	525	1025

TABLE 5B.5 Summary of Average Inventory Levels Given Different Service Levels

Calculating Fill Rate

Fill rate represents the magnitude of the stockout. If a manager wants to hold 280 units as safety stock, what will the fill rate be?[3]

The fill rate can be calculated using the following formula:

$$FR = 1 - \frac{\sigma c}{EOQ} [I(K)]$$

where
- FR = Fill rate
- σc = Combined safety stock required to consider both variability in lead-time and demand (one standard deviation)
- EOQ = Order quantity = 1000 (in this example)
- $I(K)$ = Service function magnitude factor (provided by Table 5B.6), based on desired number of standard deviations

K (the safety factor) is the safety stock the manager decides to hold divided by c. Returning to the question presented above, the manager's economic order quantity (EOQ) is 1000. The safety stock determined by the manager is 280 units. Therefore, K is equal to 280 divided by 175, or 1.60. If K = 1.60, Table 5B.6 can be used to identify $I(K)$ = 0.0236 (see last column of Table 5B.6).

Safety factor K	Stock protection (single tail)	Stockout probability F(K)	Service function (magnitude factor) partial expectation I(K)
0.00	0.5000	0.5000	0.3989
0.10	0.5394	0.4606	0.3509
0.20	0.5785	0.4215	0.3067
0.30	0.6168	0.3832	0.2664
0.40	0.6542	0.3458	0.2299
0.50	0.6901	0.3099	0.1971
0.60	0.7244	0.2756	0.1679
0.70	0.7569	0.2431	0.1421
0.80	0.7872	0.2128	0.1194
0.90	0.8152	0.1848	0.0998
1.00	0.8409	0.1591	0.0829
1.10	0.8641	0.1359	0.0684
1.20	0.8849	0.1151	0.0561
1.30	0.9033	0.0967	0.0457
1.40	0.9194	0.0806	0.0369
1.50	0.9334	0.0666	0.0297
1.60	0.9454	0.0546	0.0236
1.70	0.9556	0.0444	0.0186
1.80	0.9642	0.0358	0.0145
1.90	0.9714	0.0286	0.0113
2.00	0.9773	0.0227	0.0086
2.10	0.9822	0.0178	0.0065
2.20	0.9861	0.0139	0.0049
2.30	0.9893	0.0107	0.0036
2.40	0.9918	0.0082	0.0027
2.50	0.9938	0.0062	0.0019
2.60	0.9953	0.0047	0.0014
2.70	0.9965	0.0035	0.0010
2.80	0.9974	0.0026	0.0007
2.90	0.9981	0.0019	0.0005
3.00	0.9984	0.0014	0.0004
3.10	0.9990	0.0010	0.0003
3.20	0.9993	0.0007	0.0002
3.30	0.9995	0.0005	0.0001
3.40	0.9997	0.0003	0.0001
3.50	0.9998	0.0002	0.0001
3.60	0.9998	0.0002	
3.70	0.9999	0.0001	
3.80	0.9999	0.0001	
3.90	0.9999	0.0001	
4.00	0.9999	0.0001	

TABLE 5B.6 Inventory Safety Stock Factors

Source: Professor Jay U. Sterling, University of Alabama; adapted from Robert G. Brown, *Material Management Systems*. New York: John Wiley, 1977, p. 429.

Now the fill rate can be calculated using the following formula:

$$FR = 1 - \frac{\sigma c}{EOQ}[I(K)]$$

$$= 1 - \frac{175}{1,000}(0.0236)$$

$$= 1 - .0041$$

$$= .9959$$

Thus, the average fill rate is 99.59 per cent. That is, of every 100 units of product A demanded, 99.59 will be on hand to be sold if the manager uses 280 units of safety stock and orders 1000 units each time.[4]

If the manager wants to know how much safety stock of product A to hold to attain a 95 per cent fill rate, the same formula can be used:

$$FR = 1 - \frac{\sigma c}{EOQ}[I(K)]$$

$$I(K) = (1 - FR)\left(\frac{EOQ}{\sigma c}\right)$$

$$I(K) = (1 - 0.95)\left(\frac{1,000}{175}\right)$$

$$= (0.05)(5.714)$$

$$= 0.2857$$

Looking up the value for $I(K)$ in Table 5B.6, the corresponding K value is approximately 0.25. Since K is equal to the safety stock the manager decides to hold divided by c, the safety stock required to provide a 95 per cent fill rate is 44 units (175 × 0.25).

APPENDIX 5C   DERIVATION OF ECONOMIC ORDER QUANTITY

This appendix illustrates the mathematical derivation of EOQ in Figure 5.4.

Total annual cost $(TAC) = [1/2\ (Q) \times (V) \times (C)\ ] + [(P) \times (D/Q)\ ]$

where

Q = Average number of units in the economic order quantity during the order cycle

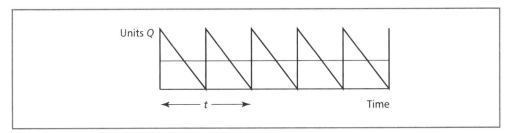

FIGURE 5C.1 Saw Tooth Diagram

Mathematical solution:

$$\frac{d\,TAC}{dQ} = \frac{VC}{2} - \frac{PD}{Q^2}$$

$$\text{Set} = \text{Zero}: \frac{VC}{2} - \frac{PD}{Q^2}$$

$$\frac{VC}{2} = \frac{PD}{Q^2}$$

$$VCQ^2 = 2PD$$

$$Q^2 = \frac{2PD}{CV}$$

$$Q = \sqrt{\frac{2PD}{CV}}$$

References

[1] Robert Hammond of McKinsey and Company, Inc., as reported in Robert Fetter and Winston C. Dalleck, *Decision Models for Inventory Management*. Burr Ridge, IL: Richard D. Irwin, 1961, pp. 105–8. For an application of the formula in a simulation model, see Walter Zinn, Howard Marmorstein and John Charnes, 'The Effect of Autocorrelated Demand on Customer Service', *Journal of Business Logistics* 13, no. 1 (1992), pp. 173–92.

[2] For this example, the average number of working days per year was assumed to be 250.

[3] This example was provided by Professor Robert L. Cook, Central Michigan University, Mount Pleasant, Michigan.

[4] If demands, lead-times, order quantities or safety stocks change significantly, the fill rate percentage will also change.

CHAPTER 6
MATERIALS MANAGEMENT

OBJECTIVES

CHAPTER OBJECTIVES

■ To identify the activities of materials management

■ To examine the concept of total quality management (TQM)

■ To identify and describe a variety of materials management philosophies and techniques, including Kanban/just-in-time (JIT), MRP, ERP and DRP

INTRODUCTION

As defined in this book, logistics management is concerned with the efficient flow of raw materials, in-process inventory, and finished goods from point of origin to point of consumption. An integral part of the logistics management process is **materials management**, which encompasses the administration of raw materials, subassemblies, manufactured parts, packing materials and in-process inventory.

Materials management is critical to the total logistics process. Although materials management does not directly interface with the final customer, decisions made in its portion of the logistics process will directly affect the level of customer service offered, the ability of the firm to compete with other companies, and the level of sales and profits the firm is able to achieve in the marketplace.

Without efficient and effective management of inbound materials flow, the manufacturing process cannot produce products at the desired price and at the time they are required for distribution to the firm's customers. For this reason, it is essential that the logistics executive understand the role of materials management and its impact on the company's cost–service mix. In a manufacturing setting, not having the proper materials when needed can cause manufacturing processes to slow down or even shut down, which can result in stockouts.

Poor materials management can result in stockouts at retail, causing customers to seek substitutes or shop elsewhere. In a service such as health care, lack of needed materials may delay testing or vital patient treatment, causing at best inconvenience and at worst a threat to patient health.

As organizations have developed and matured, the role of materials management has expanded to meet the challenges of market-driven, rather than production-driven, economies. While many things, such as the need to reduce costs and provide high levels of customer service, will remain important, materials management will be characterized by a changing set of priorities and issues: global orientation, shorter product life cycles, lower levels of inventory, electronic data processing and a market-orientated focus.

This chapter identifies the various components of materials management, and describes how to manage materials flows effectively within a manufacturing environment. We will examine specific management strategies and techniques used in the planning, implementation and control of materials flows within organizations.

Scope of Materials Management

Materials management is typically comprised of four basic activities:

1 anticipating materials requirements
2 sourcing and obtaining materials
3 introducing materials into the organization
4 monitoring the status of materials as a current asset.[1]

Functions performed by materials managers include purchasing or procurement, inventory control of raw materials and finished goods, receiving, warehousing, production scheduling and transportation. The scope of materials management used in this chapter views the activity as an organizational system with the various functions as interrelated, interactive subsystems.

The specific objectives of materials management are closely tied to the firm's main objectives of achieving an acceptable level of profitability or return on investment (ROI), and remaining competitive in an increasingly competitive marketplace.

Figure 6.1 highlights the major objectives of materials management: low costs, high levels of service, quality assurance, low levels of tied-up capital, and support of other functions. Each objective is clearly linked to overall corporate goals and objectives. Thus trade-offs among the objectives must be made using a broad perspective of materials flow throughout the total system, from source of supply to the ultimate customer.[2]

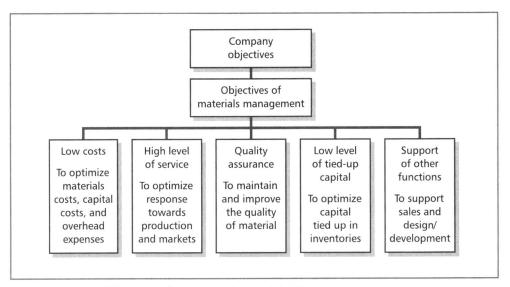

FIGURE 6.1 The Objectives of Integrated Materials Management

Source: Yunus Kathawala and Heino H. Nauo, 'Integrated Materials Management: A Conceptual Approach', *International Journal of Physical Distribution and Materials Management* 19, no. 8 (1989), p. 10.

Materials management encompasses a variety of logistics activities. The primary differences between the process of materials management and that of finished goods distribution are that the items handled in materials management are incoming finished goods, raw materials, component parts and subassemblies to be further processed or sorted before being received by the final customer. The recipient of the materials management effort is the production or manufacturing group and other internal customers, not the final customer.

Integral aspects of materials management include purchasing and procurement, production control, inbound traffic and transportation, warehousing and storage, management information system (MIS) control, inventory planning and control, and salvage and scrap disposal.

Purchasing and Procurement

The acquisition of materials has been, and will continue to be, an important aspect of materials management:

> The rapidly changing supply scene, with cycles of abundance and shortages, and varying prices, lead times, and availability, provides a continuing challenge to those organizations wishing to obtain a maximum contribution from this area.[3]

The terms *purchasing* and *procurement* are often used interchangeably, although they do differ in scope. **Purchasing** generally refers to the actual buying of materials and those activities associated with the buying process. **Procurement** is broader in scope and includes purchasing, traffic, warehousing and all activities related to receiving inbound materials. Purchasing and procurement are support activities in Porter's value chain (see Chapter 1). These topics were discussed in detail in Chapter 4, and will not be examined here.

Production Control

Production control is an activity traditionally positioned under manufacturing, although a few firms place it under logistics. Its position in the firm's organizational chart is not crucial so long as both manufacturing and logistics have input into the production planning and control activities. The production or conversion activity is also considered by Porter to be a primary activity within a firm.

Production affects the logistics process in two significant ways. First, production activity determines the quantity and type of finished goods that are produced. This in turn influences when and how the products are distributed to the firm's customers. Second, production directly determines the company's need for raw materials, subassemblies and component parts used in the manufacturing process. Therefore, production control decisions are often jointly shared by manufacturing and logistics.

Inbound Logistics

Materials management is concerned with product flows or the inbound logistics into the firm, which is also seen by Porter as a primary activity. Much like the firm's target markets, manufacturing requires satisfactory levels of customer service, which depend on the ability of materials management to effectively coordinate with a variety of functions, including traffic and transportation, warehousing and storage, and management information system (MIS) control.

One of the most important activities of materials management is working with the logistics function to manage inbound traffic and transportation. Like their counterparts responsible for finished goods, materials managers must be aware of the available transport modes and combinations, government regulations that affect the firm's transportation carriers, private versus for-hire carrier issues, leasing, evaluation of mode/carrier performance, and the cost–service trade-offs involved in the inbound movement of product.

Three major differences exist between the administration of inbound and outbound transportation. First, the market demand that generates the need for outbound movement is generally uncertain and fluctuating. The demand that concerns materials managers originates with the production scheduling activity and tends to be more predictable and stable; thus materials managers do not encounter the same types of problem as their counterparts in the outbound traffic area.

Second, the materials manager is more likely to be concerned with bulk movements of raw materials or large shipments of parts and subassemblies. In addition, raw materials and parts have different handling, loss or damage characteristics that will affect mode/carrier selection and evaluation. Third, firms may exercise less control over their inbound transportation because purchasing procedures tend to look at 'total delivered price' where the transportation cost is not identified separately. Thus, a separate analysis of inbound costs is not performed as often or in as much depth, and significant cost savings are possible.

Warehousing and Storage

Raw materials, components parts and subassemblies are placed in storage until they are required by the production process. Unlike the warehousing of finished goods, which usually occurs in the field, items awaiting use in production are usually stored on-site – that is, at the point of manufacture; or they are delivered on an 'as needed' basis by a just-in-time (JIT) supplier.

If a JIT delivery system is utilized, the need for inbound warehousing is greatly minimized or eliminated altogether. If the JIT system is not used, and warehouses are used for the storage of inbound materials, the materials manager is usually much more concerned with warehousing and inventory costs because they account for a larger percentage of product value.

In addition, the warehousing requirements for raw materials and other items are usually quite different. For example, open or outside storage is possible with many raw materials, such as iron ore, sand and gravel, coal and other unprocessed materials. Also, damage or loss due to weather, spoilage or theft is minimal with raw materials because of their unprocessed state or low value per pound.

Data and Information Systems

The materials manager needs direct access to the organization's information system to properly administer materials flow into and within the organization. The types of information needed include demand forecasts for production, names of suppliers and supplier characteristics, pricing data, inventory levels, production schedules, transportation routing and scheduling data, and other financial and marketing facts. Additionally, materials management supplies input into the firm's MIS. Data on inventory levels for materials, delivery schedules, pricing, forward buys and supplier information are a few of the inputs provided by materials management. The organization's electronic data processing (EDP) and other technology systems are considered support activities by Porter.

Integrated materials management has a multitude of data to process, a task that would not be possible without EDP-supported systems. Numerous software packages for individual functional elements of integrated materials management have been developed during the last few years. These are packages tailored for particular branches of industry and particular company sizes. One of the best-known suites of software packages for enterprise resource planning (ERP) systems comes from Germany's SAP AG and is discussed in the Technology box.

TECHNOLOGY

The Use of SAP at Sharp Electronics (Europe)

SAP AG of Germany is the world's leading and largest ERP software company. In 2003, it provided services to more than 21,600 firms and more than 12 million users worldwide, including *Fortune* 500 corporations, entrepreneurial start-ups and government agencies, and has more than 30 per cent of the total market for ERP systems. SAP R/3 Enterprise is an integrated suite of financial, manufacturing, distribution, logistics, quality control and human resources application systems.

Its software architecture consists of three main layers:

1 the SAP graphical user interface (GUI), representing the presentation layer
2 the SAP application layer, and
3 the SAP database layer.

Reported experiences in SAP-enabled supply chain reengineering applications (Al-Mashari and Zairi, 2000) have shown that effective implementation requires establishing the following five core competencies:

1 change strategy development and deployment
2 enterprise-wide project management
3 change management techniques and tools
4 business process reengineering integration with IT, and
5 strategic, architectural and technical aspects of SAP installation.

One firm that has enjoyed success with SAP is Sharp Electronics (Europe). Sharp Europe is the European operating unit of Sharp Corporation of Japan, one of the world's largest and most respected manufacturers of electronics products, including television and audio equipment, document systems, liquid crystal display (LCD) screens, mobile phones and solar cells. From its office headquarters in Hamburg, Germany, Sharp Europe coordinates sales and distribution activities throughout Europe. Sharp Europe's ERP software needs are complex: its broad product portfolio includes some of the most technologically advanced consumer electronics in the world, it deals with various retailers in more than a dozen European countries and it has a workforce that speaks numerous different languages.

Sharp Europe places a high value on standardization and ease of use for software applications required to manage all of this complexity. Its relationship with SAP began in 1983 with the implementation of SAP R/2, and Sharp Europe today uses SAP R/3 Enterprise. Sharp Europe employs about 3000 people; more than 1000 of whom are SAP users. The SAP software serves more than a dozen of Sharp Europe's sales organizations, including those in Sweden, Denmark, Norway, Germany, the Netherlands, Belgium, France, Austria, Switzerland, Poland, Italy, Spain and the United Kingdom.

'Our goal in implementing R/3 Enterprise,' Wolfram Konertz, CIO of the European IT Center explains, 'was to have one centre of operations, as well as unification of our hardware and applications. SAP has helped us meet that goal. We very much appreciate the common features of the SAP R/3 Enterprise solution, such as user adaptable screen layouts, downloads into Microsoft Excel, and decentralized reporting and printing. It helps minimize interfaces between our subsidiaries and locations, reduces the need for local development, and simplifies documentation and training. With SAP, we achieved standardization of codes, customer groups and other factors. This also enabled us to set up standardized interfaces to other IT systems and to have joint development of add-on functions.'

Operating in so many different countries, Sharp Europe has made extensive use of the multi-lingual capabilities of SAP solutions. 'Our groups work in German, English, French, Italian, Spanish, Dutch, Swedish, Norwegian and even Polish,' says Michael Helm, manager of IT Systems. 'And SAP supports them all, including the printing of the non-Latin-based Polish characters.' SAP has also helped Sharp Europe cope with the currency issues of its European operations, and supported the transition to the euro currency in 2001.

Overall, SAP has helped Sharp Europe standardize its most important business processes. 'SAP R/3 Enterprise helps us compare financial metrics for different markets,' notes Marion Backhausen, manager application development, 'so we can conduct regional sales and profit analyses. The sales and distribution component is flexible and allows us to employ very sophisticated pricing techniques and react very quickly to new requirements from our salespeople.' The SAP sales and distribution solution also gives greater flexibility to Sharp's sales force. 'Even when they're working remotely from a home office,' says Backhausen, 'our sales professionals are now able to connect through a virtual private network to get their sales figures and reports.' And with a shared view of its European inventory, the company can optimize supply levels to its various markets.

Based on the success of its European group, Sharp has implemented a variety of SAP solutions throughout its global operations. In addition, the company is currently evaluating additional SAP solutions for the future, including mySAP Customer Relationship Management and SAP Business Information Warehouse. 'We have found that for globally operating companies,' says Konertz, 'SAP solutions provide the best way to support business transactions.'

Question: **How could products from SAP or other information technology companies be used in small and medium-sized enterprises (SMEs)?**

Source: adapted from Nigel Slack, Stuart Chambers and Robert Johnson, *Operations Management*, 4th edn. Harlow, UK: FT Prentice Hall, 2004, pp. 482–3, p. 506; and Majed Al-Mashari and Mohamed Zairi, 'Supply-chain Re-engineering Using Enterprise Resource Planning (ERP) Systems: An Analysis of a SAP R/3 Implementation Case', *International Journal of Physical Distribution & Logistics Management* 30, no. 3/4 (2000), pp. 296–313.

Thus modern information technology will offer opportunities for the fast and safe transmission and processing of extensive amounts of data, both internally for users within the company and externally for suppliers and customers. Paperless communication is coming to the fore, whereby routine tasks in order processing and scheduling will be decisively facilitated. As a result, new information technology offers great opportunities for linking the planning, control and processing functions of materials management that hitherto were performed independently, thereby creating a foundation for the establishment of integrated materials management.[4]

The proliferation of computerized information systems and databases, coupled with electronic data interchange (EDI), will make this facet of materials management even more significant in the future.

Inventory Planning and Control

Inventory planning and control of raw materials, component parts, subassemblies and goods-in-process are just as important as the management of finished goods inventory. Many of the concepts discussed in Chapter 5, such as ABC analysis, inventory carrying costs and economic order quantity (EOQ), are directly applicable to materials management.

Materials Disposal

One of the most important areas of materials management that a firm often overlooks or considers minor is the disposal of scrap, surplus, recyclable or obsolete materials. During the last few years, this area, referred to as reverse logistics, has gained significant importance because of increased public awareness of the environment, more stringent government legislation and a better recognition of the opportunities it offers in return.

Many materials can be salvaged and sold to other companies but must be properly disposed of when they cannot be salvaged or sold. Almost all firms also produce surplus materials as by-products of their operations. In addition to the normal packaging materials associated with products (e.g. cartons, pallets, shrink-wrap, baling wire), this material can result from overoptimistic sales forecasts, changes in product specifications, errors in estimating materials usage, losses in processing, and overbuying due to forward buys or quantity discounts on large purchases.

The basic tasks of the disposal function include a disposal classification based on whether something can be reused and the possibility of environmental pollution it bears. As a result of legal requirements and technical conditions, the hazard represented by some waste products is being recognized by firms. Additionally, increased public awareness of the environment has highlighted the importance of reverse logistics issues.

The European Parliament has issued several directives concerning disposal. For example, the Directive on Waste Electrical and Electronic Equipment (WEEE) provides guidance for organizations on the design, production, dismantling and recovery, and reuse and recycling of various products, including large and small household appliances, information technology and telecommunications equipment, and electrical and electronic tools. This aspect of materials management is likely to become much more important in the future.

Forecasting

Predicting the future is important because it allows logistics executives to primarily be proactive rather that reactive. Every area of logistics is affected in some way by the forecasting process – that is, conducting or developing forecasts, providing information to be used in forecasting, or receiving forecasting results and implementing necessary actions. While other activities of logistics are more actively involved in the forecasting process, materials management utilizes forecasts employed in MRP and DRP (discussed later in this chapter) efforts, and is indirectly affected by the forecasts developed (and the subsequent actions taken) by others (e.g. inventory planning, purchasing and demand forecasting). **Forecasting** attempts to predict the future through quantitative or qualitative methods, or some combination of these. The essence of forecasting is to aid in logistics decision-making.

A study of the forecasting practices of a large number of companies indicated that the most widely cited reasons for engaging in forecasting included:

- increasing customer satisfaction
- reducing stockouts
- scheduling production more efficiently
- lowering safety stock requirements
- reducing product obsolescence costs
- managing shipping better
- improving pricing and promotion management
- negotiating superior terms with suppliers
- making more informed pricing decisions.[5]

Effective and efficient materials management requires many types of forecast, including the following.

- *Demand forecast*: investigation of the firm's demand for the item, to include current and projected demand, inventory status and lead-times. Also considered are competing current and projected demands by industry and product end use.
- *Supply forecast*: collection of data about current producers and suppliers, the aggregate projected supply situation, and technological and political trends that might affect supply.
- *Price forecast*: based on information gathered and analysed about demand and supply. Provides a prediction of short- and long-term prices and the underlying reasons for those trends.[6]

Additionally, forecasts can be long-term, medium-term or short-term. Typically, firms would use all three types of forecasting.

1 *Long-term forecasts* are usually several years in scope and are used for long-range planning and strategic issues. These will be performed in broad terms – that is, overall demand or throughput capacity.

2 *Medium-term forecasts* usually range from three months to one year and address budgeting issues and sales plans. Again, these might predict more than demand.

3 *Short-term forecasts* are most important for the operational logistics planning process. They project demand into the next several weeks or months. These are needed in units, by actual items to be shipped and for finite periods of time – monthly or perhaps weekly.[7]

The firm may utilize a variety of forecasting techniques, ranging from qualitative methods based on general market information (from suppliers, sales force, customers and others) to highly quantitative and sophisticated computer algorithms. The specific technique or approach a firm selects should be appropriate for the unique characteristics of the company and its markets. Just-in-time systems, materials requirements planning (MRP I), manufacturing resource planning (MRP II), enterprise resource planning (ERP), distribution requirements planning (DRP I) and distribution resource planning (DRP II) systems can also improve the efficiency of inventory planning and control. We will briefly describe these systems later in this chapter.

Total Quality Management

Total quality management (TQM) and reengineering are concepts that gained much attention and popularity in the 1980s and 1990s. TQM has been defined as:

> a philosophy of how to approach the organization of quality improvement . . . [that] stresses the 'total' of TQM . . . [and] lays particular stress on
>
> ■ meeting the needs and expectations of customers;
> ■ covering all parts of the organization;
> ■ including every person in the organization;
> ■ examining all costs which are related to quality, especially failure costs;
> ■ getting things 'right first time', i.e. designing-in quality rather than inspecting it in;
> ■ developing the systems and procedures which support quality and improvement;
> ■ developing a continuous process of improvement.[8]

TQM has particular relevance and importance to materials flow within logistics. Many leading authorities have championed the importance of quality in business, including W. Edwards Deming, Joseph M. Juran and Philip B. Crosby.[9] Additionally, awards such as the Deming Prize, the Malcolm Baldrige National Quality Award and the European Foundation for Quality Management's (EFQM) European Quality Award have helped shape corporate thinking on quality issues.[10] Traditional concepts about quality have been modified and enhanced to form the TQM approach outlined in Table 6.1.

The TQM approach stresses long-term benefits resulting from continuous improvements to systems, programmes, products and people. Improvements most often result from a combination of small innovations. A structured, disciplined operating methodology is used to maximize customer service levels. Sometimes, however, significant changes are required to bring about system improvements. If so, reengineering might have to occur.

The concept of **reengineering** deals with 'starting with a clean slate' – that is, taking systems and processes, and rethinking and redesigning them in order to create significant improvements in quality, cost, speed and service.

Traditional management	Total quality management
Looks for 'quick fix'	Adopts a new management philosophy
Firefights	Uses structured, disciplined operating methodology
Operates the same old way	Advocates 'breakthrough' thinking using small innovations
Randomly adopts improvement efforts	'Sets the example' through management action
Focuses on short term	Stresses long-term, continuous improvement
Inspects for errors	Prevents errors
Throws resources at a task	Uses people to add value
Motivated by profit	Focuses on the customer
Relies on programmes	A new way of life

TABLE 6.1 Traditional Management and TQM Comparison
Source: James H. Saylor, 'What Total Quality Management Means to the Logistician', *Logistics Spectrum* 24, no. 4 (Winter 1990), p. 20.

Table 6.2 identifies the relationships between TQM and logistics. Underlying the specific items listed in the table is the notion that quality is a philosophy of doing business. It is like the marketing concept, cost trade-off analysis and the systems approach, in that each provides a way of doing business that influences how individuals, departments and companies plan, implement and control marketing and logistics activities.

Therefore, all people involved in logistics must understand their role in delivering a level of quality to suppliers, internal operations and customers. TQM focuses on continuous improvement through employee involvement and top-level management support. Studies have shown that quality is more important than cost within materials management, especially in outsourcing and supplier selection decisions.[11]

Implementation of TQM within the materials management environment has resulted in significant benefits and improvements for many companies. One example is GSM Primographic in the United Kingdom. GSM specializes in printing onto polycarbonates, plastics and metals, employs 75 people and generates turnover of £4.5 million from customers including Creda, BAE Systems, Ford, Jaguar and Rolls-Royce. It has ISO 9000 approved status and has also adopted a 5-S quality implementation programme. The Japanese 5-S practice is considered an important base for successfully implementing TQM; 5-S is the acronym for five Japanese words *seiri*, *seiton*, *seiso*, *seiketsu* and *shitsuke*, which literally mean organization, neatness, cleanliness, standardization and discipline, respectively. GSM has passed stage 3 (i.e. 'Cleaning'). Three out of its four production teams have been trained and cost savings so far are £75,000 mainly through sorting out, cleaning up, rationalizing space, improving work flow, reducing job times and raw material savings.[12] (See the Global box for a discussion of international quality standards.)

In summary, TQM and logistics are interrelated. 'Managing logistics without incorporating the costs of quality is just as shortsighted as looking at the management of quality without considering the role of logistics.'[13] Thus, it is important that the flow of materials be administered and controlled utilizing the concepts of TQM.

TQM	Logistics
Provides a TQM management environment	Uses systematic, integrated, consistent, organization-wide perspective for satisfying the customer
Reduces chronic waste	Emphasizes 'doing it right the first time'
Involves everyone and everything	Involves almost every process
Nurtures supplier partnerships and customer relationships	Knows the importance of supply partnerships Keys to customer relations. Customer relations are directly dependent on training, documentation, maintenance, supply support, support equipment, transportation, manpower, computer resources and facilities
Creates a continuous improvement system	Uses logistics support analysis to continuously improve the system
Includes quality as an element of design	Influences design by emphasizing reliability, maintainability and supportability, using the optimum mix of manpower and technology
Provides constant training	Provides constant technical training for everyone
Leads long-term continuous improvement efforts geared to prevention	Focuses on reducing life-cycle costs by quality improvements geared to prevention
Encourages teamwork	Stresses the integrated efforts of everyone
Satisfies the customer	Places the customer first

TABLE 6.2 Direct Relationship between TQM and Logistics

Source: James H. Saylor, 'What Total Quality Management Means to the Logistician', *Logistics Spectrum* 24, no. 4 (Winter 1990), p. 22.

 GLOBAL

Certifying Quality with ISO 9000

ISO 9000? Total quality management? Quality assurance? Quality system? Quality policy? Depending upon whom you ask, these terms can conjure up many different and sometimes conflicting definitions.

Since 1987 one set of standards, the ISO 9000 series, has attempted to define a single definition for 'quality' and a 'quality system'. The ISO 9000 series is a set of five international standards that establish the minimum requirements for an organization's quality system.

The five standards were authored by the International Organization for Standardization, headquartered in Geneva, Switzerland. Contrary to popular belief, ISO is not an acronym for the International Standardization Organization. ISO is an official 'nickname', derived from *isos*, a Greek word meaning 'equal'.

The standards themselves are numbered ISO 9000, 9001, 9002, 9003 and 9004. The ISO 9000 series was adopted by the United States as the ANSI/ASQC Q90 series of standards. ANSI is the American National Standards Institute, while ASQC is the American Society for Quality Control.

Each of the five standards has a particular application, explained below.

- ISO 9000/Q90 specifies the guidelines for selection and use of the other series standards.
- ISO 9001/Q91 specifies a quality system model for use by organizations that design/develop, produce, install and service a product.
- ISO 9002/Q92 specifies a quality system model for use by organizations that produce and install a product or service.
- ISO 9003/Q93 specifies a quality system model for use by organizations that include final inspection and testing.
- ISO 9004/Q94 provides a set of guidelines for an organization to develop and implement a quality system, and interpret the standards of the other series.

When a firm becomes ISO 9000 certified, it proves to an independent assessor that it meets all the requirements of either ISO 9001/Q91, ISO 9002/Q92 or ISO 9003/Q93. Generally, ISO 9000 certification is valid for a period of three years.

Question: **What benefits should ISO certification bring to firms?**

Source: Lance L. Whitacre, *ISO 9000: Certifying Quality in Warehousing and Distribution*. Oak Brook, IL: Warehousing Education and Research Council, March 1994, pp. 5–6.

Administration and Control of Materials Flow

Like all the functions of logistics, materials management activities must be properly administered and controlled. This requires some methods to identify a firm's level of performance. Specifically, a firm must be able to *measure*, *report* and *improve* performance.

In measuring the performance of materials management, a firm should examine a number of elements, including supplier service levels, inventory, prices paid for materials, quality levels and operating costs.

Service levels can be measured using several methods, including order cycle time and fill rate for each supplier, and the number of production delays caused by materials being out of stock. This and related topics were examined in Chapter 2.

Inventory is an important aspect of materials management. It can be controlled by considering the amount of slow-moving inventory and comparing actual inventory levels and turnover with targeted and historical levels, for example. Chapter 5 examined these issues.

Materials *price level* measures include gains and losses resulting from forward buying, a comparison of prices paid for major items over several time periods, and a comparison of actual prices paid for materials with targeted prices.

Measures that can be used in the area of *quality control* are the number of product failures caused by defects in materials and the percentage of materials rejected from each shipment from each supplier.

As an overall measure of performance, management can compare the *actual budget* consumed by materials management to the *targeted budget* determined at the beginning of the operating period. This and related topics will be examined in Chapter 11.

Once the company has established performance measures for each component in the materials management process, data must be collected and results reported to individuals in decision-making positions. The major operating reports that should be developed by materials management include (1) market and economic conditions and price performance, (2) inventory investment changes, (3) purchasing operations and effectiveness, and (4) operations affecting administration and financial activities. Table 6.3 presents a summary of these reports.

Market and economic conditions and price performance

- Price trends and changes for the major materials and commodities purchased; comparisons with:
    1. standard costs where such accounting methods are used
    2. quoted market prices
    3. target costs as determined by cost analysis
- Changes in demand–supply conditions for the major items purchased; effects of labour strikes or threatened strikes
- Lead-time expectations of major items

Inventory investment changes

- Euro investment in inventories, classified by major commodity and materials groups
- Days' or months' supply, and on order, for major commodity and materials groups
- Ratio of inventory euro investment to sales euro volume
- Rates of inventory turnover for major items

Purchasing operations and effectiveness

- Cost reductions resulting from purchase research and value analysis studies
- Quality rejection rates for major items
- Percentage of on-time deliveries
- Number of out-of-stock situations that caused interruption of scheduled production
- Number of change orders issued, classified by cause
- Number of requisitions received and processed
- Number of purchase orders issued
- Employee workload and productivity
- Transportation costs

Operations affecting administration and financial activities

- Comparison of actual departmental operating costs to budget
- Cash discounts earned and cash discounts lost
- Commitments to purchase, classified by types of formal contract and by purchase orders, aged by expected delivery dates
- Changes in cash discounts allowed by suppliers

TABLE 6.3 Operating Reports that Should be Developed by Purchasing and Materials Management Functions

Source: Michiel R. Leenders, Harold E. Fearon, Anna E. Flynn and P. Fraser Johnson, *Purchasing and Supply Management*, 12th edn. Burr Ridge: IL: McGraw-Hill Irwin, 2002, pp. 527–8.

The organization must look for opportunities for reengineering and continuous improvement based on a comparison of actual with desired performance. To initiate improvements, the materials manager can address some key questions.

1. What is the means of communication between materials management and production? What issues are communicated and how often?
2. What is our suppliers' involvement in the process of materials forecasting and inventory management?

3 What sort of relationships do we have with our suppliers? Are they eager to serve us and meet our needs, even in times when supplies are allocated?

4 Who schedules production runs? On what basis are production runs scheduled?

5 How frequently is scheduling performed and updated?

6 How do the policies or procedures of materials management impact other parts of the organization?

These questions relate to how the product is produced and how inventories are controlled. Computers also are used to improve materials management performance. Systems that have gained acceptance in many firms are Kanban/just-in-time (JIT), MRP, ERP and DRP.

Kanban/Just-in-time Systems

Kanban and just-in-time systems have become much more important in manufacturing and logistics operations in recent years. **Kanban**, also known as the **Toyota production system (TPS)**, was developed by the Toyota Motor Company during the 1950s and 1960s. The philosophy of Kanban is that parts and materials should be supplied at the very moment they are needed in the factory production process. This is the optimal strategy, from both a cost and service perspective. The Kanban system can apply to any manufacturing process involving repetitive operations (see Box 6.1).

Just-in-time (JIT) systems extend Kanban, linking purchasing, manufacturing and logistics. The primary goals of JIT are to minimize inventories, improve product quality, maximize production efficiency and provide optimal customer service levels. It is basically a philosophy of doing business.

JIT has been defined in several ways, including the following.

■ As a production strategy, JIT works to reduce manufacturing costs and to improve quality markedly by waste elimination and more effective use of existing company resources.[14]

■ A philosophy based on the principle of getting the right materials to the right place at the right time.[15]

■ A programme that seeks to eliminate non-value-added activities from any operation with the objectives of producing high-quality products (i.e. 'zero defects'), high productivity levels and lower levels of inventory, and developing long-term relationships with channel members.[16]

At the heart of the JIT system is the notion that waste should be eliminated. This is in direct contrast to the traditional 'just-in-case' philosophy in which large inventories or safety stocks are held just in case they are needed. In JIT, the ideal lot size or EOQ is one unit, safety stock is considered unnecessary and any inventory should be eliminated.

Perhaps the best-known example of Kanban and JIT systems is the approach developed by Toyota. The company identified problems in supply and product quality through reduction of inventories, which forced problems into the open. Safety stocks were no longer available to overcome supplier delays and faulty components, thus forcing Toyota to eliminate 'hidden' production and supply problems.

The same type of procedure has been applied to many companies in Europe. The advantages of the system becomes evident when we see that raw materials can be reduced by 75 per cent with JIT implementation.[17] Not every component can be handled by the Kanban or JIT approaches, but the systems work very well for items that are used repetitively.

Many firms have successfully adopted the JIT approach. Companies in industries such as metal products, automobile manufacturing, electronics, and food and beverage have implemented JIT and realized a number of benefits, including:

- productivity improvements and greater control between various production stages
- diminished raw materials, work in process and finished goods inventory
- a reduction in manufacturing cycle times
- dramatically improved inventory turnover rates.

Box 6.1 Kanban Card Procedure

Work centre 1 stockpoints   Work centre 2 stockpoints

Inbound   Outbound   Inbound   Outbound

•••• Move card path. When a container of parts is selected for use from an inbound stockpoint, the move card is removed from the container and taken to the outbound stockpoint of the preceding work centre as authorization to pick another container of parts.

—— Production card path. When a container of parts is picked from an outbound stockpoint, the production card is removed and left behind as authorization to make a standard container of parts to replace the one taken.

'Kanban' literally means 'signboard' in Japanese. The system involves the use of cards (called 'kanbans') that are attached to containers that hold a standard quantity of a single part number. There are two types of kanban cards: 'move' card and 'production' cards.

When a worker starts to use a container of parts, the move card, which is attached to it, is removed and is either sent to or picked up by the preceding or feeding work centre (in many cases this is the supplier). This is the signal – or 'sign' – for that work centre to send another container of parts to replace the one now being used. This replacement container has a production card attached to it which is replaced by the 'move' card before it is sent. The production card then authorizes the producing work centre to make another container full of parts. These cards circulate respectively within or between work centres or between the supplier and the assembly plant.

For Kanban to work effectively, the following rules must be observed.

1. Only one card can be attached to a container at any one time.

2. The using (or following) work centre must initiate the movement of parts from the feeding (or preceding) work centre.
3. No fabrication of parts is allowed without a kanban production card.
4. Never move or produce other than the amount indicated by the kanban card.
5. Kanban cards must be handled on a first-in, first-out (FIFO) basis.
6. Finished parts must be placed at the location point indicated on the kanban card.

Because each kanban card represents a standard number of parts being made or used within the production process, the amount of work-in-progress inventory can easily be controlled by controlling the number of cards on the plant floor. Japanese managers, by simply removing a card or two, can test or strain the system and reveal bottle-necks. Then they have a problem they can address themselves to – an opportunity to improve productivity, the prime goal of Kanban.

Source: 'Why Everybody is Talking about 'Just-in-time', *Warehousing Review* 1, no. 1 (October 1984), p. 27. Reprinted with permission from *Warehousing Review* 1984 Charter Issue; The American Warehouse Association (publisher), 1165 N. Clark, Chicago, IL 60610.

In general, JIT produces benefits for firms in four major areas: improved inventory turns, better customer service, decreased warehouse space and improved response time. In addition, reduced distribution costs, lower transportation costs, improved quality of supplier products, and a reduced number of transportation carriers and suppliers can result from the implementation of JIT.

A specific example of a firm that achieved success through JIT is Rank Xerox Manufacturing (Netherlands). As the largest Xerox company outside the United States, Rank Xerox (a joint venture between Xerox Corporation and Britain's Rank Corporation) produces and refurbishes mid-volume copier equipment for distribution worldwide. Throughout most of the 1980s, Rank Xerox implemented a JIT programme.

As part of this JIT programme, the firm also installed an automated materials-handling system and information-processing system. Production procedures were modified at the same time. As a result of the JIT programme and other system changes, Rank Xerox realized the following specific benefits.

- Its supplier base was reduced from 3000 to 300.
- Ninety-eight per cent on-time inbound delivery was achieved, with 70 per cent of materials arriving within an hour of the time they were needed.
- Warehouse stock was reduced from a three-month to a half-month supply.
- Overall material costs were reduced by more than 40 per cent.
- Most inbound product inspection stations were eliminated because of higher-quality materials from suppliers.
- Reject levels for defective or inferior materials fell from 17.0 per cent to 0.8 per cent.
- Positions for 40 repack people were eliminated because of standardized shipment-packaging criteria.
- Inbound transportation costs were reduced by 40 per cent.
- On-time inbound delivery performance was improved by 28 per cent.[18]

While JIT offers a number of benefits, it may not be suitable for all firms. It has some inherent problems, which fall into three categories: production scheduling (plant), supplier production schedules, and supplier locations.

When levelling of the production schedule is necessary due to uneven demand, firms will require higher levels of inventory. Items can be produced during slack periods even though they may not be demanded until a later time. Finished goods inventory has a higher value because of its form utility; thus, there is a greater financial risk resulting from product obsolescence, damage or loss.

However, higher levels of inventory, coupled with a uniform production schedule, can be more advantageous than a fluctuating schedule with less inventory. In addition, when stockout costs are great because of production slowdowns or shutdowns, JIT may not be the optimal system. JIT reduces inventory levels to the point where there is little, if any, safety stock, and parts shortages can adversely affect production operations.

Supplier production schedules are a second problem with JIT. The success of a JIT system depends on suppliers' ability to provide parts in accordance with the firm's production schedule. Smaller, more frequent orders can result in higher ordering costs and must be taken into account when calculating any cost savings due to reduced inventory levels. When a large number of small lot quantities are produced, suppliers incur higher production and set-up costs. Generally, suppliers will incur higher costs unless they are able to achieve the benefits associated with implementing similar systems with their suppliers.

Supplier locations can be a third problem. As distance between the firm and its suppliers increases, delivery times may become more erratic and less predictable. Shipping costs increase as less-than-truckload (LTL) movements are made. Transit time variability can cause inventory stockouts that disrupt production scheduling; when this is combined with higher delivery costs on a per-unit basis, total costs may be greater than the savings in inventory carrying costs.

Other problem areas that can become obstacles to JIT, especially in implementation, are organizational resistance, lack of systems support, inability to define service levels, a lack of planning and a shift of inventory to suppliers. Overcoming these and the previously discussed problems requires cooperation and integration within and between companies.

JIT has numerous implications for logistics executives. First, proper implementation of JIT requires that the firm fully integrate all logistics activities. Many trade-offs are required, but without the coordination provided by integrated logistics management, JIT systems cannot be fully implemented.

Second, transportation becomes an even more vital component of logistics under a JIT system. In such an environment, the demands placed on the firm's transportation network are significant and include a need for shorter, more consistent transit times; more sophisticated communications; the use of fewer carriers with long-term relationships; a need for efficiently designed transportation and materials handling equipment; and better decision-making strategies relative to when private, common or contract carriage should be used.

Third, warehousing takes on an expanded role as it assumes the role of a consolidation facility instead of a storage facility. Since many products come into the manufacturing operation at shorter intervals, less space is required for storage, but there must be an increased capability for handling and consolidating items. Different forms of materials handling equipment may be needed to facilitate the movement of many products in smaller quantities. The location decision for warehouses serving inbound materials needs may change because suppliers are often located closer to the manufacturing facility in a JIT system.

JIT systems are usually combined with other systems that plan and control material flows into, within and out of the organization. MRP, ERP and DRP are often used to implement the JIT philosophy. They will be presented in the following sections.

JIT II and JIT III

JIT II applies JIT concepts to the purchasing function by having a representative of the supplier locate at the buying organization's facility. Developed by Bose Corporation, this approach improves mutual understanding between the buyer and supplier, reduces waste and redundancy of efforts, improves supplier responsiveness and creates a positive working environment. JIT III extends JIT II where a supplier locates its own factory close to its customer to facilitate sequential and frequent deliveries.[19] These concepts were discussed in Chapter 4.

MRP Systems

MRP has been used to signify systems called **materials requirements planning (MRP I)** and **manufacturing resource planning (MRP II)**. Introduced first, MRP I developed into MRP II with the addition of financial, marketing and purchasing aspects.

MRP I became a popular concept in the 1960s and 1970s. It is a computer-based production and inventory control system that attempts to minimize inventories while maintaining adequate materials for the production process. From a managerial perspective, MRP I requires a sufficient computer system to process information from the three main sources: the master production schedule, the bill-of-materials for each product, and inventory records.[20]

MRP I systems offer many advantages over traditional systems, including:

- improved business results (i.e. return on investment, profits)
- improved manufacturing performance results
- better manufacturing control
- more accurate and timely information
- less inventory
- time-phased ordering of materials
- less material obsolescence
- higher reliability

- more responsiveness to market demand
- reduced production costs.

MRP I does have a number of drawbacks that should be examined by any firm considering adopting the system. First, it does not tend to optimize materials acquisition costs. Because inventory levels are kept to a minimum, materials must be purchased more frequently and in smaller quantities. This results in increased ordering costs.

Higher transportation bills and higher unit costs are incurred because the firm is less likely to qualify for large volume discounts. The company must weigh the anticipated savings from reduced inventory costs against the greater acquisition costs resulting from smaller and more frequent orders.

Another disadvantage of MRP I is the potential hazard of a production slowdown or shutdown that may arise because of factors such as unforeseen delivery problems and materials shortages. The availability of safety stocks gives production some protection against stockouts of essential material. As safety stocks are reduced, this level of protection is lost.

A final disadvantage of MRP I arises from the use of standardized software packages, which may be difficult to accommodate within the unique operating situations of a given firm. Firms buying off-the-shelf software will often have to modify it so that it meets their specific needs and requirements.

The master production schedule serves as the major input into the MRP I system (see Figure 6.2A). Other inputs include the bill-of-materials file and the inventory records file. The bill-of-materials file contains the component parts of the finished product, identified by part number. The inventory records file maintains a record of all inventory on hand and on order. It also keeps track of due dates for all component parts.

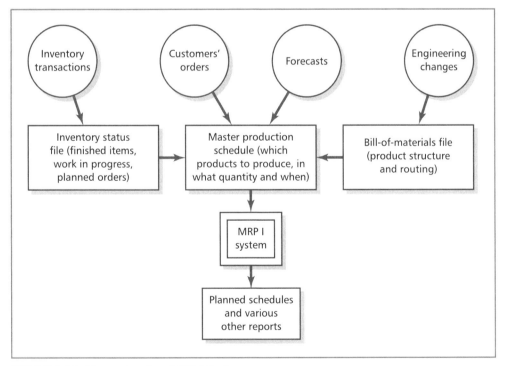

FIGURE 6.2A Elements of an MRP I System

Source: Amrik Sohal and Keith Howard, 'Trends in Materials Management', *International Journal of Physical Distribution and Materials Management* 17, no. 5 (1987), p. 11.

Reports generated from MRP I systems include timetables of operations for the master schedule, timetables of orders for materials from external suppliers, planning reports that give information for longer-term planning decisions, performance reports for identifying and determining actual system performance, and exception reports that detail unusual discrepancies requiring management action.[21]

While MRP I is still being used by many firms, it has been updated and expanded to include financial, marketing and logistics elements. This newer version is called manufacturing resource planning, or MRP II.

MRP II includes the entire set of activities involved in the planning and control of production operations. It consists of a variety of functions of modules (see Figure 6.2B), and includes production planning, resource requirements planning, master production scheduling, materials requirements planning (MRP I), shop floor control and purchasing.

The advantages of MRP II include:

- inventory reductions of one-quarter to one-third
- higher inventory turnover

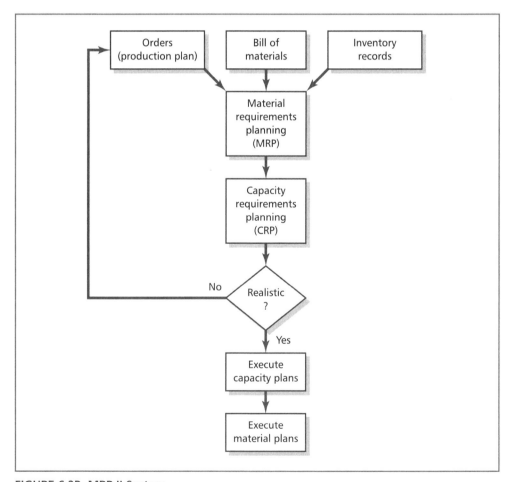

FIGURE 6.2B MRP II System
Source: Karl A. Hatt, 'What's the Big Deal about MRP II?', *Winning Manufacturing* 5, no. 2 (1994), p. 2.

- improved consistency in on-time customer delivery
- reduction in purchasing costs due to fewer expedited shipments
- minimization of workforce overtime.

These advantages typically result in savings to a firm beyond the initial costs of implementing MRP II. However MRP II also has practical implementation drawbacks. MRP II uses an iterative approach to generate an optimal solution and it is difficult to find agreement from all stakeholders who have different interests and aims. There are also difficulties in completely integrating all functions and systems due to the inflexibility of MRP systems.

ERP Systems

Enterprise resource planning (ERP) is a system that includes core finance and accounting, marketing, human resource and information technology functions, coupled with logistics functions, to manage the distribution and manufacturing components of the organization. ERP also interfaces with the organization's entire supply chain, thus ERP is essentially the newest generation of MRP systems and represents the extended application of MRP principles to the supply chain.[22]

Benefits of ERP systems include:

- visibility throughout the organization and more sophisticated communication with suppliers, customers and other business partners due to integrated software communication
- a discipline of forcing process-based changes to make the organization more efficient
- a better sense of operational control to enhance continuous improvement
- the capability of integrating the entire supply chain beyond 'tier 1' customers and suppliers.[23]

ERP depends on trust and information sharing between organizations. The Internet and e-commerce make the flow of information easy to organize, however ERP systems are very complex and have many practical problems. The modular R/3 system from SAP AG (see the Technology box on page 176) helps overcome these disadvantages. However, a common reason for firms not to invest in an ERP system is that they cannot adapt the standard modules and 'reconcile the assumptions in the software of the ERP system with their core business processes'.[24] In short, although many system suppliers may disagree, ERP is still evolving.

DRP Systems

Distribution requirements planning (DRP I) is the application of MRP principles to the flow of products out to customers.

> **Distribution resource planning (DRP II)** is an extension of distribution requirements planning (DRP I). Distribution requirements planning applies the time-phased DRP I logic to replenish inventories in multiechelon warehousing systems. Distribution resource planning extends DRP I to include the planning of key resources in a distribution system – warehouse space, manpower levels, transport capacity (e.g., trucks, railcars), and financial flows.[25]

An extension of DRP I, DRP II uses the needs of distribution to drive the master schedule, controlling the bill of materials and, ultimately, materials requirements planning. In essence, DRP I and DRP II are outgrowths of MRP I and MRP II, applied to the logistics activities of a firm.

Companies use DRP-generated information to project future inventory requirements. Specifically, the information is used to:

- coordinate the replenishment of SKUs coming from the same source (e.g. a company-owned or vendor's plant)
- select transportation modes, carriers and shipment sizes more cost efficiently
- schedule shipping and receiving labour
- develop a master production schedule for each SKU.[26]

Figure 6.3 depicts the DRP II system schematically. Although not shown in the figure, accurate forecasts are essential ingredients for successful DRP II systems.

A DRP [II] system translates the forecast of demand for each SKU at each warehouse and distribution center into a time-phased replenishment plan. If the SKU forecasts are inaccurate, the plan will not be accurate.[27]

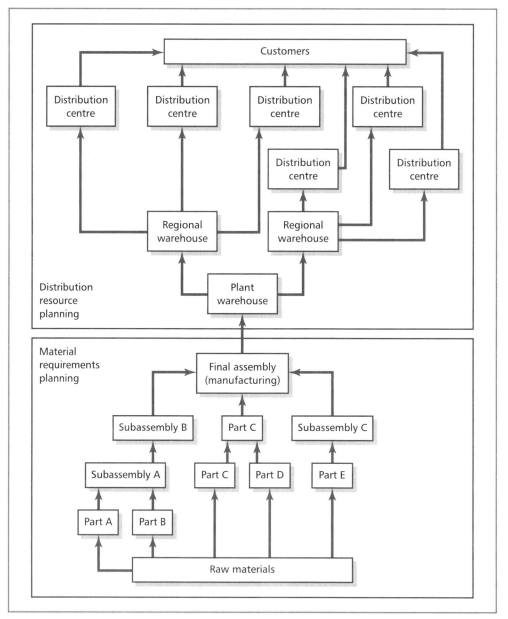

FIGURE 6.3 Distribution Resource Planning (DRP II)

Source: 'How DRP Helps Warehouses Smooth Distribution', *Modern Materials Handling* 39, no. 6 (9 April 1984), p. 53. *Modern Materials Handling*, copyright 1984 by Cahners Publishing Company, Division of Reed Holdings, Inc.

The Logistics/Manufacturing Interface

Systems such as Kanban, JIT, MRP, DRP and ERP require that the logistics and manufacturing activities of a firm work together closely. Without a cooperative effort, the full advantages of systems like JIT can never be realized. Conflicts, both real and perceived, must be minimized. This requires joint logistics/manufacturing planning and decision-making. There are a number of areas in which cooperation is necessary and great improvements can be made. The following actions can be of significant benefit.

- Logistics must reduce replenishment lead-times to increase manufacturing flexibility and reduce order fulfilment lead-times.
- Manufacturing and logistics must work together in the production scheduling area to reduce production planning cycle time. Logistics can provide input into production scheduling and system requirements.
- Manufacturing and logistics strategies, such as shortening of lead-times, set-up times and production run sizes must be used to minimize average inventory levels and stockouts.
- Logistics must develop strategies to reduce supplier lead-times for parts and supplies.
- Logistics must adopt the philosophy that slow movers (i.e. products with low inventory turnover ratios) should be produced only after orders are received, rather than held in stock.

Many other areas of logistics/manufacturing interface exist. It is important that each functional area of the firm examine its role in the JIT, MRP, DRP or ERP system, and identify how it can work individually and jointly to optimize the firm's strategic position (see Creative Solutions box).

CREATIVE SOLUTIONS

Excellence in Logistics Strategies

Sequent Computer Systems has a preferred logistics supplier programme, which concentrates its business with a small number of high-performing suppliers. A commodity team manages the programme.

'We have a formal programme for order fulfilment reduction,' says Sequent's Martha McMahon, worldwide logistics manager. It's a team effort within the organization and focuses on the areas of supply of materials and materials management.

Another area of focus is linking with customers in the field. Sequent's 'prelim' order programme allows the company to see what potential orders are coming in and anticipate its material needs.

The third area is staff education. 'It's important for everyone to understand what the goals are, why the goals have been set up and how to monitor the progress on achieving those goals to keep everyone informed.'

Sequent has achieved a significant reduction of order-to-shipment time from 35 days down to 7.5 days, which exceeds industry best in class by 10 days. Spare parts can be made available to customers within two to four hours if their computer goes down. Significant cost savings have resulted from the programme: over $350,000 for the year and an inventory reduction of $3 million. On-time delivery is averaging 98 per cent with preferred suppliers.

Question: **What cost and customer service trade-offs are at issue here?**

Source: Sarah A. Bergin, 'Recognizing Excellence in Logistics Strategies', *Transportation and Distribution* 37, no. 10 (October 1996), p. 50.

SUMMARY

This chapter examined the broad areas of materials flow. We explored the functions of purchasing and procurement, production control, inbound logistics, warehousing and storage, data and information systems, inventory planning and control, and materials disposal. The relationships between materials management and total quality management (TQM) were discussed. The TQM process was examined and some examples of its implementation presented.

The administration and control of materials flow requires that firms measure, report and improve performance. Concepts and approaches being used or developed include Kanban/just-in-time, MRP I, MRP II, ERP, DRP I and DRP II systems. Each system has been implemented by a variety of firms, with significant results. Advances in computer technology have enabled many of the systems to be implemented successfully in manufacturing, retailing and service firms. The impact on logistics has been substantial.

KEY TERMS

A full Glossary can be found at the back of the book.

QUESTIONS AND PROBLEMS

1 How does total quality management (TQM) differ from traditional management? How can TQM be applied to logistics?
2 Briefly describe the concept of just-in-time (JIT) and its relationship to logistics.
3 Discuss the role of suppliers in a JIT system. Identify areas where potential conflicts may occur.
4 Briefly discuss how forecasting can be used in materials management. Identify the general uses according to the (a) type of forecast and (b) the time frame of the forecast.
5 MRP, ERP and DRP are computer systems in materials management and manufacturing. Describe the types of situation where MRP, ERP and DRP can be effectively and/or efficiently used in a firm.

PROBLEM: MATERIALS MANAGEMENT OF SPARE PARTS

Spare parts and related support have surfaced as major challenges to many firms due to the amount of investment required for this part of the supply chain. Materials management in this area has received little attention until recently. The UK telecommunications company, BT International, is two-thirds of the way through its 'Global Spares Initiative', and estimates it is saving €35 million a year in reduced inventory and operating costs. BT believes that amount could be as high as €80 million a year once the programme is fully implemented, and that value doesn't even include savings in labour!

However, the automotive industry has its own special characteristics, which have so far prevented much action in this area. There is a customer need for rapid response to ensure consumers' vehicles are repaired in a timely manner; however, maintaining minimal stocks and shipping by air freight cannot be justified on a cost basis.

Thus, automakers typically maintain a network of distribution centres across Europe instead of one central facility, to ensure fulfilment is made to dealers for repair and service. However, inventory and handling costs are high. Given this problem, some members in the auto spares supply chain are suggesting a shared user approach where automakers share facilities and information systems, and thus costs.

The lack of suitable software and systems is also an issue as most manufacturers have MRP/ERP systems that don't work well with the uncertain demand parameters of aftermarket sales and support for automotive spares.

One suggestion is to break up automotive spare parts into categories and use different logistics strategies, and perhaps supply chains, for each. However, this is likely to entail developing new supply chains and systems.

Another suggestion is to outsource the spares supply chain to the parts manufacturers themselves instead of the automaker. However, in this case the automakers might find it difficult to give up control.

But an explosion in customer expectations has made automotive spare parts logistics and materials management important. Motorists expect to be mobile for life, and competition for their repair and servicing business is heating up as the end of the European Union 'block exemption' sees more competitors entering the marketplace.

What Is Your Solution?

Adapted from: Chris Lewis 'Pain or Gain?', *Logistics Europe* 12, no. 5 (June 2004), pp. 26–31.

SUGGESTED READING

BOOKS
Dixon, Lance and Anne Millen Porter, *JIT II: Revolution in Buying and Selling*. Newton, MA: Cahners, 1994.

Leenders, Michiel, Harold E. Fearon, Anna E. Flynn and P. Fraser Johnson, *Purchasing and Supply Management*, 12th edn. Burr Ridge, IL: McGraw-Hill Irwin, 2002.

Waters, Donald, *Inventory Control and Management*, 2nd edn. Chichester, UK: John Wiley & Sons Ltd, 2003.

JOURNALS

Daugherty, Patricia J., Dale S. Rogers and Michael S. Spencer, 'Just-in-time Functional Model: Empirical Test and Validation', *International Journal of Physical Distribution and Logistics Management* 24, no. 6 (1994), pp. 20–6.

Demmy, W. Steven and Arthur B. Petrini, 'MRP II + JIT + TQM + TOC: The Path to World Class Management', *Logistics Spectrum* 26, no. 3 (Fall 1992), pp. 8–13.

Evolution Continues in MRP II Type Systems: New Functionality for Flexible Enterprise Management', *Manufacturing Systems* 12, no. 7 (July 1994), pp. 32–5.

Garreau, Alain, Robert Lieb and Robert Millen, 'JIT and Corporate Transport: An International Comparison', *International Journal of Physical Distribution and Logistics Management* 21, no. 1 (1991), pp. 42–7.

Germain, Richard, Cornelia Droge and Nancy Spears, 'The Implications of Just-in-time for Logistics Organization Management and Performance', *Journal of Business Logistics* 17, no. 2 (1996), pp. 19–34.

Ho, Samuel K.M., 'Is the ISO 9000 Series for Total Quality Management?,' *International Journal of Quality & Reliability Management* 11, no. 9 (1994), pp. 74–89.

Masters, James M., Greg M. Allenby and Bernard J. La Londe, 'On the Adoption of DRP', *Journal of Business Logistics* 13, no. 1 (1992), pp. 47–67.

Mozeson, Mark H., 'What Your MRP II Systems Cannot Do', *Industrial Engineering* 23, no. 12 (December 1991), pp. 20–4.

Oliver, Nick, 'JIT: Issues and Items for the Research Agenda', *International Journal of Physical Distribution and Logistics Management* 20, no. 7 (1990), pp. 3–11.

Snehemay, Banejee and Damodar Y. Golhar, 'EDI Implementation: A Comparative Study of JIT and Non-JIT Manufacturing Firms', *International Journal of Physical Distribution and Logistics Management* 23, no. 7 (1993), pp. 22–31.

Sohal, Amrik S., Liz Ramsay and Danny Samson, 'JIT Manufacturing: Industry Analysis and a Methodology for Implementation', *International Journal of Physical Distribution and Logistics Management* 23, no. 7 (1993), pp. 4–21.

Swenseth, Scott R. and Frank P. Buffa, 'Just-in-time: Some Effects on the Logistics Function', *The International Journal of Logistics Management* 1, no. 2 (1990), pp. 25–34.

Von Flue, Johann L., 'The Future with Total Quality Management', *Logistics Spectrum* 24, no. 1 (Spring 1990), pp. 23–7.

Waters-Fuller, Niall, 'The Benefits and Costs of JIT Sourcing: A Study of Scottish Suppliers', *International Journal of Physical Distribution and Logistics Management* 26, no. 4 (1996), pp. 35–50.

Zipkin, Paul H., 'Does Manufacturing Need a JIT Revolution?', *Harvard Business Review* 91, no. 1 (January–February 1991), pp. 40–50.

REFERENCES

[1] Michael Leenders, Harold E. Fearon, Anna E. Flynn and P. Fraser Johnson, *Purchasing and Supply Management*, 12th edn. Burr Ridge, IL: McGraw-Hill Irwin, 2002, p. 7.

[2] Yunus Kathawala and Heino H. Nauo, 'Integrated Materials Management: A Conceptual Approach', *International Journal of Physical Distribution and Materials Management* 19, no. 8 (1989), p. 10.

3 Leenders *et al.*, *Purchasing and Supply Management*, p. 3.

4 Kathawala and Nauo, 'Integrated Materials Management', p. 14.

5 Glen Galfond, Kelly Ronayne and Christian Winkler, 'State-of-the-art Supply Chain Forecasting', *PW Review*, November 1996, p. 3.

6 Leenders *et al.*, *Purchasing and Supply Management*, pp. 701–4.

7 Donald Waters, *Inventory Control and Management*. Chichester, UK: John Wiley & Sons Ltd, 2003, p. 233. Copyright John Wiley & Sons Ltd. Reproduced with permission.

8 Nigel Slack, Stuart Chambers and Robert Johnson, *Operations Management*, 4th edn. Harlow, UK: FT Prentice Hall, 2004 pp. 722–3.

9 Ibid., pp. 719–22.

10 Ibid., pp. 740–3.

11 Arnold Maltz, 'The Relative Importance of Cost and Quality in the Outsourcing of Warehousing', *Journal of Business Logistics* 15, no. 2 (1994), pp. 45–62.

12 Stephen J. Warwood and Graeme Knowles, 'An Investigation into Japanese 5-S Practice in UK Industry', *The TQM Magazine* 16, no. 5 (2004) pp. 347–53.

13 James M. Kenderdine and Paul D. Larson, 'Quality and Logistics: A Framework for Strategic Integration', *International Journal of Physical Distribution & Materials Management* 18, no. 6 (1988), p. 9.

14 Amrik S. Sohal, Liz Ramsay and Danny Samson, 'JIT Manufacturing: Industry Analysis and a Methodology for Implementation', *International Journal of Physical Distribution & Logistics Management* 23, no. 7 (1993), pp. 4–21.

15 Snehemay Banejee and Damodar Y. Golhar, 'EDI Implementation: A Comparative Study of JIT and Non-JIT Manufacturing Firms', *International Journal of Physical Distribution & Logistics Management* 23, no. 7 (1993), pp. 22–31.

16 Larry C. Giunipero and Waik K. Law, 'Organizational Support for Just-in-time Implementation', *The International Journal of Logistics Management* 1, no. 2 (1990), pp. 35–6.

17 Sohal *et al.*, 'JIT Manufacturing', p. 13.

18 Lisa H. Harrington, 'Why Rank Xerox Turned to Just-in-time', *Traffic Management* 27, no. 10 (October 1988), pp. 82–7.

19 Lars-Erik Gadde and Håkan Håkansson, *Supply Network Strategies*. Chichester, UK: John Wiley & Sons Ltd, 2001, p. 8. Copyright John Wiley & Sons Ltd. Reproduced with permission.

20 Waters, *Inventory Control and Management*, p. 311.

21 Ibid., p. 322.

22 Ibid., p. 334.

23 Slack *et al.*, *Operations Management*, p. 506.

24 Ibid., p. 508.

25 John F. Magee, William C. Copacino and Donald B. Rosenfield, *Modern Logistics Management: Integrating Marketing, Manufacturing, and Physical Distribution*. New York: John Wiley, 1985, p. 150.

26 Alan J. Stenger, 'Distribution Resource Planning', in *The Logistics Handbook*, James F. Robeson and William C. Copacino, eds. New York: Free Press, 1994, p. 392.

27 Mary Lou Fox, 'Closing the Loop with DRP II', *Production and Inventory Management Review* 7, no. 5 (May 1987), pp. 39–41.

CHAPTER 7
TRANSPORTATION

OBJECTIVES

CHAPTER OBJECTIVES

- ■ To examine transportation's role in logistics and its relationship to marketing
- ■ To describe alternative transport modes, intermodal combination and other transportation options
- ■ To examine the impact of deregulation on carriers and shippers
- ■ To examine the issues of transportation cost and performance measurement
- ■ To examine international dimensions of transportation
- ■ To identify major transportation management activities of carriers and shippers
- ■ To identify areas where computer technology is important

INTRODUCTION

This chapter provides an overview of the transportation function and its importance to logistics. We will examine alternative transportation modes and intermodal combinations. We also describe key transportation management issues of shippers and carriers, transportation cost and performance measurement, and the role of computers.

Efficient transportation systems are the hallmark of industrialized societies. The transportation sector of most industrialized economies is so pervasive that we often fail to comprehend the magnitude of its impact on our way of life. In 2001, total expenditure on European transportation, both passenger and freight or goods, was over €60 billion. Since 1970, the European transportation sector has grown considerably. Figure 7.1 shows that since 1990 freight or goods transportation has been growing more rapidly than gross domestic product (GDP) in the 15 EU countries (prior to 1 May 2004), known hereafter as the EU15.[1]

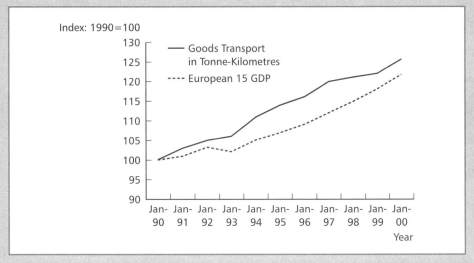

FIGURE 7.1 European Freight Transportation Growth Compared to GDP: 1999–2000
Source: Eurostat Main Transport Indicators, http://europa.eu.int/comm/eurostat/ (2005)

Time and Place Utility

Transportation physically moves products from where they are produced to where they are needed. This movement across space or distance adds value to products. This value added is often referred to as place utility.

Time utility is created by warehousing and storing products until they are needed. Transportation is also a factor in time utility; it determines how fast and how consistently a product moves from one point to another. This is known as **time-in-transit** and *consistency of service*, respectively.

If a product is not available at the precise time it is needed, there may be expensive repercussions, such as lost sales, customer dissatisfaction and production downtime, when the product is being used in the manufacturing process. Transportation service providers such as DHL Solutions, TNT Logistics, Exel, Fiege and Wincanton have achieved success as pan-European operators because they are able to provide consistent time-in-transit and thus increase the time and place utility of their customers' products.

Transportation/Logistics/Marketing Interfaces

Transportation moves products to markets that are geographically separated and provides added value to customers when the products arrive on time, undamaged and in the quantities required. In this way, transportation contributes to the level of customer service, which is one of the cornerstones of customer satisfaction: an important component of the marketing concept.

Because transportation creates place utility and contributes to time utility – both of which are necessary for successful marketing efforts – the availability, adequacy and cost of transportation impact business decisions seemingly unrelated to managing the transportation function itself – that is, what products should be produced, where should they be sold, where should facilities be located and where should materials be sourced.

Transportation is one of the largest logistics costs and may account for a significant portion of the selling price of some products. Low value-per-pound products such as basic raw materials (e.g. sand and coal) are examples. Transportation costs for computers, business machines and electronic components may be only a small percentage of the selling price. Generally, the efficient management of transportation becomes more important to a firm as inbound and outbound transportation's share of product cost increases. Even with high-value products, expenditures for transportation are important although the percentage of selling price may be low, primarily because the total cost of transportation in absolute terms is significant.

Factors Influencing Transportation Costs and Pricing

In general, factors influencing transportation costs/pricing can be grouped into two major categories: product-related factors and market-related factors.

Product-related Factors

Many factors related to a product's characteristics influence the cost/pricing of transportation. They can be grouped into the following categories:

- density
- stowability
- ease or difficulty of handling
- liability.

Density refers to a product's weight-to-volume ratio. Items such as steel, canned foods, building products and bulk paper goods have high weight-to-volume ratios; they are relatively heavy given their size. On the other hand, products such as electronics, clothing, luggage and toys have low weight-to-volume ratios and thus are relatively lightweight given their size. In general, low-density products – those with low weight-to-volume ratios – tend to cost more to transport on a per-pound (or kilo) basis than high-density products.

Stowability is the degree to which a product can fill the available space in a transport vehicle. For example, grain, ore and petroleum products in bulk have excellent stowability because they can completely fill the container (e.g. railcar, tank truck, pipeline) in which they are transported. Other items, such as automobiles, machinery, livestock and people, do not have good stowability, or cube utilization. A product's stowability depends on its size, shape, fragility and other physical characteristics.

Related to stowability is the *ease or difficulty of handling* the product. Difficult-to-handle items are more costly to transport. Products that are uniform in their physical characteristics (e.g. raw materials and items in cartons, cans or drums) or that can be manipulated with materials-handling equipment require less handling expense and are therefore less costly to transport.

Liability is an important concern. Products that have high value-to-weight ratios are easily damaged, they are subject to higher rates of theft or pilferage, and cost more to transport. Where the transportation carrier assumes greater liability (e.g. with computer, jewellery, and home entertainment products), a higher price will be charged to transport the product.

Other factors, which vary in importance depending on the product category, are the product's hazardous characteristics and the need for strong and rigid protective packaging. These factors are particularly important in the chemical and plastics industries.

Market-related Factors

In addition to product characteristics, important market-related factors affect transportation costs/pricing. The most significant are:

- degree of intramode and intermode competition
- location of markets, which determines the distance goods must be transported
- nature and extent of government regulation of transportation carriers
- balance or imbalance of freight traffic into and out of a market
- seasonality of product movements
- whether the product is transported domestically or internationally.

Each of these factors affects the costs and pricing of transportation. These topics will be examined later in this chapter. In addition, there are important service factors that need to be considered.

Customer service is a vital component of logistics management. While each activity of logistics management contributes to the level of service a company provides to its customers, the impact of transportation on customer service is one of the most significant. The most important transportation service characteristics affecting customer service levels are:

- dependability – consistency of service
- time-in-transit
- market coverage – the ability to provide door-to-door service
- flexibility – handling a variety of products and meeting the special needs of shippers
- loss and damage performance
- ability of the carrier to provide more than basic transportation service (i.e. to become part of a shipper's overall marketing and logistics programmes).

Each mode of transport – truck or road, rail, air, water and pipeline – has varying service capabilities. In the next section, we will examine each mode in terms of its economic and service characteristics.

Carrier Characteristics and Services

Any one or more of five transportation modes –truck or road, rail, air, water and pipeline – may be selected to transport products. In addition, intermodal combinations are available: rail–motor, motor–water, motor–air and rail–water. Intermodal combinations offer specialized or lower-cost services not generally available when a single transport mode is used. Other transportation options that offer a variety of services to shippers include freight forwarders, shippers' associations, intermodal marketing companies (or shippers' agents), third-party logistics service providers, parcel post and air express companies.

Truck or Road

The trucking industry transported over 11.3 million tonnes of goods throughout Europe in 2001.[2] Trucks transport over 75 per cent of the tonnage of agricultural products such as fresh and frozen meats, dairy products, bakery products, confectionery items, beverages and consumer tobacco products. Many manufactured products are transported primarily by motor carriers, including amusement, sporting and athletic goods; toys; watches and clocks; farm machinery; radios and television sets; carpets and rugs; clothing; drugs; and office equipment and furniture. Most consumer goods are transported by motor carrier. Motor carriage offers fast, reliable service with little damage or loss in transit.

Nationally and in a pan-European context, trucks compete with air for small shipments and rail for large shipments. Shipments transported by trucks are referred to as truckload (TL) or less-than-truckload (LTL). Smaller shipments transported by trucks are LTL, which is any quantity of freight weighing less than the amount required for the application of a truckload rate.

Efficient truck carriers can realize greater efficiencies in terminal, pick-up and delivery operations, which enables them to compete with air carriers on **point-to-point service** for any size of shipment if the distance involved is 500 kilometres or less. Point-to-point service refers to a single transport mode picking up products at origin and delivering them to their final destination. No additional transport modes are necessary.

Truck carriers compete directly with railroads for TL shipments that are transported 500 kilometres or more. However, rail is the dominant mode when shipment sizes exceed 25 tonnes. Truck carriers dominate the market for smaller shipments.

The average length of haul for trucks in Europe was approximately 350 kilometres in 2001.[3] Some national carriers have average hauls that are much longer, while some intracity carriers may average only a few kilometres. LTL shipments are generally shorter hauls than TL shipments, but significant variability exists.

Trucks are very flexible and versatile. The flexibility of trucks is made possible by a network of about 3.5 million kilometres of roads across the EU15 countries, including some 51,500 kilometres of motorways,[4] thus enabling them to offer point-to-point service between almost any origin–destination combination. This gives trucks the widest market coverage of any mode. Trucks are versatile because they can transport products of varying sizes and weights over any distance.

. Virtually any product can be transported by trucks, including some that require equipment modifications. Their flexibility and versatility have enabled them to become the dominant form of transport in Europe (based on the amount of freight transported as measured in tonnes) and in many other parts of the world. Many truck carriers, particularly those involved in just-in-time programmes, operate on a scheduled timetable. This results in very short and reliable transit times.

The amount of freight transported by trucks has increased since the 1980s by about 45 per cent, measured in tonne-kilometres.[5] Truck carriage has become an important part of the logistics networks of most firms because the characteristics of the trucking industry are more compatible than other transport modes with the service requirements of the firms' customers.

As long as it can provide fast, efficient service at rates between those offered by rail and air, the trucking industry will continue to prosper.

Rail

In Europe, rail is the next dominant mode of transport. Railroads carried over 880,000 tonnes of goods across Europe in 2001, however its share has declined to only 8 per cent of total European freight transported.[6]

Railroads have an average length of haul in Europe of approximately 1600 kilometres. Rail service is available in almost every major metropolitan centre in Europe and in many smaller communities; the EU15 rail network of over 156,000 kilometres is more extensive than the motorway network. The EU10 accession countries (which joined the EU on 1 May 2004) add almost another 60,000 kilometres of rail networks, but these are in need of restructuring and investment.[7]

Rail transport lacks the versatility and flexibility of motor carriers because it is limited to fixed track facilities. As a result, railroads – like air, water and pipeline transport – provide terminal-to-terminal service rather than point-to-point service unless companies have a rail siding at their facility, in which case service would be point to point. However, the rail track gauge varies between many European countries and the EC has presented a railway package to increase interoperability and open up the European railway market by January 2007.

Rail transport generally costs less (on a weight basis) than air and truck carriage. For many shipments, rail does not compare favourably with other modes in terms of loss and damage ratios. Compared to trucks, it has disadvantages in terms of transit time and frequency of service.

Many trains travel on timetable schedules, but depart less frequently than trucks. If a shipper has strict arrival and departure requirements, trucks usually have a competitive advantage over railroads. Some of this rail disadvantage may be overcome through the use of trailer-on-flatcar (TOFC) or container-on-flatcar (COFC) services, which offer the economy of rail or water movements combined with the flexibility of trucking. TOFC and COFC eliminate much of the inventory penalty associated with rail transportation. Most logistics executives refer to TOFC and COFC as a 'piggyback' service; in Germany it is known as a *Huckepack*.

Truck trailers or containers are delivered to the rail terminals, where they are loaded on flatbed wagons. At the destination terminal, they are offloaded and delivered to the consignee, the customer who receives the shipment. We will examine these services in greater detail later in this chapter.

Railroads suffer in comparison to trucks in terms of equipment availability. Railroads use their own as well as each other's wagons, and at times this equipment may not be located where it is most needed. Wagons may be unavailable because they are being loaded, unloaded, moved within railroad sorting yards or undergoing repair. Other wagons may be standing idle or lost within the vast rail network.

A number of developments in the rail industry have helped to overcome some of these utilization problems. Advances have included computer routing and scheduling; the upgrading of equipment, railtrack beds and terminals; improvements in wagon identification systems; wagons owned or leased by the shipper; and the use of **block trains** or dedicated through-train services between major metropolitan areas (i.e. non-stop shipments of one or a few shippers' products). Block trains operate on agreed train paths between specific origin–destination pairs (e.g. Hamburg–Vienna) without intermediate stops and on set days and times. Locomotive traction is provided from different railway enterprises. The train is operated at the risk of the service provider with the wagon mix driven by market requirements.

Railroads for the most part own their rolling stock, with the remainder owned by companies who lease them to the rail operating companies, particularly in the UK.

Air

Air carriers transport less than 1 per cent of tonne-kilometre traffic in Europe. Revenues of scheduled air carriers from movement of freight within Europe were about €46 billion in 1998, but this represented only about 4 per cent of total European turnover.[8]

Although increasing numbers of shippers are using air freight for regular service, most view air transport as a premium, emergency service because of its higher cost. However, when an item must be delivered to a distant location quickly, air freight offers the quickest time-in-transit of any transport mode. For most shippers, however, these time-sensitive shipments are relatively few in number or frequency.

Modern aircraft have cruising speeds of 800 to 1000 kilometres per hour and are able to travel internationally. The average length of haul within Europe is very small, at about 800 kilometres, however international movements may be thousands of kilometres.

To a great extent, air freight competes directly with trucks, and to a much lesser degree with rail. Where countries are separated by large expanses of water, the major competitor for international air freight is water carriage.

Air carriers generally handle high-value products – for example, Rolex watches or other jewellery. Air freight cannot usually be cost-justified for low-value items, because the high price of air freight would represent too large a percentage of the product cost. Customer service considerations may influence the choice of transport, but only if service issues are more important than cost issues.

Air transport provides frequent and reliable service and rapid time-in-transit, but terminal and delivery delays and congestion may appreciably reduce some of this advantage. On a point-to-point basis over short distances, motor transport often matches or outperforms the total transit time of air freight. It is the *total* transit time that is important to the shipper rather than the transit time from terminal to terminal.

Despite the limitations of air carriers, the volume of air freight has grown slightly in Europe over the years, but it continues to have a negligible impact on the amount of tonne-kilometres moved in Europe.[9] Undoubtedly, as customers demand higher levels of service and as global shipments increase, air freight will have a potentially greater role in the distribution plans of many firms.

Water

Water transportation can be broken down into several distinct categories: (1) inland waterways, such as rivers and canals, (2) coastal and intercoastal oceans, and (3) international deep sea. In Europe, water carriage competes primarily with rail and pipeline, since the majority of commodities carried by water are semiprocessed or raw materials transported in bulk. It is concentrated in low-value items (e.g. iron ore, grains, pulpwood products, coal, limestone and petroleum) where speed is not critical.

Other than in ocean transport, water carriers are limited in their movement by the availability of rivers, canals or intercoastal waterways. Reliance on water carriage depends to a greater or lesser degree on the geography of the particular location. In northern and central Europe, water carriage is much more important because of the vast system of navigable waterways, the accessibility to major population centres provided by water routes, and the relatively shorter distances between origins and destinations. For example, the approximately 675,000 tonnes of freight moved by inland waterways in 2001 was based around major European rivers such as the Rhone, Rhine, Elbe and Danube.[10]

The average length of haul varies tremendously depending on the type of water transport. For international ocean movements, the length of haul can be many thousands of kilometres. Generally, water is the dominant mode in international shipping. Within Europe, movements are of shorter length, depending on the length of navigable waterways.

Water carriage is perhaps the most inexpensive method of shipping high-bulk, low-value commodities. However, because of the inherent limitations of water carriers, it is unlikely that water transport will gain a larger role in domestic commerce, although international developments have made marine shipping increasingly important.

The development of **very large crude carriers (VLCCs)**, or supertankers, has enabled marine shipping to assume a vital role in the transport of petroleum between oil-producing and oil-consuming countries. Because of the importance of energy resources to industrialized nations, water carriage will continue to play a significant role in the transportation of energy resources. In addition, container ships have greatly expanded the use of water transport for many products.

Many domestic and most international shipments involve the use of containers. Containers are typically 8 feet high, 8 feet wide and of various lengths (e.g. 53 ft, 48 ft, 45 ft, 40 ft, 20 ft) and are compatible with conventional truck or rail equipment. A common transport statistic is the TEU, a 20-foot container equivalent.

The shipper in one country places cargo into an owned or leased container at its facility or at point of origin. Then the container is transported by rail or motor carriage to a water port for loading on to a container ship. After arrival at the destination port, it is unloaded and tendered to a rail or motor carrier in that country, and subsequently delivered to the customer or consignee. The shipment leaves the shipper and arrives at the customer's location with no or minimal handling of the items within the container.

The use of containers in intermodal logistics reduces staffing needs, minimizes in-transit damage and pilferage, shortens time in transit because of reduced port turnaround time, and allows the shipper to take advantage of volume shipping rates.

The largest ocean water carriers include Maersk Sealand, Evergreen Marine Corp., Hanjin Shipping, APL Limited and Orient Overseas Container Line Ltd (OOCL) of Hong Kong. These companies utilize both container and general cargo ships. The container ships are very large; in 2003 OOCL launched the world's largest container ship, the 323-metre-long and 8063 TEU OOCL *Shenzhen*, which displaces 34,300 tonnes when empty and is capable of carrying 99,500 tonnes of cargo, fuel and stores, giving a full load displacement of 133,800 tonnes. The *Shenzhen* has a crew of only 19, as computer automation means the entire ship can be controlled and adjusted from a single Windows-based computer terminal. When the *Shenzhen* arrives in port, five cranes work on it simultaneously. Travelling at a full-load speed of 25 knots the ship takes only 56 days to do a round-trip from Shanghai to Hamburg, calling in at 12 other ports on the way. The *Shenzhen* has since been joined by five other ships of this SX class and size in OOCL's fleet with the last, OOCL *Atlanta*, being launched in 2005.[11]

Often, a carrier will form alliances with other ocean carriers to maximize market coverage and customer service levels. Such networks are connected through large intermodal hubs such as Hamburg or Rotterdam. See the Global box for a discussion of the port of Hamburg's role as a major intermodal hub.

GLOBAL

The Port of Hamburg
The Port of Hamburg in Germany has developed into one of Europe's leading logistics and distribution centres. Situated at the mouth of the River Elbe on a world-renowned harbour, Hamburg also has an airport with a large cargo centre and adjoining canal systems that extend as far as the Czech Republic. Hamburg is thus a gateway to the rest of Germany and beyond, which also reflects its Hanseatic merchant past.

Over 12,000 ships a year pass through Hamburg, including luxury passenger liners and the world's largest container ship, the 323-metre-long and 8063 TEU OOCL *Shenzhen*. The port handled 110 million tonnes of freight in 2004 from 6.1 million TEU of container traffic. Over 140,000 jobs are directly or indirectly dependent on port business.

An average of 160 intermodal trains are loaded and unloaded in the port everyday while seaborne feeder traffic accounts for an average of 25 departures daily to other destinations in the Baltic region. The majority of German and Austrian industrial centres can be reached within 24 hours by train, offering next-morning delivery in most cases.

Hamburg's main distribution centres are located close to the E45, E22 and E26 motorways. More than 1700 road transport companies have branches in the city.

The city's inland waterway connections complete the multimodal options of Hamburg. The Elbe-Seintenkanal serves Dresden, Halle, Leipzig, Madeburg and Prague.

Dr Olaf Mager, media director for the Hamburg Port and Warehousing Corporation, notes, 'The Port of Hamburg is a true transport hub that fulfils all the obligations attached to this.'

Question: **Identify other major ports in the European Union 25 (EU25) member states and the markets they serve.**

Source: 'Using Total Transport', *Logistics Business Magazine* (May/June 2004), pp. 10–14.

Pipeline

Pipelines are able to transport only a limited number of products, including natural gas, crude oil, petroleum products, water, chemicals and slurry products. Natural gas and crude oil account for the majority of pipeline traffic. Oil pipelines transport approximately 10 per cent of all intra-European intercity freight traffic measured in millions of tonnes.[12]

There are over 20,500 kilometres of oil pipelines, and over 1.2 million kilometres of natural gas transmission and distribution pipelines in the EU15 countries. Pipelines offer the shipper an extremely high level of service dependability at a relatively low cost. Pipelines are able to deliver their product on time because of the following factors.

■ The flows of products within the pipeline are monitored and controlled by computer.
■ Losses and damage due to pipeline leaks or breaks are extremely rare.
■ Climatic conditions have minimal effects on products moving in pipelines.
■ Pipelines are not labour-intensive; therefore, strikes or employee absences have little effect on their operations.

The advantages in cost and dependability that pipelines have over other transport modes have stimulated shipper interest in moving other products by pipeline. Certainly, if a product is, or can be, in liquid, gas or slurry form, it can be transported by pipeline. As the costs of other modes increase, shippers may give additional consideration to pipelines as a mode of transport for non-traditional products.

Each mode transports a large amount of freight in Europe. The particular mode a shipper selects depends on the characteristics of the mode coupled with the needs of the company and its customers. Table 7.1 summarizes the economic and service characteristics of the five basic modes of transport.

Third Parties

Third parties are companies similar to channel intermediaries that provide linkages between shippers and carriers. Often, third parties do not own transportation equipment themselves; instead, they partner with a number of carriers who provide the necessary equipment to trans-

	Motor	Rail	Air	Water	Pipeline
Economic characteristics					
Cost	Moderate	Low	High	Low	Low
Market coverage	Point- to point	Terminal- to-terminal	Terminal- to-terminal	Terminal- to-terminal	Terminal- to-terminal
Degree of competition (number of competitors)	Many	Few	Moderate	Few	Few
Predominant traffic	All types	Low–moderate value, moderate high density	High value, low–moderate density	Low value, high density	Low value, high density
Average length of haul (in kilometres)	350	1600	330	376 to 1,367	276 to 343
Equipment capacity (tonnes)	10 to 25	50 to 1200	5 to 125	1000 to 60,000	30,000 to 2,500,000
Service characteristics					
Speed (time-in-transit)	Moderate to fast	Moderate	Fast	Slow	Slow
Availability	High	Moderate	Moderate	Low	Low
Consistency (deliver time variability)	High	Moderate	High	Low to moderate	High
Loss and damage	Low	Moderate	Low	Low to moderate	Low
Flexibility (adjustment to shipper's needs)	High	Moderate	Moderate to high	Low to moderate	Low

TABLE 7.1 Comparison of Transportation Modes

port their shipments. There are several types of third party, including transportation brokers, freight forwarders (domestic and foreign), shippers' associations or cooperatives, intermodal marketing companies (shippers' agents) and third-party logistics service providers.

Transportation Brokers
Transportation brokers are companies that provide services to both shippers and carriers by arranging and coordinating the transportation of products. They charge a fee to do so, which is usually taken as a percentage of the revenue collected by the broker from the shipper. The broker in turn pays the carrier.

Shippers with minimal traffic support, or no traffic department at all, can use brokers to negotiate rates, oversee shipments and do many of the things the shipper may not be able to do because of personnel or resource constraints. In these instances, the broker partially replaces some of the firm's own traffic department. The broker does not completely replace the traffic function; it merely assumes some of the transportation functions. Small and medium-sized shippers are the major users of transportation brokers, although larger firms utilize them in smaller markets.

Freight Forwarders

Freight forwarders purchase transport services from various carriers, although in some instances they own the equipment themselves. For example, the most successful air freight forwarders typically purchase and operate their own equipment, rather than relying on other air carriers. Freight forwarders consolidate small shipments from a number of shippers into large shipments moving into a certain region at a lower rate. Because of consolidation efficiencies, these companies can offer shippers lower rates than the shippers could obtain directly from the carrier. Often, the freight forwarder can provide faster and more complete service because they are able to tender larger volumes to the carrier.

Freight forwarders can be classified as domestic or international, depending on whether they specialize in shipments within a country or between countries. They can be surface or air freight forwarders. If they are involved in international shipments, freight forwarders will provide documentation services, which is especially vital for firms with limited international marketing experience.

Often, freight forwarders and transportation brokers are viewed similarly, but there are important differences, as outlined below.

- A forwarder is the shipper to a carrier and the carrier to a shipper.
- A broker is neither shipper nor carrier, but an intermediary between the two.
- A forwarder can arrange for transportation of freight by any mode.
- A broker can arrange for freight transportation only by a truck carrier.
- A forwarder is primarily liable to a shipper for cargo loss and damage.
- A broker is not usually liable for cargo loss and damage, although many do provide this coverage.[13]

Shippers' Associations

In their operations, shippers' associations are much like freight forwarders, but they differ in terms of perception by regulatory authorities. A shippers' association can be defined as a non-profit cooperative that consolidates small shipments into truckload freight for member companies.

Shippers' associations primarily use truck and rail carriers for transport. Because small shipments are much more expensive to transport (on a per-tonne or per-unit basis) than large shipments, companies band together to lower their transportation costs through consolidation of many small shipments into one or more larger shipments. The members of the shippers' association thus realize service improvements.

Shippers' associations also can handle truckload shipments by purchasing large blocks of flatbed rail wagons at discount rates. They then fill the available wagons with the trailers on flatcars (TOFCs) of member companies. Both parties benefit as a result. Shippers are charged lower rates than they could get by themselves (shipping in smaller quantities), while the railroads realize better equipment utilization and the economies of large, direct-route piggyback trains.

Intermodal Marketing Companies (or Shippers' Agents)

Intermodal marketing companies (IMCs), or shippers' agents, act much like shippers' associations or cooperatives. They specialize in providing piggyback services to shippers and are an important intermodal link between shippers and carriers. As the use of intermodal transportation increases in the future, shippers' agents will grow in importance as they purchase large quantities of TOFC/COFC services at a discount and resell them in smaller quantities. An example is the International Union of combined Road–Rail transport companies (UIRR) founded in 1970 with the objective of shifting as much freight transport as possible from road to rail. To achieve this goal, UIRR members offer different products, often developed in collaboration with their clients, such as **direct trains** that operate non-stop between destinations with variable capacity and composition of wagon loads all destined for the same location, or **shuttle trains** that have fixed composition of wagon loads.[14]

Third-party Logistics Service Providers

This sector is growing very rapidly. The number of European companies that have turned to third-party logistics (3PL) service providers has nearly doubled, from 28 per cent to 53 per cent since 2000.[15]

With an increasing emphasis on supply chain management, more companies are exploring the third-party option. For some firms, dealing with one third-party firm who will handle all or most of their freight offers a number of advantages, including the management of information by the third party, freeing the company from day-to-day interactions with carriers, and having the third party oversee hundreds or even thousands of shipments. Activities such as freight payment and dedicated contract carriage have been administered by third parties for many years. However, additional transportation and logistics activities are being outsourced. In some instances, some companies have outsourced large parts of their logistics operations to third parties.

Brokers, freight forwarders, shippers' associations, intermodal marketing companies and third-party logistics service providers can be viable shipping options for a firm in the same way that the five basic modes and intermodal combinations can. The logistics executive must determine the optimal combination of transport alternatives for his or her company.

Small-package Carriers

In addition to the preceding alternatives, many companies find that other transport forms can be used to distribute their products. Small-package carriers such as Federal Express (FedEx) and DHL are important transporters of many time-sensitive products. These entities use a combination of transport modes, especially air.

For companies such as electronics firms, catalogue merchandisers, cosmetic companies and textbook distributors, small-package carriers can be important transportation options. Growth rates for this transport sector are expected to average 10 to 15 per cent per year.

Parcel Post

Various European postal services, such as the UK's Royal Mail and Switzerland's Express Post, provide parcel post services to companies shipping small packages. The advantages of parcel post are low cost and wide geographical coverage, both domestically and within Europe. Disadvantages include specific size and weight limitations, variability in transit time, higher loss and damage ratios than other forms of shipment, and inconvenience because packages must be prepaid and deposited at a postal facility. Mail-order houses are probably the most extensive users of parcel post service.

Air Express Companies

Characterized by high levels of customer service, the air express industry has significantly expanded since its inception in 1973. DHL, owned by Deutsche Post World Net, is one of the best-known examples of an air express company that illustrates how the concept of supplying rapid transit with very high consistency has paid off. In 2004, Deutsche Post World Net had worldwide revenues of €44 billion and profits of €1.6 billion. DHL offers expertise in express, air and ocean freight, overland transport and logistics solutions, combined with worldwide coverage and an in-depth understanding of local markets. Its DHL Express Division covers over 4000 offices and more than 120,000 destinations worldwide, and has over 160,000 employees to provide fast and reliable services to meet customer expectations.[16] Because some firms need to transport certain products quickly, the air express industry is able to offer overnight (or second-day) delivery of small parcels to many locations throughout the world.

Many carriers have experienced a drop in order quantities as their customers aim to reduce inventory by ordering more frequently and in smaller quantities. This has increased the demand for air express-type services. Competition is fierce among the 'giants' of the industry, including FedEx, UPS, TNT and DHL. As long as there is a need to transport products quickly and with very high levels of consistency, the air express companies will continue to provide a valuable service for many shippers.

Intermodal Services
In addition to the five basic modes of transport, a number of intermodal combinations are available to the shipper. The more popular combinations are trailer on flatcar (TOFC), container on flatcar (COFC) and rolling road trains. Intermodal movements combine the cost and/or service advantages of two or more modes in a single product movement.

Piggyback (TOFC/COFC)
In piggyback service, a truck trailer or a container is placed on a rail flatcar and transported from one terminal to another. Axles can be placed under the containers so that they can be delivered by a truck. At the terminal facilities, motor carriers perform the pick-up and delivery functions. As shown in Figure 7.2, piggyback service thus combines the low cost of long-haul rail movement with the flexibility and convenience of truck movement.

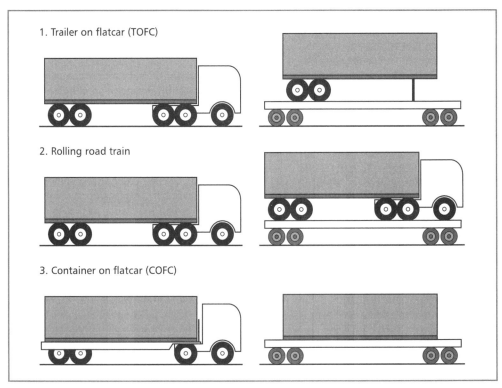

1. Trailer on flatcar (TOFC)

2. Rolling road train

3. Container on flatcar (COFC)

FIGURE 7.2 Selected Forms of Intermodal Transportation

Rolling Road Train

An innovative intermodal concept was introduced in Europe in the late 1990s. Rolling road trains, or accompanied traffic as they are sometimes called, combine truck and rail transport. As shown in Figure 7.2, the whole truck and trailer unit is driven up a ramp and on to a special low wagon. The trains of the 'rolling road' can be rapidly loaded and unloaded; this technique is particularly suitable when the transport has to be quick. Also, the transporters do not need any special equipment, and nothing in particular is needed at the terminals.

A major disadvantage is that the rolling road also has to carry a lot of dead weight (i.e. the truck unit). Also, in some west and southern European countries the railway gauge is not sufficient to transport 4-metre-high trucks on the rolling road.

Miscellaneous Intermodal Issues

Many other intermodal combinations are possible. In international commerce, for example, the dominant modes of transportation are air and water. Both include intermodal movements through the use of containers and truck trailers. Combinations of air–sea, air–rail, truck–sea, and rail–sea are used globally.

As an example:

> By shipping cargo by ocean from . . . Japan to Seattle, then transferring it to a direct flight to Europe from Seattle-Tacoma Airport, Asian exporters reap substantial benefits. They can cut their transit times from 30 days for all-water service to about 14 days, and slash freight costs by up to 50 per cent compared with all-air service.[17]

Between 1990 and 1996, intermodal freight movements in Europe increased steadily from 117 to 199 billion tonnes-kilometre.[18] Intermodal movements by carriers and intermodal marketing companies (IMCs) continue to be important means of transporting products domestically and internationally. While overall industry growth may have stabilized, many shippers and carriers are exploring expanded usage of this form of transport.

Global Issues

International freight transportation can involve any of the five basic modes of transportation, although air and water carriage are perhaps the most important. Truck and rail carriage are the most important freight movements *within* nations in Europe.

Managers of firms involved in international markets must be aware of the services, costs and availability of transport modes within and between the countries where their products are distributed. For example, air and water transportation compete directly for transoceanic shipments. Management must consider many factors when it compares the two alternatives.

Within countries, differences can exist because of taxes, subsidies, regulations, government ownership of carriers, geography and other factors. Because of government ownership or subsidies to railroads in Europe, rail service benefits from newer or better-maintained equipment, track and facilities. Japan and Europe utilize water carriage to a much larger degree than the United States or Canada due to the length and favourable characteristics of coastlines and inland waterways.

In general, international transportation costs represent a much higher fraction of merchandise value than domestic transportation costs. This is primarily due to the longer distances involved, administrative requirements, and related paperwork that must accompany international shipments.

Intermodal transportation is much more common in international movements. Even though rehandling costs are higher than for single-mode movements, cost savings and service improvements can result. One form of international intermodal distribution is a **landbridge**, where a foreign cargo crosses a country en route to another country. For example, goods from Finland are sent to China across a landbridge in Russia.

In making traffic and transportation decisions, the logistics manager must know and understand the differences between the domestic and international marketplace. Modal availability, rates, regulatory restrictions, service levels and other aspects of the transportation mix may vary significantly from one market to another. Most international terms of trade have been set under internationally accepted international commercial terms, known as **INCOTERMS**, which are discussed further in Chapter 12.

Regulatory Issues

There are two major areas of transportation regulation: economic and safety. All freight movements are subject to safety regulation, but not all are subject to economic regulation. The regulation of the transportation sector has had an enormous impact on the logistics activities of carriers and shippers. We will briefly describe economic and safety regulation, legal forms of transportation, and the impact of deregulation on shippers and carriers.

Forms of Regulation

Historically, transportation regulation has developed along two lines. The first, and perhaps the most publicized in recent years, is *economic* regulation. Economic regulation affects business decisions such as mode/carrier selection, rates charged by carriers, service levels, and routing and scheduling. *Safety* regulation deals with labour standards, working conditions for transportation employees, shipment of hazardous materials, vehicle maintenance, insurance and other elements relating to public safety.

The 1970s, 1980s and 1990s were periods of **deregulation** in North America, Europe and elsewhere throughout the world. At the same time, safety regulation has been increasing in terms of its scope and breadth. In Europe the European Conference of Ministers of Transport (ECMT) is an intergovernmental forum to cooperate on transport policy. The role of the ECMT is to help create an integrated transport system in Europe that is 'economically and technically efficient, meets the highest possible safety and environmental standards, and takes full account of the social dimension' and that 'build[s] a bridge between the European Union and the rest of the continent at a political level'.[19]

The European Council (EC) has developed a White Paper on 'European transport policy for 2010' to help ECMT fulfil its role. The overall objective is to develop a European transport system that shifts the balance between modes of transport, revitalizing the railways, promoting transport by sea and inland waterway, and controlling the growth in air transport. The 60-plus proposed measures, not all of which have yet been set in legislation, are meant to fit in with the sustainable development strategy adopted by the EC in Gothenburg in 2001.[20]

Legally Defined Forms of Transportation

In addition to classifying alternative forms of transportation by mode, carriers can be classified on the basis of the four legal forms: common, contract, exempt and private carriers. The first three forms are for-hire carriers and the last is owned by a shipper. **For-hire carriers** transport freight belonging to others and are subject to various government statutes and regulations. For the most part, **own account carriers** transport their own goods and supplies in their own equipment and are exempt from most regulations, except for those dealing with safety and taxation.

Deregulation has reshaped how logistics executives view the transport modes, particularly the legal forms of transportation. In principle, these legal designations no longer exist because of deregulation. However, the terms are used within the industry and do provide some guidance with respect to transportation type.

Common Carriers

Common carriers offer their services to any shipper to transport products, at published rates, between designated points. To operate legally, they must be granted authority from the appropriate regulatory agency. With deregulation, common carriers have significant flexibility with respect to market entry, routing and pricing. Common carriers must offer their services to the general public on a non-discriminatory basis – that is, they must serve all shippers of the commodities that their equipment can feasibly carry. A significant problem facing common carriers is that the number of customers cannot be predicted with certainty in advance. Thus, future demand is uncertain. The result has been that many common carriers have entered into contract carriage.

Contract Carriers

A **contract carrier** is a for-hire carrier that does not hold itself out to serve the general public; instead, it serves a limited number of shippers under specific contractual arrangements. The contract between the shipper and the carrier requires the carrier to provide a specified transportation service at a specified cost. In most instances, contract rates are lower than common carrier rates because the carrier is transporting commodities it prefers to carry for cost and efficiency reasons. An advantage is that transport demand is known in advance.

Own Account Carriers

An own account carrier is generally not for hire and is not subject to government economic regulation. Own account carriage means that a firm is providing transportation primarily for its own products. As a result, the company must own or lease the transport equipment and operate its own facilities. From a legal standpoint, the most important factor distinguishing own account carriage from for-hire carriers is the restriction that the transportation activity must be incidental to the primary business of the firm.

Own account carriage has had an advantage over other carriers because of its flexibility and economy. The major advantages of private carriage have been related to cost and service. With deregulation, common and contract carriage can often provide excellent service levels at reasonable costs. Later in this chapter, we will examine the own account versus for-hire transportation decision, and discuss more fully the pros and cons of own account carriage.

Impact of Deregulation

The degree to which the transportation sector has been regulated has varied over the years but in 1993 the EU began the deregulation of intra-EU transport. Throughout the next decade road transport increased dramatically while other inland modes remained static. For example, **cabotage**, where a carrier from one EU country is allowed to perform domestic transport in another EU country, represented 10.5 billion tonne-kilometres during the period 1990–98 and increased by a factor of six during that time.[21]

Deregulation of the major transportation modes has had a significant impact on motor, rail, air and water carriers, and the shippers who use their services. Freight transportation has moved into a new age. The next decade promises to be an exciting time for carriers and shippers.

Carrier Pricing and Related Issues

Several pricing issues are important in transportation. They involve how rates are developed in general and how specific rates are determined by a carrier to transport a shipment between an origin and destination point. Rates and rate determination will be overviewed, followed by a description of the issues relating to the specific rates carriers charge. The most significant approach is **free-on-board (FOB)** pricing. Also relevant are the availability of quantity discounts and allowances provided to the buyer by the carrier or shipper.

Rates and Rate Determination

Two forms or methods of transportation pricing can be utilized: cost of service and value of service. **Cost-of-service pricing** establishes transportation rates at levels that cover a carrier's fixed and variable costs, plus allowance for some profit. Transportation costs can vary within the cost-of-service pricing approach because of two major factors: distance and volume. Naturally, this approach is appealing because it establishes the lower limit of rates. However, it has some inherent difficulties.

First, a carrier must be able to identify its fixed and variable costs. This involves a recognition of the relevant cost components and an ability to measure these costs accurately. Many carrier firms are unable to measure costs precisely. Second, this approach requires that fixed costs be allocated to each freight movement (shipment). As the number of shipments increase, the fixed costs are spread over a larger number of movements, and thus the fixed cost per unit becomes smaller. As the number of shipments decreases, the fixed cost per unit becomes larger. As a result, the allocation of fixed costs changes the price based on the volume of shipments. Clearly, this method creates problems.

A second method of transportation pricing is **value-of-service pricing**. This approach is based on charging what the market will bear, and on the demand for transportation services and the competitive situation. This approach establishes the upper limit on rates. The rates set will maximize the difference between revenues received and the variable cost incurred for carrying a shipment. In most instances, competition will determine the price charged. An example of online freight exchanges that determine free market prices is contained in the Creative Solutions box.

CREATIVE SOLUTIONS

Freight Traders

Freight Traders is a web-based transport procurement service provider that designs and manages freight tenders for major manufacturers and retailers. Using Internet technology and bespoke software it provides a truly market-driven solution for the freight purchasing needs of firms. Formed in May 2000 by managing director Garry Mansell, Freight Traders is owned by Mars, Incorporated, the well-known confectionery company.

Freight Traders has provided its customers with, on average, reduced transport costs of 5 per cent per tender using Internet technology to remove inefficiencies from the marketplace; this process has benefited carriers as well. Freight Traders works closely with shippers and carriers through the four distinct phases of tendering: tender design and specification; carrier engagement; collection, monitoring and analysis of offers through the online bidding process; and decision support to aid evaluation of the bids.

Website transactions during 2004 increased 28 per cent on 2003, while revenue grew by 30 per cent during the same period to over £750,000 or €1 billion. The number of companies using Freight Traders grew as well, with more than 240 shippers and 2300 carriers in its web community. The combined number of users is approximately 20,000.

Question: **Can Internet exchanges be used for other logistics activities, such as warehousing and inventory management?**

Source: adapted from Freight Traders, http://www.freight-traders.com/home/ (2005).

FOB Pricing

The FOB pricing terms that are offered by sellers to buyers have a significant impact on logistics generally and transportation specifically. For example, if a seller quotes a delivered price to the buyer's retail store location, the total price includes not only the cost of the product, but the cost of moving the product to the retail store. This rather simple illustration highlights a number of important considerations for the buyer or consignee (i.e. the recipient of the product being distributed).

- The buyer knows the final delivered price prior to the purchase.
- The buyer does not have to manage the transportation activity involved in getting the product from the seller's location to the buyer's.
- The buyer typically will not control the transportation decision, so it is possible that a mode or carrier could be selected by the seller that might be disadvantageous to the buyer (e.g. due to poor service levels provided by the mode/carrier).

While it is easier from a management perspective to purchase products FOB destination, the lack of control of the transportation function can cause problems for the purchaser (e.g. the carrier selected by the shipper might provide poor service in your area, or only make deliveries at certain times, which may not correspond to your ideal time to receive shipments). Buyers should always know the specifics about all shipments that include delivery to ensure that the best decisions are being made on their behalf.

Factory gate pricing (FGP) has emerged as an alternative to FOB pricing for grocery retailers in the UK. The major supermarket chains are now arranging their own account or 3PL transportation of goods from suppliers, rather than letting suppliers handle transportation arrangements. In this way, the supermarkets are negotiating new pricing arrangements whereby the price excludes transport (i.e. the price at the factory gate). The UK Institute of Grocery Distribution (IGD) considers the move to

> FGP is establishing a principle which could potentially be applied to all costs involved in bringing a product to market. The first stage of transparency is to share existing information such as open book costing, quality and service measures. The second stage will be to apply measures to restructure the entire chain or value stream.[22]

Delivered Pricing

In a **delivered pricing system**, buyers are given a price that includes delivery of the product. As mentioned in the discussion of FOB pricing, this form of pricing is, in essence, FOB destination. The seller secures the transportation mode/carrier and delivers the product to the buyer. This option can be advantageous to one or both parties of the transaction, depending on which variation of delivered pricing is used by the seller.

For example, assume that two manufacturers compete for business in a market area. Manufacturer A is located in the market area, sells its product for €2.50 a unit and earns a contribution of €0.50 per unit. If manufacturer B incurs the same costs exclusive of transportation costs, but is located 400 miles from the market, €0.50 per unit represents the maximum that manufacturer B can afford to pay for transportation to the market. If two forms of transportation available to manufacturer B are equal in terms of performance characteristics, the higher-priced service would have to meet the lower rate to be competitive.

Zone Pricing

Zone pricing is a method that categorizes geographic areas into zones. Each zone will have a particular delivery cost associated with it. The closer the zone to the seller, the lower the delivery cost; the further away, the higher the delivery charge. Depending upon the buyer's location in a particular zone, some buyers will be paying more for delivery on a per-kilometre basis than others.

Quantity Discounts

Quantity discounts can be cumulative or non-cumulative. **Cumulative quantity discounts** provide price reductions to the buyer based on the amount of purchases over some prescribed period of time. **Non-cumulative quantity discounts** are applied to each order and do not accumulate over a time period.

From a transportation perspective, buyers purchasing products under a cumulative quantity discount system can order smaller quantities, paying the higher transportation costs for smaller shipments, and still gain a cost advantage (i.e. the additional cost of transportation is less than the cost savings resulting from the quantity discount).

On the other hand, if a non-cumulative quantity discount is applied, buyers must purchase sufficient quantities in order to obtain truckload (TL) or carload (CL) rates for larger shipments. While transportation costs on a per-item or per-tonne basis will be less for larger shipments, buyers will incur additional costs (e.g. warehousing and inventory carrying costs), which must be considered when they place larger, but fewer, orders with sellers. These issues will be addressed in other chapters.

In today's business environment, managers must be very responsive to customer demand in the marketplace. The trend is for companies to purchase smaller quantities more often and as quickly as possible, so it is more advantageous for buyers to have a cumulative quantity discount applied.

Allowances

Sometimes, sellers will provide price reductions to buyers that perform some of the delivery function. For example, when using a delivered pricing system, the seller assumes all costs of delivery and adds those costs to the price of the product. If the buyer is willing to assume some of the delivery functions, the seller will often provide some **allowances**, or price reductions, to the buyer.

The most common allowances are provided for customer pick-up of the product or unloading of the carrier vehicle upon delivery at the customer's location. These services cost the seller money and if the buyer is willing to perform these functions, the seller can provide a price concession.

The important element in making the right decisions about taking advantage of allowances is to know the costs associated with each delivery function. The allowance should be equal to, or greater than, the costs to the buyer for assuming these responsibilities.

Pricing and Negotiation

Shippers are concentrating more business with fewer carriers and placing greater emphasis on negotiated pricing. The goal of the negotiation process is to develop an agreement that is mutually beneficial, recognizes the needs of the parties involved, and motivates them to perform. Because most negotiations are based on cost-of-service pricing, carriers should have precise measures of their costs. Only when all costs are considered can carriers and shippers work together to reduce the carriers' cost base.

Logistics and Traffic Management

The strategies of carriers and shippers are inextricably interrelated. Transportation is an integral component of logistics strategy. Carriers must understand the role of transportation in a firm's overall logistics system, and firms must understand how carriers aid them in satisfying customer needs at a profit.

Some of the logistics issues relating to transportation will be described in the following section. The administration of transportation activities is referred to as **traffic management**, and includes major issues such as inbound and outbound transportation, carrier–shipper contracts, strategic partnerships and alliances, private carriage/leasing, mode/carrier selection, routing and scheduling, service offerings, and computer technology.

Inbound and Outbound Transportation
Transportation is one of the most significant areas of logistics management because of its impact on customer service levels and the firm's cost structure. Inbound and outbound transportation costs can account for as much as 10 per cent, 20 per cent, or more of the product's price. Firms in medium- and high-cost business sectors will be especially conscious of the transportation activity. Effective traffic management can achieve significant improvements in profitability.

To be effective, the traffic function must interface with other departments within and outside of the logistics area. Areas of interface include accounting (freight bills); engineering (packaging, transportation equipment); legal (warehouse and carrier contracts); manufacturing (just-in-time deliveries); purchasing (expediting, supplier selection); marketing/sales (customer service standards); receiving (claims, documentation); and warehousing (equipment supply, scheduling).

The transportation executive has many and varied duties, including selecting the best modes of transportation, choosing specific carriers, routing and scheduling of company-owned transport equipment, consolidation of shipments, filing claims with carriers and negotiating with carriers.

The carrier–shipper relationship is an important one; it directly affects the transportation executive's ability to manage successfully.

Carrier–shipper Contracts
The advantages of outsourcing or contracting out to 3PLs are numerous. Contracts permit the shipper to exercise greater control over the transportation activity, typically at a lower cost. They assure predictability and guard against fluctuation in rates. In addition, contracting provides the shipper with service-level guarantees and allows the shipper to use transportation to gain a competitive advantage.

Carrier–shipper contracts can prove valuable to both parties, but it is important that the contract include all of the relevant elements that apply to the shipping agreement. Because the transportation contract is a legal document and therefore binding, it should not be entered into casually. The exact format for each carrier–shipper contract will vary, depending on factors such as: the mode/carrier involved, the type of shipping firm, the products to be transported and the level of competition.

Carrier–shipper Alliances
An effective logistics network requires a cooperative relationship between shippers and carriers on both a strategic and an operational level. When this cooperation takes place, the shipper and carrier may become part of a partnership or alliance. Companies that have implemented the concept include Black and Decker, Procter & Gamble, Xerox and 3M.

In many instances, shippers and carriers do not act in concert because of differences in perceptions, practice or philosophy. Sometimes the notion that 'we never did it that way before' impedes cooperation and synergism. Such differences result in inefficiencies in the transportation system and conflicts between shippers and carriers.

In essence, a successful alliance is more than simply a set of plans, programmes and methods. Like the marketing concept and like customer service, a willingness to form a partnership is a philosophy that permeates the entire organization. It is a way of life that becomes a part of the way a firm conducts its business.

Own Account Carriage
An own account carrier is any transportation entity that moves products for the manufacturing or merchandising firm that owns it. While the equipment may transport products of other firms in some cases, own account carriers were established primarily to haul the products of their own companies.

Own account carriage should not be viewed strictly as a transportation decision; it is also a financial decision. There are two stages in evaluating the financial considerations of own account carriage. The first involves a comparison of current cost and service data of the firm's for-hire carriers with that of an own account operation. The second is to devise a plan of implementation and procedure for system control.

The feasibility study should begin with an evaluation of the current transportation situation, along with corporate objectives regarding potential future market expansion. Objectives should include a statement outlining past, current and desired service levels, as well as a consideration of the business environment, including legal restrictions and the general economic trends.

A firm must perform a cost–benefit analysis to determine whether it should use own account carriage. Any financial analysis should consider the time value of money. The company must calculate the net cash inflows (cash inflows minus cash outflows) for the life of the investment decision and discount them, using the company's minimum acceptable rate of return on new investments. The sum of these discounted cash flows must be compared with the initial capital requirement to determine if the investment is financially sound.

If the company makes the decision to engage in own account carriage, its next step is to devise a plan of implementation and a procedure for system control. Implementation begins with a review of the structure of the organization or group responsible for operating the own account fleet. Management assigns the activities to be performed to groups or individuals, and formulates a timetable for phasing in the project. Because of the risk involved, most firms begin with a low level of activity, followed by intermediate reviews of results and subsequent modification of the plan. The process is repeated until the firm achieves full implementation.

Control of own account carriage should emphasize the measurement of performance against standards, with the ability to identify specific problem areas. If management desires to use a total cost approach in order to charge cost against the product and customer, it can calculate a cost per kilometre, identifying the fixed costs associated with distribution of the product and then adding the variable costs per kilometre. This information may be useful to compare budgeted with actual expenditures, or to compare the figures for the own account fleet operation with those of common or contract carriers.

The future of own account trucking can hold tremendous promise for shippers that are able to take advantage of its benefits. For others, for-hire carriers offer the most opportunities. The most likely scenario is that shippers will utilize a mixture of own account and for-hire options.

Mode/Carrier Selection Decision Process

Economic and resource constraints, competitive pressures and customer requirements mandate that firms make the most efficient and productive mode/carrier choice decisions possible. Because transportation affects customer service, time-in-transit, consistency of service, inventories, packaging, warehousing, energy consumption, pollution caused by transportation, and other factors, traffic managers must develop the best possible mode/carrier strategies.

Four separate and distinct decision stages occur in the mode/carrier selection decision: (1) problem recognition, (2) search process, (3) choice process, and (4) post-choice evaluation.

Problem Recognition

The **problem recognition** stage of the mode/carrier choice process is triggered by a variety of factors, such as customer requirements, dissatisfaction with an existing mode, and changes in the distribution patterns of a firm. Typically, the most significant factors are related to service. In circumstances where the customer does not specify the transport mode, a search is undertaken for feasible alternatives.

Search Process

In the **search process**, transportation executives scan a variety of information sources to aid them in making optimal mode/carrier decisions. Possible sources include past experience, carrier sales calls, company shipping records and customers of the firm. Once the desired information is gathered, the decision becomes one of using the information obtained to select the optimal mode or carrier alternative.

Choice Process

The **choice process** involves the selection of an option from the several modes and carriers available. Using information previously gathered in the selection process, the transportation executive determines which of the available options best meets the firm's customer service requirements at an acceptable cost. Generally, service-related factors are the major determinants of mode/carrier choice. For example, Table 7.2 identifies a number of selection criteria used in the evaluation of truck carriers.

Attribute description	Importance mean*
Honesty of dispatch personnel	6.5
On-time pick-ups	6.5
On-time deliveries	6.5
Competitive rates	6.5
Accurate billing	6.4
Assistance from carrier in handling loss and damage claims	6.4
Prompt action on complaints related to carrier service	6.4
Honesty of drivers	6.4
Prompt response to claims	6.4
Carrier's general attitude toward problems and complaints	6.3
Prompt availability of status information on delivery	6.3
Consistent (reliable) transit times	6.3

TABLE 7.2 The Most Important Attributes Considered in the Selection and Evaluation of LTL Motor Carriers

* Respondents were asked to indicate on a seven-point scale how important the attribute was in selecting an LTL motor carrier. The scale ranged from 1 (not important) to 7 (very important).

Source: adapted from Douglas M. Lambert, M. Christine Lewis and James R. Stock, 'How Shippers Select and Evaluate General Commodities LTL Motor Carriers', *Journal of Business Logistics* 14, no. 1 (1993), p. 135.

There are similarities across modes in terms of the most important attributes used to select and evaluate carriers. Attributes such as on-time pick-ups and deliveries, prompt response to customer enquiries, consistent transit times and competitive rates seem to be important irrespective of the mode or carrier being considered.

The transportation executive selects the mode or carrier that best satisfies the decision criteria, and the shipment is sent by that option. When a similar decision may arise in the future, such as a repeat order from a customer, management may establish an order routine, so that the same choice process will not have to be repeated. Order routines eliminate the inefficiencies associated with making the same decision repeatedly.

Post-choice Evaluation

Once management has made its choice of mode or carrier, it must institute some evaluation procedure to determine the performance level of the mode/carrier. Depending on the individual firm, the **post-choice evaluation** process may be extremely detailed or there may be no evaluation at all. For the majority of firms, the degree of post-choice evaluation lies somewhere between the two extremes. It is rare that a company does not at least respond to customer complaints about its carriers, and this is one form of post-choice evaluation. Many firms use other techniques, such as cost studies, audits and reviews of on-time pick-ups and delivery performance. Some will statistically analyse the quality of carrier service attributes, such as on-time performance and loss–damage ratios.

Mode and carrier selection is becoming much more important as shippers reduce the number of carriers with whom they do business and develop core carriers. By leveraging freight volumes to get bigger discounts and higher levels of service, shippers are able to reduce their transportation costs. At the same time, carriers benefit by having to deal with fewer shippers, each shipping larger volumes of product consistently over longer periods of time.

Routing and Scheduling

Considering the significant capital investments in equipment and facilities, along with operating expenses, carriers recognize the importance of good routing and scheduling in achieving acceptable levels of company profit and customer service. In recent years, these areas have become much more significant because of increased competition and deregulation, and a number of economic factors (e.g. fuel, labour and equipment).

Carriers can achieve sizeable benefits by optimizing their routing and scheduling activities. For example, by prescheduling shipments into specific market areas while simultaneously reducing the frequency of delivery, a vehicle's load factor can be increased. The result is a cost savings to the carrier. A reduction in the frequency of pick-ups and deliveries can result in a reduced level of transportation necessary to deliver the same amount of goods. Thus, the cost of transportation is reduced and productivity is increased.

Other examples include the use of fixed routes instead of variable routes for some shipments, and changing customer delivery hours. If customers can accept shipments at off-peak hours, the carrier will have a larger delivery time window and thus can improve vehicle utilization and reduce equipment cost on a per-delivery basis.

In general, the benefits to a carrier of improved routing and scheduling include greater vehicle utilization, higher levels of customer service, lower transportation costs, reduced capital investment in equipment and better management decision-making.

Advances in technology have also helped routing and scheduling decision-making. The UK company Paragon Software Systems provides 'software solutions for operations ranging from 10 vehicles based at a single site to hundreds of vehicles running out of several sites, and applications include home delivery, complex multi-site distribution planning and real-time vehicle tracking and monitoring'.[23]

Service Offerings

In the traditional areas of service – pick-up and delivery, claims, equipment availability, time-in-transit and consistency of service – competitive pressures from the marketplace have worked to improve service levels. Carriers have had to develop customer service packages that meet the needs of increasingly demanding customers. Such improvements have benefited shippers and have required carriers to maximize their efficiency and productivity in order to remain profitable.

Examples of carriers providing higher levels of traditional transportation services can be found in all modes. In 1991 the international ocean marine carriers A.P. Moller-Maersk Line and

Sea-Land Service (a division of CSX) began a vessel-sharing plan in order to improve service levels. The plan reduced the number of ships deployed and increased the number of trans-Pacific sailings to five per week. At the same time, Maersk and Sea-Land continued to compete against each other. However, they merged in 1999 to become Maersk Sealand.[24]

Carriers have begun to expand into non-traditional areas such as warehousing, logistics consulting, import–export operations and facility location analysis. In effect, the transportation carrier has become a logistics service firm. The trend will continue as carriers expand their traditional and non-traditional service offerings. In addition, competitive pressures will force overall carrier service levels to improve.

Computer Technology

The use of computers has become widespread in logistics, especially in the area of traffic management. Generally, computerized transportation activities can be categorized into four groups: transportation analysis, traffic routing and scheduling, freight rate maintenance and auditing, and vehicle maintenance.

Transportation Analysis

This software allows management to monitor costs and service by providing historical reporting of key performance indicators such as carrier performance, shipping modes, traffic lane utilization, premium freight usage and backhauls.

Traffic Routing and Scheduling

Software in this area provide features such as the sequence and timing of vehicle stops, route determination, shipping paperwork preparation and vehicle availability.

Freight Rate Maintenance and Auditing

These software systems maintain a database of freight rates used to rate shipments or to perform freight bill auditing. They compare actual freight bills with charges computed from the lowest applicable rates in the database. The systems can then pay, authorize payment or report on exceptions.

Vehicle Maintenance

Features commonly provided by these packages include vehicle maintenance scheduling and reporting.

The degree and scope of computer usage will vary from firm to firm and by transportation activity. Despite these variations, it is clear that the computer has had a significant impact on traffic management and will have an even greater impact in the future. The Technology box describes how the use of **vehicle telematics** enhances traffic management.

TECHNOLOGY

Vehicle Telematics Systems

Vehicles Telematics Systems (VTSs) can be used as a source of key performance indicator (KPI) data for the measurement of road transport efficiency. However, VTSs can also be used for better traffic management, improved road safety, and road user charging through electronic toll collection (ETC).

VTSs use global positioning system (GPS) satellites to track truck movement through onboard units (OBUs) in the truck cab. There are now many suppliers and users of VTS in the road freight sector in Europe. The main focus appears to be on vehicle time utilization from the VTS suppliers' point of view, while users understandably focus on the main cost drivers for the operation of the vehicle: fuel consumption and drivers' hours. Governments, on the other hand, are concerned with using VTSs to assist their ETC initiatives.

However, VTSs currently have an inability to monitor vehicle loading either directly through the use of sensors or indirectly through interfacing with other company IT systems, which is a major constraint on the wider adoption of performance measures used in any transport KPI initiative, particularly energy-intensity indices.

Further, it has taken 10 years for the ETC industry to reach agreement on a European standard, but this standard still has issues of interoperability with other technologies such as GPS. Europe also has a poor record on the interoperability of ETC services between different countries as each nation has developed its own system for toll collection.

Question: **Do you believe vehicle telematics have any consumer logistics uses, and if so what are they?**

Sources: adapted from David McClelland and Alan C. McKinnon, 'Use of Vehicle Telematics Systems for the Collection of Key Performance Indicator Data in Road Freight Transport', Heriot-Watt University Logistics Research Centre, http://www.sml.hw.ac.uk/logistics/pdf/vehicletelematics.pdf (April 2004); Robin Meczes, 'Public Policy to Drive Telematics Take-up?', *Logistics & Transport Focus* 5, no. 7 (September 2003), pp. 38–40.

Transportation Productivity Issues

Both shippers and carriers are concerned with improving transportation productivity. Such improvements are absolutely vital to the success of the logistics system.

Areas in which transportation productivity can be improved can be categorized into three groups:

1 improvements in the transportation system's design, and its methods, equipment and procedures
2 improvements in the utilization of labour and equipment
3 improvements in the performance of labour and equipment.[25]

Each of the following groups offers examples of possibilities for productivity improvement.

- Group 1: inbound consolidation, company operates over-the-road trucking, local pick-up and delivery operations, and purchases for-hire transportation.
- Group 2: breakbulk operations, backhaul use of fleet, routing and scheduling systems, tracing and monitoring systems, customer delivery hours, shipment consolidation and pooling, and driver utilization.
- Group 3: standards for driver activity, first-line management improvements, establishment of a transportation database, incentive compensation to encourage higher productivity and safety, and programmes to increase fuel efficiency.[26]

From a shipper's perspective, some of the most common types of data that measure carrier effectiveness and efficiency include claims and damage ratios, transit time variability, on-time pick-up and delivery percentages, cost per tonne-kilometre, billing accuracy, and customer complaint frequency. In many firms, the data do not appear on a formal report and carrier performance is therefore examined informally.

Carriers employ similar measures, although they view them from the perspective of a provider rather than a receiver of services. Some carriers measure euro contributions by traffic lane, shipper, salesperson or terminal. These measures are used primarily for internal performance evaluations, but they may be provided to customers in special situations such as rate negotiations or partnership arrangements. The exact format for data collection is not as important as the need to have the information available in some form.

SUMMARY

In this chapter, we examined the role of transportation in logistics. The five basic transport modes were described, including intermodal combinations, third-party providers and small-package carriers.

Transportation regulation and deregulation issues in Europe were briefly presented in order to gain a perspective on today's competitive environment. The regulatory environment is fast becoming one of open competition, with shippers and carriers having significant freedom to negotiate prices, service levels and other transportation aspects.

Transportation pricing issues were examined, including rates and rate determination, categories of rates, FOB pricing, delivered pricing, quantity discounts, allowances and pricing-negotiation issues.

We described transportation management in terms of the importance of the carrier–shipper interface and some of the specific functions of management. Important perspectives included inbound and outbound transportation, carrier–shipper contracts, carrier–shipper alliances, private carriage, mode/carrier selection, routing and scheduling, service offerings, computer technology and transportation productivity.

KEY TERMS

A full Glossary can be found at the back of the book.

QUESTIONS AND PROBLEMS

1 What are time utility and place utility? How does the transportation function add utility to products?
2 Briefly discuss the five transport modes, based on their economic and service characteristics.
3 Discuss the role and functions of brokers in the transportation system.
4 Trailer-on-flatcar (TOFC) and container-on-flatcar (COFC) intermodal combinations are referred to as piggyback movements. Describe piggyback movements from the perspectives of cost, service and availability, and identify their major strengths and weaknesses.
5 What is the difference between cost-of-service and value-of-service pricing in transportation? How does each affect the rates charged by carriers?
6 Most transportation executives believe that service factors are generally more important than cost factors in causing firms to switch from one transport mode to another. Under what circumstances would service factors be more important than cost factors?
7 In the evaluation of transportation modes, *consistency of service* is significantly more important to shippers than *time-in-transit*. What is the difference between the two terms? Identify some reasons why shippers consider consistency of service more important. In your answer, be sure to consider the impact on inventory levels.
8 Own account carriage should not be viewed strictly as a transportation decision – it is also a financial decision. Briefly explain the reasons that underlie this statement.

THE LOGISTICS CHALLENGE!

PROBLEM: EU ENLARGEMENT

In May 2004 the European Union grew from 15 member states (EU15) with the addition of 10 central and eastern European countries (EU10) that now make up the EU25. The growth in the EU is regarded as a sensible response to changing economic environments, as competition has become fiercer than ever in the global economy. Consequently, the efficient management of the flow of materials, goods and information between EU countries is fast becoming an important factor for success.

The subsequent growth in trade between the EU25 countries has created coincident growth for the demand of logistics services. Twenty per cent of respondents to a European Logistics Association (ELA) survey in 2003 had plans to significantly increase business in the EU10 countries within one year, while 16 per cent intended to do so in three to five years.

Many respondents believed local logistics service providers for transport and warehousing would provide cost advantages and relatively few (31 per cent) were planning to restructure their logistics infrastructures in the new member countries. Notwithstanding, firms may still have to redesign their logistics networks, especially the location of distribution centres. Almost 60 per cent of the distribution centres responsible for European distribution are situated in the Netherlands. As a result of the enlargement three regions will gain more importance: the Baltic Sea, the continental EU countries adjacent to the former east–west border (Austria, Czech Republic, Germany and Poland) and the Mediterranean Sea.

Companies and logistics service providers expanding their business towards the EU10 will have to take this shift into consideration and rethink their logistical distribution concepts. A better service level towards the customer at an affordable price can be achieved by an adequate centralization of the system in combination with cross-docking strategies. To achieve this, one possibility is to cooperate with local logistics providers. In this way, companies can benefit from the existing knowledge of local partners and thus enjoy a lower investment risk. The challenge here is to find the right partner, with whom a trustful collaboration is possible, and to design adequate, flexible contracts with them.

A more expensive way would be to re-adapt the existing logistical topology and install hubs and distribution centres in the EU10 member countries. This allows for maximum flexibility and control alongside customized solutions. For some companies an alternative might be to establish a production plant on-site, hence profiting from low wages and distributions costs.

However, many EU15 logistics service providers are not concerned about competition from eastern logistics companies. This is due to the fact that special regulations in the enlargement treaty ensure that the new members are integrated gradually. During the first years following enlargement there will be no completely liberalized competition between logistics service providers from the new and old EU countries. The main concern was that in a free competition situation EU15 companies would not be able to compete with EU10 wages. Another disadvantage of EU10 logistics service providers is their out-to-date transport assets, which will soon have to be replaced by more environmentally friendly carriers.

If you were the managing director of an EU15 manufacturer looking to expand into the EU10 countries, which option would you choose? An alliance with an EU10 logistics service provider, or the establishment of your own transport and distribution centres in the EU10 region?

What Is Your Solution?

Sources: adapted from Lauri Ojala, 'The Logistical Impact of EU Enlargement', *Economic Trends* (April 2004), p. 29–33; Alfred Angerer, Daniel Corsten, Frank Straube and Philippe Tufinkgi,'The EU Enlargement – Influence on European Logistics', St Gallen, CH: Kuehne-Institute for Logistics, University of St Gallen (September 2003).

SUGGESTED READING

BOOKS

European Council of Ministers for Transport, 'Regulatory Reform in Road Freight Transport', *Proceedings of the International Seminar February 2001*. Paris: OECD Publications, 2001.

Mohring, Herbert (ed.), *The Economics of Transport*, vols 1 and 2. Brookfield, VT: Edward Elgar, 1994.

Ojala, Lauri, Tapio Naula and Torsten Hoffman, *Trade and Transport Facilitation Audit of The Baltic States (TTFBS): On a Fast Track to Economic Development*. Washington, DC: The World Bank, 2005.

Schary Philip B. and Tage Skjøtt-Larsen, *Managing the Global Supply Chain*, 2nd edn. Copenhagen: Copenhagen Business School Press, 2001.

JOURNALS

Buxbaum, Peter A., 'Winning Together', *Transportation and Distribution* 36, no. 4 (April 1995), pp. 47–50.

Carter, Joseph L., Bruce G. Ferrin and Craig R. Carter, 'The Effect of Less-than-truckload Rates on the Purchase', *Transportation Journal* 34, no. 3 (Spring 1995), pp. 35–47.

Cooke, James Aaron, '3PLs Look toward More Realistic Growth', *Distribution* 35, no. 12 (December 1996), pp. 29–32.

Daugherty, Patricia J. and Michael S. Spencer, 'Just-in-time Concepts: Applicability to Logistics/Transportation', *International Journal of Physical Distribution and Logistics Management* 20, no. 7 (1990), pp. 12–18.

D'Este, Glen, 'An Event-based Approach to Modelling Intermodal Freight Systems', *International Journal of Physical Distribution and Logistics Management* 26, no. 6 (1996), pp. 5–16.

Garreau, Alain, Robert Leib and Robert Millen, 'JIT and Corporate Transport: An International Comparison', *International Journal of Physical Distribution and Logistics Management* 21, no. 1 (1991), pp. 42–7.

Gooley, Toby B., 'How to Choose a Third Party', *Traffic Management* 32, no. 10 (October 1993), pp. 85A–87A.

Kasarda, John D., 'Transportation Infrastructure for Competitive Success', *Transportation Quarterly* 50, no. 1 (Winter 1996), pp. 35–50.

McConville, Daniel J., 'Private Fleet Leasing: Two Sides to Teamwork', *Distribution* 96, no. 3 (March 1997), pp. 40–6.

Muller, E.J., 'A New Paradigm for Partnerships', *Distribution* 93, no. 1 (January 1994), pp. 45–8.

Richardson, Barbara C., 'Transportation Ethics', *Transportation Quarterly* 49, no. 2 (Spring 1995), pp. 117–26.

Richardson, Helen L., 'Will Shrinking Shipments Shrink Profits?', *Transportation and Distribution* 36, no. 3 (March 1995), pp. 45–50.

Spizziri, Martha, 'Intermodal Overcomes the Obstacles', *Distribution* 33, no. 4 (April 1994), pp. 39–42.

Theurmer, Karen, 'Multimodal Carriers at Your Service', *Intermodal Shipping* 29, no. 7 (September 1994), pp. 28–31.

Trunick, Perry A., 'Air Cargo Needs Positive Action', *Transportation and Distribution* 35, no. 9 (September 1994), pp. 53–8.

Vantuono, William C., 'Roadrailer Hits the Big Time', *Railway Age* 195, no. 10 (October 1994), pp. 49–52.

Waters, C.D.J., 'Computer Use in the Trucking Industry', *International Journal of Physical Distribution and Logistics Management* 20, no. 9 (1990), pp. 24–31.

REFERENCES

[1] Eurostat Main Transport Indicators, http://europa.eu.int/comm/eurostat/ (2005).

[2] Ibid.

[3] Ibid.

[4] Ibid.

[5] Philip B. Schary and Tage Skjøtt-Larsen, *Managing the Global Supply Chain*, 2nd edn. Copenhagen: Copenhagen Business School Press, 2001, p. 221.

[6] Eurostat Main Transport Indicators, http://europa.eu.int/comm/eurostat/ (2005).

[7] Ibid.

[8] European Commission, *EU Transport in Figures: Statistical Pocket Book 2000*. Luxembourg: European Communities, 2000, p. 16.

[9] Schary and Skjøtt-Larsen, *Managing the Global Supply Chain*, pp. 221–2.

[10] Eurostat Main Transport Indicators, http://europa.eu.int/comm/eurostat/ (2005).

[11] Orient Overseas Container Line Ltd, http://www.oocl.com/ (2005).

[12] European Commission, *EU Transport in Figures: Statistical Pocket Book 2000*. Luxembourg: European Communities, 2000, p. 16.

[13] Mitchell E. MacDonald, 'Broker vs Forwarder', *Traffic Management* 31, no. 6 (June 1992), p. 62.

[14] International Union of combined Road-Rail transport companies, http://www.uirr.com/ (2005).

[15] Charles Davis, 'Countering the Costs', *Logistics Europe* 12 no. 7, (September 2004), p. 27.

[16] DHL, http://www.dhl.com (2005).

[17] Toby B. Gooley, 'Air Freight Hits the Rails', *Logistics Management* 35, no. 3 (March 1996), p. 112A.

[18] European Commission, *EU Transport in Figures: Statistical Pocket Book 2000*. Luxembourg: European Communities, 2000, p. 78.

[19] European Council of Ministers for Transport, 'Regulatory Reform in Road Freight Transport', *Proceedings of the International Seminar February 2001*. Paris: OECD Publications, 2001, p. 2.

[20] European Council, 'White Paper: European Transport Policy for 2010', http://europa.eu.int/scadplus/leg/en/lvb/l24007.htm (2005).

[21] European Council of Ministers for Transport, 'Regulatory Reform in Road Freight Transport', p. 36.

[22] Institute of Grocery Distribution, 'Factory Gate, Open Book & Beyond', http://www.igd.org.uk/cir.asp?cirid=642&search=1 (2005).

[23] Paragon Software Systems, http://www.paragonrouting.com/ (2005).

[24] Elizabeth Canna, 'The Maersk/Sea-Land Deal', *American Shipper* 33, no. 5 (May 1991), p. 43.

[25] A.T. Kearney, Inc., *Measuring and Improving Productivity in Physical Distribution Management*. Chicago: National Council of Physical Distribution Management, 1984, pp. 174.

[26] Ibid., pp. 174–85.

CHAPTER 8
WAREHOUSING

OBJECTIVES

CHAPTER OBJECTIVES

- ■ To show why warehousing is important in the logistics system
- ■ To identify the major types or forms of warehousing
- ■ To examine the primary functions of warehousing
- ■ To identify the key decision criteria affecting the type of warehousing
- ■ To identify the factors that affect the size and number of warehouses
- ■ To examine the warehouse site-selection decision from macro- and microperspectives
- ■ To describe the factors that affect warehouse layout and design
- ■ To provide an overview of the importance of productivity and accounting/control issues in warehouse management.

INTRODUCTION

Warehousing is an integral part of every logistics system. There are almost one million warehouse facilities worldwide, including state-of-the-art, professionally managed warehouses, as well as company stockrooms, garages, self-store facilities and even garden sheds. Warehousing plays a vital role in providing a desired level of customer service at the lowest possible total cost. Warehousing activity is an important link between the producer and the customer. Over the years, warehousing has developed from a relatively minor facet of a firm's logistics system to one of its most important functions.

We can define **warehousing** as that part of a firm's logistics system that stores products (raw materials, parts, goods-in-process, finished goods) at and between point of origin and point of consumption, and provides information to management on the status, condition and disposition of items being stored. The term **distribution centre (DC)** is sometimes used, but the terms are not identical. **Warehouse** is the more generic term.

> Warehouses store all products, DCs hold minimum inventories and predominantly high-demand items. Warehouses handle most products in four cycles [receive, store, ship and pick], DCs handle most products in two: receive and ship. Warehouses perform a minimum of value-added activity, DCs perform a high percentage of value adding, including possible final assembly. Warehouses collect data in batches, DCs collect data in real-time. Warehouses focus on minimizing the operating cost to meet shipping requirements, DCs focus on maximizing the profit impact of meeting customer delivery requirements.[1]

With an increasing interest in improving inventory turns and reducing time to market, the role of distribution increasingly focuses on filling orders rapidly and efficiently.

Effective warehouse management involves a thorough understanding of the functions of warehousing, the merits of public versus private warehousing, and the financial and service aspects of warehousing decisions. Managers need knowledge of the methods that can improve warehousing performance and a strategy for locating warehousing facilities at optimal locations.

Warehousing decisions may be strategic or operational. *Strategic* decisions deal with the allocation of logistics resources over an extended time in a manner consistent and supportive of overall enterprise policies and objectives. They can take either long-range or project-type forms.

An example of a long-range strategic decision is the choice of logistics system design. A project-type decision might deal with consolidation of branch warehouses into a regional distribution centre. Other examples of typical strategic questions include the following.

- Should warehousing be owned, leased or rented, or some combination of these?
- Should the warehousing functions be outsourced – that is, contracted out to a third-party provider?
- Should the company install new materials handling equipment or continue to hire more labour?

Operational decisions are used to manage or control logistics performance. Typically, these decisions are routine in nature and involve time spans of one year or less. They relate to the coordination and performance of the logistics system. For example, a warehouse manager would be concerned with how best to utilize labour in the shipping department. Due to the short time horizon involved, these decisions have more certainty than strategic decisions.

The Nature and Importance of Warehousing

Warehousing has traditionally provided storage of products (referred to as inventory) during all phases of the logistics process. Two basic types of inventory can be placed into storage: (1) raw materials, components and parts (physical supply); and (2) finished goods (physical distribution). Also, there may be inventories of goods-in-process and materials to be disposed of or recycled, although in most firms these constitute only a small portion of total inventories.

Why do companies hold inventories in storage? Traditionally, the warehousing of products has occurred for one or more of the following reasons – to:

- achieve transportation economies
- achieve production economies
- take advantage of quantity purchase discounts and forward buys
- maintain a source of supply
- support the firm's customer service policies
- meet changing market conditions (e.g. seasonality, demand fluctuations, competition)
- overcome the time and space differentials that exist between producers and consumers
- accomplish least total cost logistics commensurate with a desired level of customer service
- support the just-in-time programmes of suppliers and customers
- provide customers with a mix of products instead of a single product on each order
- provide temporary storage of materials to be disposed of or recycled (i.e. reverse logistics).

Several Uses of Warehousing

Figure 8.1 identifies some of the uses of warehousing in both the physical supply and physical distribution systems. Warehouses can be used to support manufacturing, to mix products from multiple production facilities for shipment to a single customer, to breakbulk or subdivide a large shipment of product into many smaller shipments to satisfy the needs of many customers, and to combine or consolidate a number of small shipments into a single higher-volume shipment.

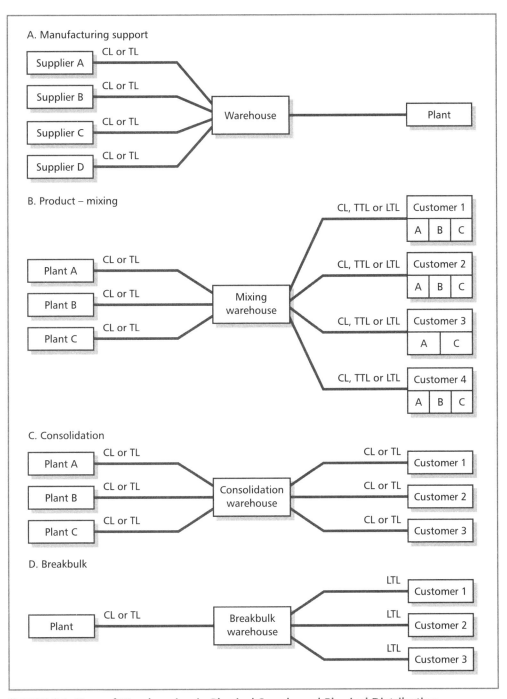

FIGURE 8.1 Uses of Warehousing in Physical Supply and Physical Distribution

Warehousing is used increasingly as a 'flow-through' point rather than a 'holding' point, or even bypassed (e.g. scheduled deliveries direct to customers), as organizations increasingly substitute information for inventory, purchase smaller quantities and use warehouses as 'consolidation points' to receive purchased transportation rates and service levels.

> The traditional method [of distribution] is a push system. Production plans are based on capabilities and capacities of the plant, and product is produced in the expectation that it will sell. When it is produced faster than it can be sold, it is stockpiled at plant warehouses. If sales cannot be accelerated, then the plant will be slowed down until supply moves into balance with demand. In this system, warehousing serves to absorb excess production. Today's pull system depends on information. It is based on a constant monitoring of demand. . . . With a pull system, there is no need for a reservoir. Instead, the warehouse serves as a flow-through center, offering improved service by positioning inventory closer to the customer.[2]

Manufacturing Support

In supporting manufacturing operations, warehouses often play the important role of inbound consolidation points for the receipt of shipments from suppliers. As shown in Figure 8.1, part A, firms order raw materials, parts, components or supplies from various suppliers, who ship truckload (TL) or carload (CL) quantities to a warehouse located in close proximity to the plant. Items are transferred from the warehouse to the manufacturing plant(s).

Product Mixing

From a physical distribution or outbound perspective, warehouses can be used for product mixing, outbound consolidation or breakbulk. Product mixing (see Figure 8.1, part B) often involves multiple plant locations (e.g. plant A, plant B and plant C) that ship products (e.g. products A, B and C) to a central warehouse. Each plant manufactures only a portion of the total product offering of the firm. Shipments are usually made in large quantities (TL or CL) to the central warehouse, where customer orders for multiple products are combined or mixed for shipment.

Consolidation

When a warehouse is used for outbound consolidation (see Figure 8.1, part C), TL or CL shipments are made to a central facility from a number of manufacturing locations. The warehouse consolidates or combines products from the various plants into a single shipment to the customer.

Breakbulk

Breakbulk warehouses (see Figure 8.1, part D) are facilities that receive large shipments of product from manufacturing plants. Several customer orders are combined into a single shipment from the plants to the breakbulk warehouse. When the shipment is received at the warehouse, it is broken down into smaller LTL shipments, which are sent to customers in the geographical area served by the warehouse.

Warehousing and Transportation

Transportation economies are possible for both the physical supply system and the physical distribution system. In the case of physical supply, small orders from a number of suppliers may be shipped to a consolidation warehouse near the source of supply; in this way, the producer can achieve a TL or CL shipment to the plant, which normally is situated at a considerably greater distance from the warehouse. The warehouse is located near the sources of supply so that the LTL rates apply only to a short haul, and the volume rate is used for the long haul from the warehouse to the plant.

Warehouses are used to achieve similar transportation savings in the physical distribution system. In the packaged goods industry, manufacturers often have multiple plant locations, with each plant manufacturing only a portion of the company's product line. Such plants are often referred to as *focused factories*.

Usually, these companies maintain a number of **field warehouse** locations from which mixed shipments of the entire product line can be made to customers. Shipments from plants to field warehouses are frequently made by rail in full carload quantities of the products manufactured at each plant. Orders for customers, comprised of various items in the product line, are shipped by truck at TL or LTL rates. The use of field warehouses results in lower transportation costs than direct shipments to customers. Savings are often significantly larger than the increased costs resulting from warehousing and the associated increase in inventory carrying costs.

Warehousing and Production

Short production runs minimize the amount of inventory held throughout the logistics system by producing quantities near to current demand, but they carry increased costs of set-ups and line changes. If a plant is operating near or at capacity, frequent line changes may leave the manufacturer unable to meet product demand. If so, the cost of lost sales – the lost contribution to profit on unrealized sales – could be substantial.

On the other hand, the production of large quantities of product for each line change results in a lower per unit cost on a full-cost basis and more units for a given plant capacity. However, long production runs lead to larger inventories and increased warehouse requirements. Consequently, production cost savings must be balanced with increased logistics costs in order to achieve least total cost.

Traditionally, warehousing was necessary if a company was to take advantage of quantity purchase discounts on raw materials or other products. Not only is the per unit price lower as a result of the discount, but if the company pays the freight, transportation costs will be less on a volume purchase because of transportation economies. Similar discounts and savings can accrue to manufacturers, retailers and wholesalers. Once again, however, these savings must be weighed against the added inventory costs incurred as a result of larger inventories.

Increasingly, companies operating with a JIT manufacturing philosophy are negotiating with their suppliers to receive cumulative quantity discounts. Thus, they receive the lower rate based on total yearly order volume rather than individual order size.

Holding inventories in warehouses may be necessary to maintain a source of supply. For example, the timing and quantity of purchases is important in retaining suppliers, especially during periods of shortages. It may be necessary to hold an inventory of items that are in short supply as a result of damage in transit, vendor stockouts or a strike against one of the company's suppliers.

Warehousing and Customer Service

Customer service policies, such as a 24-hour delivery standard, may require a number of field warehouses in order to minimize total costs while achieving the standard. Changing market conditions may make it necessary to warehouse product in the field, primarily because companies are unable to accurately predict consumer demand and the timing of retailer or wholesaler orders. By keeping some excess inventory in field warehouse locations, companies can respond quickly to meet unexpected demand. In addition, excess inventory allows manufacturers to fill customer orders when shipments to restock the field warehouses arrive late.

Warehousing and Least Total Cost Logistics

The majority of firms utilize warehousing to accomplish least total cost logistics at some prescribed level of customer service, considering the cost and logistics functional area trade-offs discussed in Chapter 1. Factors that influence a firm's warehousing policies include:

- the industry
- the firm's philosophy
- capital availability
- product characteristics such as size, perishability, product lines, substitutability and obsolescence rates
- economic conditions
- competition
- seasonality of demand
- use of just-in-time programmes
- production process in use.

Types of Warehousing

In general, firms have a number of warehousing alternatives. Some companies may market products directly to retail customers (called **direct store delivery**), thereby eliminating warehousing in the field. Mail-order catalogue companies, for example, utilize warehousing only at a point of origin, such as sales headquarters or plant.

Cross-docking

Another alternative is to utilize cross-docking concepts, whereby warehouses serve primarily as 'distribution mixing centres'. Product arrives in bulk and is immediately broken down and mixed in the proper range and quantity of products for customer shipment. In essence, the product never enters the warehouse. This topic will be described more fully in the next section.

Cross-docking is becoming popular among retailers, who can order TL, then remix and immediately ship to individual store locations. Products usually come boxed for individual stores from the supplier's location. For example, a supplier will ticket merchandise, place it on hangers and box it up for individual retail stores to replace items sold. The trailer load of boxes arrives at the retailer's DC where products are cross-docked to trucks for stores. At stores, the boxes are opened and garments are immediately ready to hang on display racks.

Most firms warehouse products at some intermediate point between plant and customers. When a firm decides to store product in the field, it faces two warehousing options: rented facilities, called *public warehousing*, or owned or leased facilities, called *private warehousing*.

Contract Warehousing

Another option exists, termed **contract warehousing**, which is a variation of public warehousing. Contract warehousing is an arrangement between the user and provider of the warehousing service, which in some cases is a third-party logistics service provider (3PL). It has been defined as:

> . . . a long-term mutually beneficial arrangement which provides unique and specially tailored warehousing and logistics services exclusively to one client, where vendor and client share the risks associate with the operation. [There is a] focus on productivity, service and efficiency, not the fee and rate structure itself.[3]

Firms must examine important customer service and financial considerations to choose between public and private warehousing. For example, operating costs for a public warehouse tend to be higher because the warehouse will attempt to operate at a profit; it may also have selling and advertising costs. However, a firm makes no initial investment in facilities when it uses public warehousing. From a customer service perspective, private warehousing can generally provide higher service levels because of its more specialized facilities and equipment, and its better familiarity with the firm's products, customers and markets.

The two options must be examined closely. In some instances, innovative public warehouses can provide higher levels of service owing to their expertise and strong competitive drive to serve the customer.

Public Warehouses

There are many types of *public warehouse*, including: (1) general merchandise warehouses for manufactured goods, (2) refrigerated or cold storage warehouses, (3) bonded warehouses, (4) household goods and furniture warehouses, (5) special commodity warehouses, and (6) bulk storage warehouses. Each type provides users with a broad range of specialized services.

General Merchandise Warehouse

The *general merchandise warehouse* is probably the most common form. It is designed to be used by manufacturers, distributors and customers for storing almost any kind of product.

Refrigerated Warehouses

Refrigerated, or cold storage, warehouses provide a temperature-controlled storage environment. They tend to be used for preserving perishable items such as fruit and vegetables. However, a number of other items (e.g. frozen food products, some pharmaceuticals, and photographic paper and film) require this type of facility.

Bonded Warehouses

Some general merchandise or special commodity warehouses are known as **bonded warehouses**. These warehouses undertake surety bonds from the government and place their premises under the custody of a government agent. Goods such as imported tobacco and alcoholic beverages are stored in this type of warehouse, although the government retains control of the goods until they are distributed to the marketplace. At that time, the importer must pay customs duties to the appropriate government agency. The advantage of the bonded warehouse is that import duties and excise taxes need not be paid until the merchandise is sold, so that the importer has the funds on hand to pay these fees.

Household Goods Warehouses

Household goods warehouses are used for storage of personal property rather than merchandise. The property is typically stored for an extended period as a temporary layover option. Within this category of warehouses, there are several types of storage alternatives. One is the open storage concept. The goods are stored on a cubic-foot basis per month on the open floor of the warehouse. Household goods are typically confined to this type of storage. A second kind of storage is private room or vault storage, where users are provided with a private room or vault to lock in and secure goods. A third kind, container storage, provides users with a container into which they can pack goods. Container storage affords better protection of the product than open storage.

Special Commodity Warehouses

Special commodity warehouses are used for particular agricultural products, such as grains, wool and cotton. Ordinarily, each of these warehouses handles one kind of product and offers special services specific to that product.

Bulk Storage Warehouses

Bulk storage warehouses provide tank storage of liquids, and open or sheltered storage of dry products such as coal, sand and chemicals. These warehouses may provide services such as filling drums from bulk or mixing various types of chemicals with others to produce new compounds or mixtures.

Warehousing Operations: Three Functions

Warehousing serves an important role in a firm's logistics system. In combination with other activities, it provides the firm's customers with an acceptable level of service. The obvious role of warehousing is to store products, but warehousing also provides breakbulk, consolidation and information services. These latter activities emphasize product flow rather than storage.

Fast and efficient movement of large quantities of raw materials, component parts and finished goods through the warehouse, coupled with timely and accurate information about the products being stored, are the goals of every logistics system. These goals have received increasing attention from the top management of many organizations.

Warehousing has three basic functions: movement, storage and information transfer. Recently, the movement function has been receiving the most attention as organizations focus on improving inventory turns and speeding orders from manufacturing to final delivery (see Figure 8.2).

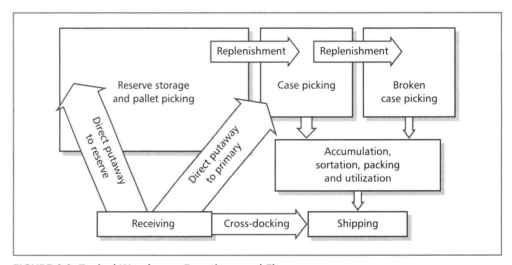

FIGURE 8.2 Typical Warehouse Functions and Flows
Source: James A. Tompkins *et al.*, *Facilities Planning*, 2nd edn. New York: John Wiley, 1996, p. 392.

Movement

The movement function can be further divided into several activities, including:

- receiving
- transfer or putaway
- order picking/selection
- cross-docking
- shipping.[4]

The *receiving* activity includes the actual unloading of products from the transportation carrier, the updating of warehouse inventory records, inspection for damage, and verification of the merchandise count against orders and shipping records.

Transfer or *putaway* involves the physical movement of the product into the warehouse for storage, movement to areas for specialized services such as consolidation, and movement to outbound shipment. Customer *order selection* or *order picking* is the major movement activity and involves regrouping products into the assortments customers desire. Packing slips are made up at this point.

Cross-docking bypasses the storage activity by transferring items directly from the receiving dock to the shipping dock (see Figure 8.3). A pure cross-docking operation would avoid putaway, storage and order picking. Information transfer would become paramount because shipments require close coordination.

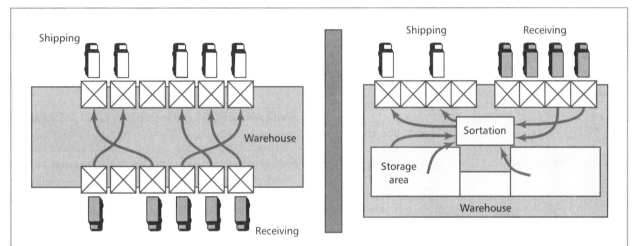

Under a cross-docking system, pallet loads can be moved directly across the warehouse floor from receiving to shipping (left). Boxes, however, must first pass through a sortation system (right).

FIGURE 8.3 Two Examples of Cross-docking
Source: James Aaron Cooke, 'Cross-docking Rediscovered', *Traffic Management* 33, no. 11 (November 1994), p. 51.

Cross-docking has become commonplace in warehousing because of its impact on costs and customer service. For example, approximately 75 per cent of food distribution involves the cross-docking of products from supplier to retail food stores.[5] Eliminating the transfer or putaway of products can reduce costs by as much as 30 per cent at a typical DC,[6] and the time goods remain at the warehouse, thus improving customer service levels.

Cross-docking should be considered as an option by firms meeting two or more of the following criteria:
- inventory destination is known when received
- customer is ready to receive inventory immediately
- shipment to fewer than 200 locations daily
- daily throughput exceeds 2000 cartons
- more than 70 per cent of the inventory is conveyable
- large quantities of individual items received by firm
- inventory arrives at firm's docks pre-labelled
- some inventory is time sensitive
- firm's distribution centre is near capacity
- some of the inventory is pre-priced.[7]

Shipping, the last movement activity, consists of product staging and physically moving the assembled orders onto carrier equipment, adjusting inventory records and checking orders to be shipped. It can consist of sortation and packaging of items for specific customers. Products are placed in boxes, cartons or other containers, placed on pallets or shrinkwrapped (i.e. the process of wrapping products in a plastic film), and are marked with information necessary for shipment, such as origin, destination, shipper, consignee and package contents.

Storage

Storage, the second function of warehousing, can be performed on a temporary or semi-permanent basis. *Temporary storage* emphasizes the movement function of the warehouse and includes only the storage of product necessary for basic inventory replenishment. Temporary storage is required regardless of the actual inventory turnover. The extent of temporary inventory storage depends on the design of the logistics system and the variability experienced in lead-time and demand. A goal of cross-docking is to utilize only the temporary storage function of the warehouse.

Semi-permanent storage is the storage of inventory in excess of that required for normal replenishment. This inventory is referred to as buffer or safety stock. The most common conditions leading to semi-permanent storage are (1) seasonal demand, (2) erratic demand, (3) conditioning of products such as fruits and meats, (4) speculation or forward buying, and (5) special deals such as quantity discounts. Semi-permanent storage also includes products that require ageing to reach their usable potential. For example, Scotch whisky may be stored for 10 years or more before being bottled, while Dutch or French cheeses may be stored to mature from several weeks for soft cheeses to several years for hard cheeses.

Information Transfer

Information transfer, the third major function of warehousing, occurs simultaneously with the movement and storage functions. Management always needs timely and accurate information as it attempts to administer the warehousing activity. Information on inventory levels, throughput levels (i.e. the amount of product moving through the warehouse), stockkeeping locations, inbound and outbound shipments, customer data, facility space utilization and personnel is vital to the successful operation of a warehouse. Organizations are relying increasingly on computerized information transfer utilizing electronic data interchange (EDI) and bar coding to improve both the speed and accuracy of information transfer.

In spite of numerous attempts by firms to reduce the flow of paperwork, the amount of paperwork is still significant. For this reason and many others, management in many firms has attempted to automate the clerical function whenever possible. The developments in electronic communications have been instrumental in reducing the clerical activities in all aspects of warehousing.

Successful completion of all of the warehousing activities already mentioned eliminates the need for *checking*. However, errors and mistakes do occur within any warehouse operation, usually making it necessary to conduct a check of previous activities. In some instances, this activity can be minimized in operations where employees are empowered to perform quality control at their respective levels within the warehouse. This activity may be performed by teams instead of individuals.

It is important to eliminate any inefficiencies in movement, storage and information transfer within the warehouse. These can occur in a variety of forms:

■ redundant or excessive handling
■ poor utilization of floor space and height availability, also known as 'cube'
■ excessive maintenance costs and downtime due to obsolete equipment
■ dated receiving and shipping dock conditions
■ obsolete computerized information handling of routine transactions.

The competitive marketplace demands more precise and accurate handling, storage and retrieval systems, as well as improved packaging and shipping systems. It is vital for a warehouse operation to have the optimal mix of manual and automated handling systems. These issues are presented in more depth in Chapter 9. The next section compares and contrasts warehousing decision criteria.

Decision Criteria for Warehousing Types

One of the most important warehousing decisions a company makes is whether to use public (rented) or private (owned or leased) facilities. Contract warehousing is a variant of public warehousing in which the organization has a contractual relationship to utilize a certain amount of space and services in a facility or facilities over a set time period. This arrangement gives the warehouser more stability and certainty in making investments and planning for the future. Companies may also outsource their warehousing requirement to third party logistics (3PL) service providers who provide storage space on a rented or leased basis. There are three primary criteria affecting this decision: financial criteria, storage space criteria and operating criteria.

Financial Criteria

Use of Capital

One of the major advantages of public (or contracted) warehouses is that they require no capital investment from the user. The user avoids the investment in buildings, land and materials handling equipment, as well as the costs of starting up the operation and hiring and training personnel.

Because of the high costs involved, many firms are simply unable to generate enough capital to build or buy a warehouse. A warehouse is a long-term, often risky, investment (which later may be difficult to sell because of its customized design). The hiring and training of employees, and the purchase of materials handling equipment makes start-up a costly and time-consuming process. And, depending on the nature of the firm, return on investment may be greater if funds are channelled into other profit-generating opportunities.

A further consideration in the decision is the rate of return that the private warehouse alternative will provide. At a minimum, the investment in a corporate-owned warehouse should generate the same rate of return as the firm's other investments. Most companies find it advantageous to use some combination of public and private warehousing. Private warehouses are used to handle the basic inventory levels required for least cost logistics in markets where the volume justifies ownership. Public warehouses are used where volume is insufficient to justify ownership or to store peak requirements.

Storage and Handling Costs

When a company uses a public warehouse, it knows the exact storage and handling costs because it receives a bill each month. The user can therefore forecast costs for different levels of activity because these are known in advance. Firms that operate their own facilities often find it extremely difficult to determine precisely the fixed and variable costs of warehousing.

Public warehouses typically charge on the basis of cases, pallets or hundredweight stored or handled. When the volume of activity is sufficiently large, public warehousing charges exceed the cost of a private facility, making ownership more attractive.

Private warehousing can be less costly over the long term. Operating costs can be 15 to 25 per cent lower if the company achieves sufficient throughput or utilization. The generally accepted industry norm for the utilization rate is 75 to 80 per cent. If a firm cannot achieve at least 75 per cent utilization, it generally would be more appropriate to use public warehousing.

Tax Benefits

A company can also realize tax benefits when it owns its warehouses. Depreciation allowances on buildings and equipment reduce taxes payable.

In some countries, a firm can have an advantage if it does not own property in the country. Ownership means that the firm is doing business in the country and is thus subject to various national *taxes*. These taxes can be substantial. If the company does not currently own property in a country, it may find it advantageous to use a public warehouse.

Some countries do not charge property or value-added taxes (VAT) on inventories in certain types of warehouse or fulfilment centres; this tax shelter applies both to regular warehouse inventories and storage-in-transit inventories. For example, Amazon.co.uk and Tesco.com own outsourced order-fulfilment DCs in the Channel Islands (in the English Channel) for online consumer sales as they are exempt from the normal UK 17.5 per cent VAT.[8]

A **free-port** provision enacted in some countries allows inventory to be held for up to one year, tax-free. The manufacturer pays no real estate tax. The public warehouse pays real estate taxes and includes this cost in its warehouse rates, but the cost is smaller on a per unit throughput basis because the cost is allocated among all of the clients using the public warehouse.

Economies of Scale

Public warehouses are able to achieve economies of scale that may not be possible for some firms. Because public warehouses handle the requirements of a number of firms, their volume allows the employment of a full-time warehousing staff. In addition, building costs are non-linear and a firm pays a premium to build a small facility. Additional economies of scale can be provided by using more expensive, but more efficient, materials handling equipment and by providing administrative and other expertise.

Economies of scale also result from the consolidation of small shipments with those of non-competitors who use the same public warehouse. The public warehouse consolidates orders of specific customers from the products of a number of different manufacturers in a single shipment. This results in lower shipping costs and reduced congestion at the customer's receiving dock. Customers who pick up their orders at the public warehouse are able to obtain the products of several manufacturers in one go, if the manufacturers all use the same facility.

Storage Space Criteria

Space to Meet Peak Requirements

If a firm's operations are subject to seasonality, the public warehouse option allows the user to rent as much storage space as needed to meet peak requirements. A private warehouse, on the other hand, has a constraint on the maximum amount of product that can be stored because it cannot be expanded in the short term. Also, it is likely to be underutilized during a portion of each year. Since most firms experience variations in inventory levels because of seasonality in demand or production, sales promotions or other factors, public warehousing offers the distinct advantage of allowing storage costs to vary directly with volume.

Shortage of Space

Public warehousing space may not be available when and where a firm wants it. Shortages of space do occur periodically in selected markets, which can have an adverse affect on the logistics and marketing strategies of a firm.

Operating Criteria

Control

In private warehousing, the company that owns the goods can exercise a greater degree of control. The firm has direct control of and responsibility for the product until the customer takes possession or delivery, which allows the firm to integrate the warehousing function more easily into its total logistics system.

With warehouse control comes a greater degree of flexibility to design and operate the warehouse to fit the needs of customers and the characteristics of the product. Companies with products requiring special handling or storage may not find public warehousing feasible. The firm must utilize private warehousing or ship the product directly to customers. The warehouse can be modified through expansion or renovation to facilitate product changes, or it can be converted to a manufacturing plant or branch office location.

Flexibility

Another major advantage offered by public warehouses is flexibility. Owning or holding a long-term lease on a warehouse can become a burden if business conditions necessitate changes in location. Public warehouses require only a short-term contract and, thus, short-term commitments. Short-term contracts available from public warehouses make it easy for firms to change field warehouse locations because of changes in the marketplace (e.g. population shifts), the relative cost of various transport modes, volume of a product sold or the company's financial position.

In addition, a firm that uses public warehouses does not have to hire or lay off employees as the volume of business changes. A public warehouse provides the personnel required for extra services when they are necessary, without having to hire them on a full-time basis.

A public warehouse may be very flexible and adaptable in terms of meeting an organization's special requirements or providing the specialized services that are discussed below. However, the space or specialized services desired may not always be available in a specific location. Many public warehouse facilities provide only local service and are of limited use to a firm that distributes regionally or nationally. A manufacturer that wants to use public warehouses for national distribution may find it necessary to deal with several different operators and monitor several contractual agreements.

Many experts feel that the major drawback of private warehousing is the same as one of its main advantages: flexibility. A private warehouse may be too costly because of its fixed size and costs. Regardless of the level of demand the firm experiences, the size of the private warehouse is restricted in the short term. A private facility cannot expand and contract to meet increases or decreases in demand. When demand is low, the firm must still assume the fixed costs as well as the lower productivity linked to unused warehouse space. The disadvantage can be minimized if the firm is able to rent out part of its space.

If a firm uses only private warehouses, it loses flexibility in its strategic location options. If a company cannot adapt to rapid changes in market size, location and preferences it may lose a valuable business opportunity. Customer service and sales could fall if a private warehouse cannot adapt to changes in the firm's product mix.

Time Risk

Companies normally plan for a distribution facility to have a life span of 20 to 40 years. By investing in a private warehouse, management assumes the risk that changes in technology or in the volume of business will make the facility obsolete. With public warehousing, the user firm can switch to another facility in a short period of time, often within 30 days.

Human Resources

The courts have ruled that a labour union does not have the right to picket a public warehouse when the union is involved in a labour dispute with one of the customers of that warehouse. Thus, using a public warehouse has the advantage of insulating the manufacturer's distribution system from labour disputes.

Indeed, by employing private warehousing, a firm can also make better use of its human resources. There is greater care in handling and storage when the firm's own workforce operates the warehouse. Some public warehouses allow their clients to use their own employees in the handling and storage of products. The company can utilize the expertise of its technical specialists.

Communication Problems

Effective communication may be a problem with public warehouses because not all computer terminals and systems are compatible. A warehouse operator may hesitate to add another terminal for only one customer. In addition, the lack of standardization in contractual agreements makes communication regarding contractual obligations difficult.

Specialized Services

Public warehouses can often offer a number of specialized services more economically than a private warehouse. These specialized services include the following:

- broken-case handling, which is breaking down manufacturers' case quantities to enable orders for less-than-full-case quantities to be filled
- packaging of manufacturers' products for shipping
- consolidation of damaged and recalled products for shipment to the manufacturer in carload or truckload quantities; in addition to the documentation and pre-packing that may be necessary, the public warehouse can perform the *reworking* (repair, refurbishing) of damaged product
- equipment maintenance and service
- stock spotting of product for manufacturers with limited or highly seasonal product lines; **stock spotting** involves shipping a consolidated carload of inventory to a public warehouse just prior to a period of maximum seasonal sales
- a breakbulk service whereby the manufacturer combines the orders of different customers in a particular market, and ships them at the carload or truckload rate to the public warehouse; there the individual orders are separated and local delivery is provided.

Intangible Benefits

There may be certain intangible benefits associated with warehouse ownership. When a firm distributes its products through a private warehouse, this can give the customer a sense of permanence and continuity of business operations. The customer sees the company as a stable, dependable and lasting supplier of products. However, customers are more concerned with on-time delivery of products and remote warehousing sites can provide similar service levels if managed properly.

Facility Development

Another important warehousing decision a logistics executive faces is how to develop an optimal warehouse network for the firm's products and customers. Such a decision encompasses a number of significant elements. Management must determine the size and number of warehouses, and ascertain their location. Each warehouse must be laid out and designed properly in order to maximize efficiency and productivity.

The Creative Solutions box describes how SPAR Austria achieved better distribution and thus better service for its retail stores through its decision to build one new distribution centre.

CREATIVE SOLUTIONS

One Warehouse, Many Benefits

The grocery retailer SPAR Group has 1500 stores throughout Austria, of various sizes and types, that present a variety of large, medium and small orders. At the end of the 1990s SPAR Austria was operating six regional warehouses that were inefficient and outdated. Also, out-of-stocks at the retail stores were high and unacceptable.

Rather than build new distribution centres or refurbish existing ones for different categories of products, SPAR Austria chose to build a new 13,020-square-metre facility in Wels, near good road and rail links.

To ensure error-free picking and store sequencing, and to deal with issues between slow- and fast-moving SKUs, SPAR chose the Witron's Dynamic Picking System (DPS) as its warehouse management system (WMS). DPS is based on the Pareto principle, or 80/20 rule, and is applicable to high picking levels and large numbers of SKUs.

The SPAR system has 93,000 bin locations and 12,500 pallet locations in 11 aisles on 18 levels, which are served by five large cranes. Fast-moving SKUs are placed in bins, while slower-moving SKUs are placed in high-bay pallet locations.

The selection of the DPS system has been the key to SPAR's successful implementation of its facility and has brought many benefits for SPAR's retail operations. The WMS is very efficient and accurate: SPAR claims an error rate of only one case in every 27,000.

SPAR ships 24,000 bins in 35 trucks each day; these are tracked by bar codes and the WMS, which ensures proper store sequencing. SPAR is also able to meet last-minute orders. Orders received by 7 pm are shipped by noon the following day.

Lastly, out-of-stock situations at SPAR stores have been reduced by 25 per cent. Thus, by using advanced automated warehouse systems, SPAR Austria has been able to improve its warehouse operations and store order fulfilment in one location.

Question: **How could SPAR use this type of system for Internet or online grocery sales and delivery?**

Source: 'Automating for Efficiency', *Logistics Manager* 11, no. 5 (June 2004), pp. 16–18.

Size and Number of Warehouses

Two issues that must be addressed are the size and number of warehouse facilities. These are interrelated decisions because they typically have an inverse relationship – that is, as the *number* of warehouses increases, the average *size* of a warehouse decreases.

Size of a Warehouse

Many factors influence how large a warehouse should be. First, it is necessary to define how size is measured. In general, size can be defined in terms of square metres or cubic space.

Unfortunately, square-metre measures ignore the capability of modern warehouses to store merchandise vertically; hence, the cubic space measure was developed. Cubic space refers to the total volume of space available *within* a facility. It is a much more realistic size estimate than square footage because it considers more of the available usable space in a warehouse. Some of the most important factors affecting the size of a warehouse are:

- customer service levels
- size of market or markets served
- number of products marketed

- size of the product or products
- materials handling system used
- throughput rate
- production lead-time
- economies of scale
- stock layout
- aisle requirements
- office area in warehouse
- types of racks and shelves used
- level and pattern of demand.

As a company's service levels increase, it typically requires more warehousing space to provide storage for higher levels of inventory. As the market served by a warehouse increases in number or size, additional space is required. When a firm has multiple products or product groupings, especially if they are diverse, it needs larger warehouses to maintain at least minimal inventory levels of all products. In general, greater space requirements are necessary when: products are large; production lead-time is long; manual materials handling systems are used; the warehouse contains office, sales or computer activities; demand is erratic and unpredictable.

To illustrate, consider the relation of warehouse size to the type of materials handling equipment used. As Figure 8.4 shows, the type of forklift truck a warehouse employs can significantly affect the amount of storage area necessary to store product. Because of different capabilities of

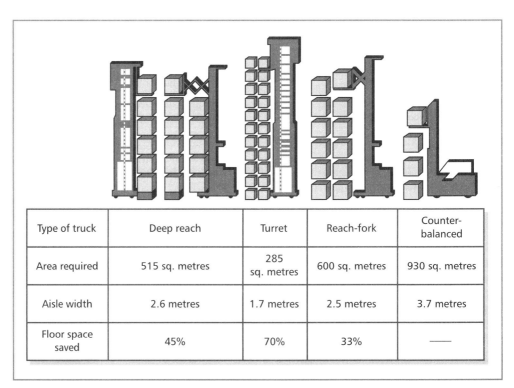

Type of truck	Deep reach	Turret	Reach-fork	Counter-balanced
Area required	515 sq. metres	285 sq. metres	600 sq. metres	930 sq. metres
Aisle width	2.6 metres	1.7 metres	2.5 metres	3.7 metres
Floor space saved	45%	70%	33%	——

FIGURE 8.4 Narrow-aisle Trucks can Reduce Floor Space

Source: James Aaron Cooke, 'When to Choose a Narrow-aisle Lift Truck', *Traffic Management* 28, no. 12 (December 1989), p. 55.

244

forklift trucks, a firm can justify the acquisition of more expensive units when it is able to bring about more effective utilization of space. The four examples in Figure 8.4 show that warehouse layout and warehouse handling systems, one of the topics described in Chapter 9, are interwined.

The simplest type of forklift truck, the counter-balanced truck, requires aisles that are 3 to 4 metres wide. At €25,000, it is the least expensive forklift. The turret truck requires aisles only 5 to 7 feet wide to handle the same amount of product, but it costs €50,000 or more. The warehouse decision-maker must examine the cost trade-offs for each of the available systems, and determine which alternative is most advantageous from a cost–service perspective.

Demand also has an impact on warehouse size. Whenever demand fluctuates significantly or is unpredictable, inventory levels generally must be higher. This results in a need for more space and thus a larger warehouse. All the warehousing space need not be private. Many firms utilize a combination of private and public warehousing. Figure 8.5 shows the relationship between demand and warehouse size.

The hypothetical firm depicted in Figure 8.5 utilizes private warehousing to store 36,000 units of inventory. This results in full utilization of its facilities all year, with the exception of July and August. For months when inventory requirements exceed private warehousing space, the firm rents short-term storage space from one or more public warehouses. In essence, the firm develops private facilities to accommodate a maximum level of inventory of 36,000 units.

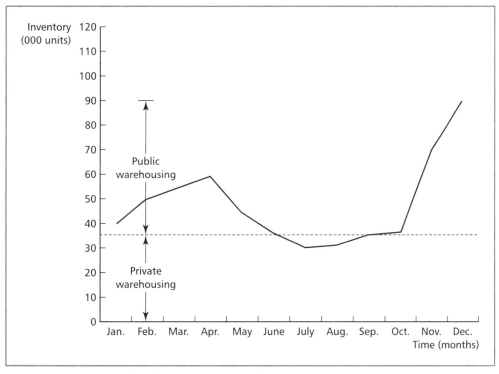

FIGURE 8.5 The Relationship of Demand to Warehouse Size

Number of Warehouses

Four factors are significant in deciding on the number of warehousing facilities: cost of lost sales, inventory costs, warehousing costs, and transportation costs. Figure 8.6 depicts these cost areas (except for cost of lost sales).

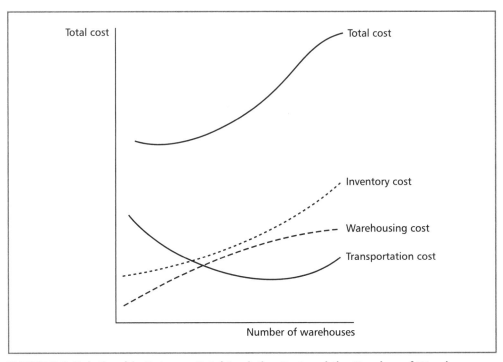

FIGURE 8.6 Relationship Between Total Logistics Cost and the Number of Warehouses

Cost of Lost Sales

Although lost sales are extremely important to a firm, they are the most difficult to calculate and predict, and they vary by company and industry. If the **cost of lost sales** appeared in Figure 8.6, it would generally slope down and to the right. The degree of slope, however, would vary by industry, company, product and customer.

The remaining components of Figure 8.6 are more consistent across firms and industries.

Inventory Costs

Inventory costs increase with the number of facilities because firms usually stock a minimum amount (e.g. safety stock) of all products at every location, although some companies have specific warehouses dedicated to a particular product or product grouping. This means that both slow- and fast-turnover items are stocked, thus more total space is required.

Warehousing Costs

Warehousing costs increase, because more warehouses mean more space to be owned, leased or rented, but they decrease after a number of warehouses are brought online, particularly if the firm leases or rents space. Public and contract warehouses often offer quantity discounts when firms acquire space in multiple locations.

Transportation Costs

Transportation costs initially decline as the number of warehouses increases, but they eventually curve upwards if too many facilities are employed, owing to the combination of inbound and outbound transportation costs. A firm must be concerned with the total delivered cost of its products, not simply the cost of moving products to warehouse locations. In general, the use of fewer facilities means lower inbound transport costs due to bulk shipments from the manufacturer or supplier.

After the number of warehouses increases to a certain point, the firm may not be able to ship its products in such large quantities and may have to pay a higher rate to the transportation carrier. Local transportation costs for delivery of products from warehouses to customers may increase because of minimum charges that apply to local cartage.

If the cost of lost sales is not included, the slopes shown in Figure 8.6, taken together, indicate that fewer warehouses are better than many warehouses. However, customer service is a critical element of a firm's marketing and logistics systems. In general, if the cost of lost sales is very high, a firm may wish to expand its number of warehouses or use scheduled deliveries. There are always cost–service trade-offs. Management must determine the optimal number of warehouses given the desired customer-service level.

Value of Computers

Computers can help minimize the firm's number of warehouses by improving warehouse layout and design, inventory control, shipping and receiving, and the dissemination of information. Coupled with more efficient warehouses, the substitution of information for inventories tends to reduce the number of warehouses needed to service a firm's customers. In essence, the more responsive the logistics system, the less need there is for warehousing.

Location Analysis

Where would be the best place to build a warehouse to service European consumers? Belgium is the best on average across a number of different measures such as land supply and prices, labour costs, transport density and congestion.[9] If a firm wished to locate facilities closest to its potential customers, using one or more warehouses in their logistics network, a number of sites would be possible.

The site-selection decision can be approached from macro and micro perspectives. The macro perspective examines the issue of where to locate warehouses geographically within a general area so as to improve the sourcing of materials and the firm's market offering (improve service and/or reduce cost). The micro perspective examines factors that pinpoint specific locations within the large geographic areas.

Macro Approaches

In one of the best-known macro approaches to warehouse location, Edgar M. Hoover, an American location theorist, identified three types of location strategy: (1) market positioned, (2) production positioned, and (3) intermediately positioned.[10] The **market-positioned strategy** locates warehouses nearest to the final customer. This maximizes customer service levels and enables a firm to utilize transportation economies – TL and CL shipments – from plants or sources to each warehouse location. The factors that influence the placement of warehouses near the market areas served include transportation costs, order cycle time, the sensitivity of the product, order size, local transportation availability and levels of customer service offered.

The **production-positioned strategy** locates warehouses in close proximity to sources of supply or production facilities. These warehouses generally cannot provide the same level of customer service as market-positioned warehouses; instead, they serve as collection points or mixing facilities for products manufactured at a number of different plants.

For multiproduct companies, transportation economies result from consolidation of shipments into TL or CL quantities. The factors that influence the placement of warehouses close to the point of production are perishability of raw materials, number of products in the firm's product mix, assortment of products ordered by customers, and transportation consolidation rates.

The **intermediately positioned strategy** places warehouses at a midpoint location between the final customer and the producer. Customer service levels are typically higher for intermediately positioned warehouses than they are for the production-positioned facilities and lower than for market-positioned facilities. A firm often follows this strategy if it must offer high customer service levels and if it has a varied product offering manufactured at several plant locations.

Another macro approach includes the combined theories of a number of well-known economic geographers. Many of these theories are based on distance and cost considerations. Johan Heinrich von Thünen (1783–1850), a German agriculturalist, called for a strategy of facility location based on cost minimization. Specifically, when locating points of agricultural production, he argued that transportation costs should be minimized to result in maximum profits for farmers. Von Thünen's model assumed that market price and production costs would be identical (or nearly so) for any point of production. Since farmer profits equal market price less production costs and transportation costs, the optimal location would have to be the one that minimized transportation expenditures.

Alfred Weber, a German economist, also developed a model of facility location based on cost minimization. According to Weber, the optimal site was one that 'minimised total transport costs, assuming that these costs varied in direct proportion to the weight of goods transported multiplied by the distance moved (traditionally expressed as *ton mileage*)'.[11] Weber classified raw materials into two categories according to their effect on transportation costs: location and processing characteristics. Location referred to the geographical availability of the raw materials. Few constraints would exist on facility locations for items that had wide availability.

Processing characteristics were concerned with whether the raw material increased, remained the same or decreased in weight as it was processed. If the processed raw material decreased in weight, facilities should be located near the raw material source because transportation costs of finished goods would be less with lower weights. Conversely, if processing resulted in heavier finished goods, facilities should be located closer to the final customers. If processing resulted in no change in weight, a location close to raw material sources or to markets for finished goods would be equivalent.

Other economic geographers included the factors of demand and profitability in the location decision. **Hoover's model** considered both cost and demand elements, and stressed cost minimization in determining an optimal location. In addition, Hoover identified that transportation rates and distance were not linearly related – that is, rates increased with distance, but at a decreasing rate. The tapering of rates over greater distances supported placement of warehouses at the end points of the channel of distribution, rather that at some intermediate location. In that regard, Hoover did not fully agree with Weber's location choices.

Another approach, the **centre-of-gravity approach**, is simplistic in scope, and locates facilities based on transportation costs. This approach locates a warehouse or distribution centre at a point that minimizes transportation costs for products moving between a manufacturing plant and the markets.

Envision two pieces of rope tied together with a knot and stretched across a circular piece of board, with unequal weights attached to each end of the rope. Initially, the knot would be located in the centre of the circle. On the release of the weights, the rope would shift to the point where the weights would be in balance. Adding ropes with varying weights would result in the same shifting of the knot (assuming the knots were all in the same place). If the weights represented transportation costs, then the position where the knot would come to rest after releasing the weights would represent the centre of gravity, or position where transportation costs would be minimized. The approach provides general answers to the warehouse location problem, but it must be modified to take into account factors such as geography, time and customer service levels.

Micro Approaches

From a micro perspective, more specific site-selection factors must be examined. If a firm wants to use private warehousing, it must consider:
- quality and variety of transportation carriers serving the site
- quality and quantity of available labour
- labour rates
- cost and quality of industrial land
- potential for expansion
- tax structure
- building codes
- nature of the community environment
- costs of construction
- cost and availability of utilities
- cost of money locally
- local government tax allowances and inducements to build.

If a firm wants to use public warehousing, it will be necessary to consider:
- facility characteristics
- warehouse services
- availability and proximity to motor carrier terminals
- availability of local cartage
- other companies using the facility
- availability of computer services and communications
- type and frequency of inventory reports.

The site-selection process is interactive, progressing from the general to the specific. It may be formalized or informal, centralized at the corporate level, decentralized at the divisional or functional level, or some combination of each. It is important that management follow some type of logical process that recognizes many trade-offs when making a location decision.

Many non-quantitative and political factors may take on great importance in the warehouse locating decision. For example, several European countries do not allow high-bay warehouses near certain areas as they are considered to be visual intrusions.

Related to the location of facilities is the decision to design an optimal structure that maximizes efficiency and effectiveness. This is the warehouse layout and design decision.

Warehouse Layout and Design

Where should products/materials be located in the logistics system and, more particularly, within the warehouse? With an average warehouse containing about 22,000 stockkeeping units (SKUs), this consideration has a critical effect on system efficiency and productivity.[12] A good warehouse layout can (1) increase output, (2) improve product flow, (3) reduce costs, (4) improve service to customers, and (5) provide better employee working conditions.[13]

The optimal warehouse layout and design for a firm will vary by the type of product being stored, the company's financial resources, the competitive environment and the needs of customers. In addition, the warehouse manager must consider cost trade-offs between labour, equipment, space and information.

For example (see Figure 8.4), the purchase of more expensive, yet more efficient, materials handling equipment can affect the optimal size of a warehouse facility. Installation of an expensive conveyor system to reduce labour costs and raise productivity can affect the configuration of a warehouse. Considering all of the possible factors and their combinations, it is imperative that the firm uses a logical and consistent decision strategy to develop an optimal warehousing system for itself. Whatever layout the company finally selects for its warehouse, it is vital that all available space be utilized as fully and efficiently as possible.

Randomized Storage

Randomized and dedicated storage are two examples of how products can be located and arranged. **Randomized storage**, or **floating slot storage**, places items in the closest available slot, bin or rack. The items are retrieved on a first-in, first-out (FIFO) basis. This approach maximizes space utilization, although it necessitates longer travel times between order-picking locations. Randomized systems often employ a computerized automatic storage and retrieval system (AS/RS), which minimizes labour and handling costs.

Dedicated Storage

Another example is **dedicated storage**, or **fixed-slot storage**. In this approach, products are stored in permanent locations within the warehouse. This tends to be common in manual labour situations where employee performance improves as employees learn each product's location. Three methods can be used to implement the dedicated storage approach, including storing items by (1) part number sequence, (2) usage rates, or (3) activity levels (e.g. grouping products into classes, or families, based on how fast products move in and out of storage).

In terms of overall warehouse layout, products may be grouped according to their compatibility, complementarity or popularity. **Compatibility** refers to whether products can be stored together harmoniously. For example, many years ago, before the development of newer paints, it was discovered that automobile tyres and consumer appliances could not be stored together. Apparently, chemical vapours given off by the tyres reacted with the pigments in the appliance paint, resulting in slight colour changes. Appliances had to be repainted or sold at a discount.

Complementarity refers to how often products are ordered together and therefore stored together. Computer disk drives and monitors, pens and pencils, and desks and chairs are examples of complementary products that are usually stored in close proximity.

Popularity relates to the different inventory turnover rates or demand rates of products. Another term used for this turnover rate is *velocity*. Items that are in greatest demand should be stored closest to shipping and receiving docks. Slow-moving items should be stored elsewhere. In a food wholesaler's warehouse, for example, non-refrigerated basic food items are stored close to the outbound shipping area, whereas slow movers are located in more remote areas of the warehouse.

Using the computer, it is possible to group products within a warehouse, so that the following objectives are met.

■ Fast movers are placed nearest the outbound truck docks. This minimizes the distances travelled daily by materials handling equipment.
■ Slow movers are located at points furthest from outbound shipping docks. This ensures that lengthy horizontal moves by materials handling equipment are minimized.
■ Remaining areas in the warehouse are reserved for products received in periodic batches, those requiring rework before shipping, those that are compatible with fast-moving products, and back-up overflow from fast-moving areas.

- Aisles are redesigned to facilitate the most efficient flow of products to and from dock areas.
- Storage areas are configured to match the velocity and dimensions of each major product, rather than designing all storage bins, racks and floor storage areas in the same dimensions. This facilitates the maximum use of available cubic space, because products are not only matched to the width of each storage slot, but also to the depth and height of each slot.

The entire area of facilities development – size and number of warehouses, location analysis, warehouse layout and design – is an important, yet complex, part of warehouse management. In recent years, computers have played a much more significant role as logistics executives attempt to optimize warehouse operations. The Technology box shows how computer modelling and simulation can assist facilities development in an era of high costs.

TECHNOLOGY

Simulating a Warehouse

The high cost of warehouse construction and operations has led to a wide range of logistics software packages being developed to design warehouses and select locations. However, one aspect that is not as prominent is computer modelling and simulation. Modelling allows a company to perform sensitivity analyses or 'what-if' scenarios on various warehouse criteria in order to determine feasible solutions to the problem at hand.

One logistics service company, Unipart, was able to win a tender to provide a warehouse solution for a manufacturer by doing just that. Unipart is 'one of Europe's leading providers of outsourced, aftermarket logistics and distribution services. It specializes in managing complete supply chains from manufacturers through to retail distributors, and seeks to add value at every level of the chain'.

Unipart modelled an optimum warehouse design by creating a three-dimensional, animated computer presentation. This use of technology captured important client issues and presented them in dramatic visual scenarios. The client was confident enough in Unipart's abilities to choose it as its warehouse designer.

Once the tender was won, Unipart continued to use modelling to develop richer simulations that provided numerous 'what-if' scenarios for testing different options for racking, picking, packing and transportation.

Simulation of warehousing operations can also be used for risk management, as well as investigating proposed changes in the ongoing management and development of existing and new facilities.

Question: Where else in its logistics systems could Unipart use computer simulation and modelling?

Source: Robin Vega, 'Thinking Outside the Box', *Logistics & Transport Focus* 6, no. 2 (March 2004), pp. 16–21.

Warehouse Productivity Measurement

To obtain maximum logistics efficiency, each component of the logistics system must operate at optimal levels. This means that high levels of productivity must be achieved, especially in the warehousing area. Productivity gains in warehousing are important to the firm in terms of reduced costs and to its customers in terms of improved customer service levels (see the Global box).

GLOBAL

Warehouse Outsourcing at Michelin

TNT Logistics (TNT) is a division of TPG, a global leader in logistics, mail and express operating in 39 countries, and based in the Netherlands. In 2002, TNT assumed operations of Michelin Group's entire North American network of tyre distribution centres, totalling 650,000 square metres in the United States and Canada. Michelin is a leading global manufacturer of tyres for the automotive, trucking and other industries.

The contract was the latest in TNT's relationship with Michelin, which began in 1995 in Europe. Since that time TNT has expanded its relationship by providing a combination of warehousing, inbound and outbound distribution services in Germany, Italy, France, Turkey, Malaysia, Australia and North America.

Michelin turned to TNT's expertise in warehouse management as part of an initiative to reduce annual operational costs by €154 million, as well as become more cost competitive and improve services to its customers.

TNT assumed the operation of 12 Michelin distribution centres in the United States and six in Canada. TNT also transitioned nearly 650 Michelin employees to its operations. In addition, TNT's internal logistics design team redesigned the distribution network and optimized material flow throughout the United States and Canada, resulting in greater efficiencies and cost savings in outbound distribution to Michelin's retail outlets throughout North America. Going forward, the plan calls for the implementation of three 'super distribution centres' to further improve distribution efficiency in North America.

Within the first year of its operation, TNT initiated the phased implementation of a new warehouse management system that allows for more efficient processes and increased visibility to replace Michelin's legacy system. After one year of operation, all Michelin distribution centres were declared 'green', meaning that they were meeting and exceeding operational goals set by Michelin. Efficiency in distribution to retail outlets has also increased by a significant percentage.

Question: **What are the control versus efficiency trade-offs in outsourcing logistics networks?**

Source: 'Case Study: Michelin Tyres', *TNT Logistics*, http://www.tntlogistics.com/en/sectors/case_studies/na_michelin_distribution.asp (2005).

Productivity has been defined in many ways, but most definitions include the notions of real outputs and real inputs, utilization and warehouse performance. One study defined these elements as follows.

- *Productivity is the ratio of real output to real input*. Examples are cases handled per labour-hour and lines selected per equipment-hour.
- *Utilization is the ratio of capacity used to available capacity*. Examples are the percentage of pallet spaces filled in a warehouse and employee-hours worked versus employee-hours available.
- *Performance is the ratio of actual output to standard output (or standard hours earned to actual hours)*. Examples are cases picked per hour versus standard rate planned per hour, and actual return on assets employed versus budgeted return on assets employed.[14]

Any working definition of productivity probably includes all three components because they are interrelated. Most firms utilize a variety of measures. Firms tend to use more sophisticated productivity measures over time.

A multitude of warehouse productivity measures are used, although they can be grouped into major categories such as labour cost per unit handled, amount of space needed to store each unit, and frequency of errors. Performance data must be available and used as the basis for corrective action and proactive improvement.

The general management notion that 'you can't manage what you don't measure' is an important warehousing performance concept. Some of the most important areas of measurement that highlight problems or opportunities include customer service (e.g. shipping performance, error rates, order cycle time), inventory accuracy (e.g. the quantity of each SKU is correct at all warehouse locations), space utilization (e.g. having the right inventory, square metre or cube utilization of facilities), and labour productivity (e.g. throughput rates).

It is not enough merely to identify problem areas; rather, it is vital that the firm take appropriate action to improve poor performance whenever possible. A company should develop decision strategies to handle most problem areas before the problems develop. This is the essence of contingency planning. Once issues are pinpointed, the firm can institute various controls or corrective actions to improve warehouse productivity.

Improving Warehouse Productivity

Because warehousing is such a significant component of the logistics process in terms of its cost and service impacts, logistics executives are acutely aware of the need to improve warehouse productivity. Productivity can be improved in many ways, including methods-related, equipment-related, systems-related and training/motivation-related programmes.[15]

Methods-related Programmes

Methods-related programmes consider alternative processes for achieving desired results. They include those involving warehouse cube utilization, warehouse layout and design, methods and procedures analysis, batch picking of small orders, combined putaway/picking, wrap packaging, inventory cycle counting, product line obsolescence, standardized packaging and warehouse consolidation.

Equipment-related Programmes

Equipment-related programmes include the use of new technology such as optical scanners, automatic labelling devices, computer-generated putaway and pick lists, automated materials handling equipment, communications devices, computers and automated storage/retrieval systems (AS/RSs), carousels and conveyors.

Systems-related Programmes

Systems-related programmes include the use of router/location systems, geographic or zone picking, and random location of products in the warehouse. These are systems related because they directly affect the way that different components of the logistics system interact.

Training/Motivation-related Programmes

Training/motivation-related programmes include employee training, management development programmes, work teams, incentive systems and awards recognition. These programmes can improve warehouse productivity by empowering those closest to the activity to make improvements in operations.

The preceding approaches can be implemented individually or in combination. Most firms utilize several methods simultaneously to improve warehouse productivity.

Financial Dimensions of Warehousing

Financial control of warehousing is closely tied to logistics productivity and corporate profitability.[16] Before the various activities of warehousing can be properly integrated into a single unified system, management must be aware of the risks and costs of each activity.

Many warehouse decisions involve risk. The risks can be of many types, but all will eventually result in some impact on costs or revenues. For example, making a capital investment in automated storage and retrieval systems increases both risk and the level of expected return on investment.[17] Firms must be able to justify such investments financially. The faster the cost of the equipment can be recovered, the less risk associated with the decision. Financial accounting and control techniques are very important in assessing the risks and rewards associated with warehousing decisions.

Activity-based Costing

One approach that has proven successful in the financial control of warehousing activities is *activity-based costing (ABC)*. Accurate and timely financial data allow warehouse executives to properly plan, administer and control warehousing activities. Traditional costing systems, in place at many firms, often do not provide financial data in the proper form for use in making warehousing decisions. Frequently, it is difficult to identify how warehousing costs impact overall corporate profitability and how changes in costs in one area affect costs in another. Some companies are implementing ABC in order to have better warehousing cost information.

With ABC, costs are determined by specific products, services or customers (see Chapter 11). It utilizes a two-stage process. The first stage assigns resource costs according to the amount of each resource consumed in performing specific warehousing activities. The second stage assigns warehousing activity costs to the products, services or customers consuming the activities. Proponents of ABC state that it unbundles traditional cost accounts and shows how resources are consumed.[18]

Levels of Sophistication in Warehouse Accounting and Control

Companies are often at various levels of sophistication in terms of warehouse accounting and control. Four levels have been identified, as follows.

Level I: warehouse costs are allocated in total, using a single allocation base.

Level II: warehouse costs are aggregated by major warehouse function (e.g. handling, storage and administration) and are assigned using a separate allocation base for each function.

Level III: warehouse costs are aggregated by major activity within each function (e.g. receiving, putaway, order pick) and are allocated using a separate base for each activity.

Level IV: costs are categorized in matrix form, reflecting each major activity, natural expense and type of cost behaviour; separate allocations are developed for each cost category, using bases that reflect the key differences in warehousing characteristics among cost objectives.[19]

Accounting and control require having the right kind of financial data available when and where they are needed, and in a form that is usable by as many functional areas of the firm as possible. Ultimately, these data are essential to making the necessary cost–service trade-offs within the warehousing activity and between other logistics functions.

In this chapter, we described the importance of warehousing and distribution in the logistics system. Economies of scale, costs and customer service are the most important considerations. The types of option available to a firm include public or rented and privately owned or leased warehousing.

The major functions of warehousing are movement, storage and information transfer. Firms may choose to perform these functions utilizing public or private warehousing. Each option has advantages and disadvantages, which must be understood so that optimal warehousing decisions are made.

Facility development is a large part of warehouse management. Decisions relating to the size and number of warehouses, the location of the facilities, and layout and design have significant impact on a firm's ability to satisfy its customers and make a profit. We described various methods, techniques and approaches relative to each decision area. This led us to explore some important management issues relating to warehouse productivity, accounting and control.

In the next chapter, we will examine the issues of materials handling and computerization in warehousing, packaging and reverse logistics. With a knowledge of key warehousing decisions, it will be possible to understand more fully the role of these parameters in the logistics system.

SUMMARY

KEY TERMS

A full Glossary can be found at the back of the book.

QUESTIONS AND PROBLEMS

1 Warehousing is used for the storage of inventories during all phases of the logistics process. Since inventory carrying costs can be so high, why is it necessary for a firm to store inventories of any kind?

2 What is meant by a cost trade-off analysis within the context of warehousing? Give two examples of the cost trade-offs involved in a firm's decision to use a combination of public and private warehousing rather than public *or* private warehousing alone.

3 What are the three basic functions of warehousing? Briefly describe each.

4 Identify and describe some of the more important factors that affect the specific size of a firm's warehouse or warehouses.

5 What are the differences between the following types of facility location strategies: (a) market positioned, (b) production positioned, and (c) intermediately positioned?

6 How can layout and design affect warehouse efficiency and productivity?

7 Productivity has been defined as the ratio of real output to real input. In terms of the warehousing function, how could a firm measure the productivity level of its storage facilities?

8 Discuss the reasoning behind the following statement: 'Financial control of warehousing is closely tied to logistics productivity and corporate profitability.'

THE LOGISTICS CHALLENGE!

PROBLEM: HOME DELIVERY FULFILMENT OF ONLINE GROCERIES

Le Shop.ch was the first online grocery service in Switzerland. The service is targeted to working people, especially women, who have to manage both family and work, to make their weekly household shopping as easy as possible. The company started with seven employees in 1998 and in its first year of operations generated revenues of 4 million Swiss francs.

According to Christian Wanner, Le Shop's head of marketing, 'Logistics are crucial for an online shop to be successful, yet they are often overlooked. Instead of trying to compete on price as many others are doing, we've implemented a new model based on speed and reliability.' The logistics for the entire distribution were handed over to the Swiss Post Office, which is a partner in the venture.

Not only does the Post Office deliver the parcels, it also packs them. Le Shop did not own a warehouse, nor did it manage its stock. Once online orders are logged at Le Shop's headquarters, they are validated against a customer's purchasing record, then transmitted electronically to a centre in Lausanne where postal employees fill yellow plastic boxes and put them directly into the traditional postal circuit, which is just a floor away.

Despite downsizing, the Swiss Post Office still remains one of the fastest and most reliable postal services around. If an order is passed before 4.30 pm, the products will be delivered the next morning. If the customer is not at home, he or she can specify another address for delivery – that of a neighbour or friend, for example. Customers are charged 10 Swiss francs for this service.

However, Le Shop's business grew substantially and, by 2001, it had 40 employees and revenues of 11.5 million Swiss francs. It could no longer remain removed from distribution. LeShop.ch determined that it needed its own home delivery fulfilment centre to meet this growth. The challenge was how to set up and operate such a facility.

One option was to do everything itself, including delivery, from one central location. Another option was to outsource the entire process to a large 3PL such as TNT or Exel Logistics. A third option was to continue to partner with the Swiss Post Office and add a transport 3PL to distribute deliveries from a large central DC to Swiss Post Office's Express Post postal distribution centres. Express Post would then handle final delivery to the customer's home.

How should Le Shop evaluate these options and come to a decision?

What Is Your Solution?

Source: 'Design with the Home in Mind', *Distribution Business* 15, no. 8 (October 2002), p. 40.

SUGGESTED READING

BOOKS

Frazelle, Edward H., *World-class Warehousing and Materials Handling*. New York: McGraw-Hill, 2002.

McKinnon, Alan C., *Physical Distribution Systems*. London: Routledge, 1989.

Stock, James R., 'Strategic Warehousing: Bringing the "Storage Game" to Life', *WERC Research Paper*. Oak Brook, IL: Warehousing Education and Research Council, 1988.

Tompkins, James A. and Dale Harmelink (eds), *The Distribution Management Handbook*. New York: McGraw-Hill, 1994.

JOURNALS

Ackerman, Kenneth B., 'The Deming Management Message: It Can Work in Your Warehouse!', *Warehousing Forum* 11, no. 4 (March 1996), pp. 1–2.

Ackerman, Kenneth B., 'Leadership in the 21st Century Warehouse', *Warehousing Forum* 7, no. 6 (May 1992), pp. 1–3.

Ballou, Ronald H. and James M. Masters, 'Commercial Software for Locating Warehouses and Other Facilities', *Journal of Business Logistics* 14, no. 2 (1993), pp. 71–107.

Bancroft, Tony, 'Strategic Role of the Distribution Center: How to Turn Your Warehouse into a DC', *International Journal of Physical Distribution and Logistics Management* 21, no. 4 (1991), pp. 45–7.

Copacino, William C., 'How Warehousing Provides a Competitive Edge', *Warehousing Forum* 6, no. 10 (September 1991), pp. 1–4.

De Koster, René, 'Recent Developments in Warehousing', paper presented at the European Logistics Association Educators' Conference, Lisbon, Portugal, September 1998.

Gooley, Toby B., 'ISO 9000 is Coming!', *Traffic Management* 34, no. 10 (October 1995), Warehousing and Distribution Supplement, pp. 77–80.

Ho, Peng-Kuan and Jossef Perl, 'Warehouse Location under Service-sensitive Demand', *Journal of Business Logistics* 16, no. 1 (1995), pp. 133–62.

Lin, Binshan, James Collins and Robert K. Su, 'Supply Chain Costing: An Activity-based Perspective', *International Journal of Physical Distribution and Logistics Management* 31, no. 10 (2001), pp. 702–13.

Maltz, Arnold B., 'The Relative Importance of Cost and Quality in the Outsourcing of Warehousing', *Journal of Business Logistics* 15, no. 2 (1994), pp. 45–62.

Murphy, Paul R. and Richard F. Poist, 'In Search of Warehousing Excellence: A Multivariate Analysis of HRM Practices', *Journal of Business Logistics* 14, no. 2 (1993), pp. 145–64.

Rogers, Dale S. and Patricia J. Daugherty, 'Warehousing Firms: The Impact of Alliance Involvement', *Journal of Business Logistics* 16, no. 2 (1995), pp. 249–69.

Stock, James R., 'Managing Computer, Communication and Information Technology Strategically: Opportunities and Challenges for Warehousing', *Logistics and Transportation Review* 26, no. 2 (June 1990), pp. 32–54.

Van Oudheusden, Dirk L. and Peter Boey, 'Design of an Automated Warehouse for Air Cargo: The Case of the Thai Air Cargo Terminal', *Journal of Business Logistics* 15, no. 1 (1994), pp. 261–85.

'Warehousing: Coping with the Challenge of Change', *Modern Materials Handling* 50, no. 6 (May 1995), pp. 12–13.

REFERENCES

[1] Richard L. Dawe, 'Reengineer Warehousing', *Transportation and Distribution* 36, no. 1 (January 1995), p. 102.

[2] Kenneth B. Ackerman, 'Push Versus Pull', *Warehousing Forum* 11, no. 7 (June 1996), p. 3.

[3] Kenneth B. Ackerman, 'Contract Warehousing – Better Mousetrap, or Smoke and Mirrors?', *Warehousing Forum* 8, no. 9 (August 1993), p. 1.

[4] Edward H. Frazelle, *World-class Warehousing and Material Handling*. New York: McGraw-Hill, 2002, pp. 8–11.

[5] 'Grocery Warehouses Turn to Cross-Docking', *Traffic Management* 34, no. 2 (February 1995), p. 77-S.

[6] Phil Whiteoak, 'Rethinking Efficient Replenishment in the Grocery Sector', in *Logistics and Retail Management*, ed. John Fernie and Leigh Sparks. London: Kogan Page, 2004, p. 144.

[7] 'Receiving Is Where Efficiency Starts', *Modern Materials Handling* 50, no. 5 (March–April 1995), p. 9.

[8] Peter Rowlands, 'You Can Save if You do it in Jersey (But Don't Tell Anyone)', *fulfilment & e.logistics* 35, (February 2005), pp. 16–18.

[9] Cushman & Wakefield Healey & Baker, *European Distribution Report*. London, 2003.

[10] Edgar M. Hoover, *The Location of Economic Activity*. New York: McGraw-Hill, 1948, p. 11.

[11] Alan C. McKinnon, *Physical Distribution Systems*. London: Routledge, 1989, p. 129.

[12] Philippe R. Hebert, 'Manage Inventory? Better Find it First!', *Transportation and Distribution*, July 1995, Buyers' Guide Issue, p. 8.

[13] Greg Owens and Robert Mann, 'Materials Handling System Design', in *The Distribution Handbook*, ed. James F. Robeson and William C. Copacino. New York: Free Press, 1994, pp. 519–45.

[14] A.T. Kearney, Inc., *Measuring and Improving Productivity in Physical Distribution*. Oak Brook, IL: National Council of Physical Distribution Management, 1984, p. 188; see also Douglas Lambert, 'Logistics Cost, Productivity, and Performance Analysis', in *The Logistics Handbook*, ed. James F. Robeson and William C. Copacino. New York: Free Press, 1994, pp. 289–91.

[15] This material was paraphrased from A.T. Kearney, Inc., *Measuring and Improving Productivity in Physical Distribution*, pp. 227–34. Used with permission of the Council of Supply Chain Management Professionals, the successor to the Council of Logistics Management and the National Council of Physical Distribution Management.

[16] Morton T. Yeomans, 'Using Warehouse Information', in *The Logistics Handbook*, ed. James F. Robeson and William C. Copacino. New York: Free Press, 1994, pp. 632–43.

[17] James D. Krasner, 'Satisfying the Chief Financial Officer', *Warehousing Forum* 10, no. 7 (June 1995), pp. 1–2.

[18] Terrance L. Pohlen, 'Activity Based Costing for Warehouse Managers', *Warehousing Forum* 9, no. 5 (May 1994), pp. 1–3; and Binshan Lin, James Collins and Robert K. Su, 'Supply Chain Costing: An Activity-based Perspective', *International Journal of Physical Distribution and Logistics Management* 31, no. 10 (2001), pp. 702–13.

[19] Ernst and Whinney, *Warehouse Accounting and Control: Guidelines for Distribution and Financial Managers*. Oak Brook, IL: National Council of Physical Distribution Management, 1985, p. 50. Used with permission of the Council of Supply Chain Management Professionals, the successor to the Council of Logistics Management and the National Council of Physical Distribution Management.

CHAPTER 9
MATERIALS HANDLING, PACKAGING AND REVERSE LOGISTICS

OBJECTIVES

CHAPTER OBJECTIVES

■ To provide an overview of the various types of automated and non-automated materials handling systems

■ To examine the role of warehousing in a just-in-time (JIT) environment

■ To demonstrate the important role of computer technology in materials management

■ To identify the role of packaging in the warehouse operation

■ To examine the importance of reverse logistics for warehousing and other logistical functions

INTRODUCTION

If one views warehousing as a means of achieving a competitive advantage, traditional perspectives of warehousing as merely storing and managing inventories are replaced by new paradigms that include information flows as well as inventories. In the new mission of warehousing, computerization, information and automation become essential ingredients in logistics success.

This chapter integrates some of the key components that affect warehousing decisions. Some companies tend to view the warehousing decisions described in Chapter 8 (private versus public, location, size and number) as separate from decisions about warehouse automation, materials handling systems, warehouse computerization, packaging and reverse logistics. Yet, these decisions are very intertwined, as demonstrated in this chapter.

Materials Handling Equipment

The Materials Handling Institute, a US industry trade association for manufacturers of materials handling equipment and systems, has estimated that:

> The hardware and software used to move, store, control, contain, and unitize materials in factories and warehouses exceeds $50 billion US annually. Much of the growth in size and variety of the market is fueled by major changes in the requirements of warehouse and distribution operations (e.g. reduced order cycle times, reduced inventory levels, reduced order sizes, SKU proliferation).[1]

Materials handling equipment and systems often represent a major capital outlay for an organization. Like the decisions related to the number, size and location of warehouses, materials handling can affect many aspects of the firm's operations.

Manual or Non-automated Materials Handling Systems

Manual or non-automated materials handling equipment has been the mainstay of the traditional warehouse and will likely continue to be important even with the move towards automated warehousing. Such equipment can be categorized according to the functions performed – that is, storage and order picking, transportation and sorting, and shipping.

Storage and Order-picking Equipment

Storage and order-picking equipment includes racks, shelving, drawers and operator-controlled devices (e.g. forklift trucks). Manual systems provide a great deal of flexibility in order picking, because they use the most flexible handling system (i.e. people).

Table 9.1 explains the many types of racks, shelving and drawers most often used in a warehouse. Storage racks normally store palletized or unitized loads. In most instances, some type of operator-controlled device places the load into the storage rack. Table 9.1 presents the types of material stored, the benefits, and other information about each item. Figures 9.1 to 9.5 illustrate what these items look like.

Equipment	Types of material	Benefits	Other considerations
Racking: conventional pallet racks	Pallet loads	Good storage density, good product security	Storage density can be increased further by strong loads two deep
Drive in racks	Pallet loads	Fork trucks can access loads, good storage density	Fork truck access is from one direction only
Drive-through racks	Pallet loads	Same as above	Fork truck access is from two directions
High-rise racks	Pallet loads	Very high storage density	Often used in AS/R systems, may offer tax advantages when used in rack-supported building
Cantilever racks	Long loads or rolls	Designed to store difficult shapes	Each different SKU can be stored on a separate shelf
Pallet stacking frames	Odd-shaped or crushable parts	Allow otherwise unstackable loads to be stacked, saving floor space	Can be disassembled when not in use
Stacking racks	Odd-shaped or crushable parts	Same as above	Can be stacked flat when not in use
Gravity-flow racks	Unit loads gravity moves loads	High-density storage,	FIFO or LIFO flow of loads
Shelving	Small, loose loads and cases	Inexpensive	Can be combined with drawers for flexibility
Drawers	Small parts and tools	All parts are easily accessed, good security	Can be compartmentalized for many SKUs
Mobile racking or shelving	Pallet loads, loose materials and cases	Can reduce required floor space by half	Come equipped with safety devices

(The equipment rows above are bracketed by the vertical label **MANUAL**.)

TABLE 9.1 Storage Guidelines for the Warehouse

	Equipment	Types of material	Benefits	Other considerations
A U T O M A T E D	Unit load AS/RS	Pallet loads, and a wide variety of sizes and shapes	Very high storage density computer controlled	May offer tax advantages when rack-supported
	Car-in-lane	Pallet loads, other unit loads	High storage density	Best used where there are large quantities of only a few SKUs
	Miniload AS/RS	Small parts	High storage density, computer controlled	For flexibility, can be installed in several different configurations
	Horizontal carousels	Small parts	Easy access to parts, relatively inexpensive	Can be stacked on top of each other
	Vertical carousels	Small parts and tools	High storage density	Can serve dual role as storage and delivery system in multifloor facilities
	Man-ride machines	Small parts	Very flexible	Can be used with high-rise shelving or modular drawers

TABLE 9.1 Continued

This table is a general guide to the types of available storage equipment and where each is best used in the warehouse. Each individual storage application should be studied in detail with the equipment supplier before any equipment is specified.

Source: 'Storage Equipment for the Warehouse', *Modern Materials Handling*, *1985 Warehousing Guidebook* 40, no. 4. (Spring 1985), p. 53. *Modern Materials Handling*, copyright 1985 by Cahners Publishing Company, Division of Reed Holdings, Inc.

The storage racks illustrated in Figure 9.1 are found in most warehouse facilities as either permanent or temporary fixtures for storage of products. They would be considered 'standard' or 'basic' components of a warehouse. All these storage racks are easily accessible by materials handling equipment such as forklift trucks.

Gravity flow storage racks (see Figure 9.2) are often used to store high-demand items. Products that are of uniform size and shape are well suited to this type of storage system. Items are loaded into the racks from the back, flow to the front of the racks, which are sloped forwards, and are then picked from the front of the system by order-picking personnel.

For small parts, bin shelving systems are useful. Figure 9.3 illustrates a typical bin configuration. Items are handpicked, so the height of the system must be within the physical reach of employees. Typically, the full cube of each bin cannot be used, so some wasted space exists. Bin shelving systems are relatively inexpensive compared with other storage systems, but they have limited usefulness beyond storage of small parts.

The modular storage drawers and cabinets shown in Figure 9.4 are used for small parts. Similar in function to bin shelving systems, they require less physical space and allow items to be concentrated into areas that are easily accessed by employees. The drawers are pulled out and items are selected. Fasteners, nuts and bolts, and other small parts and components, are often stored in this manner. By design, modular storage drawers must be low to the floor and often less than five feet in height to allow access by employees picking items from the drawers.

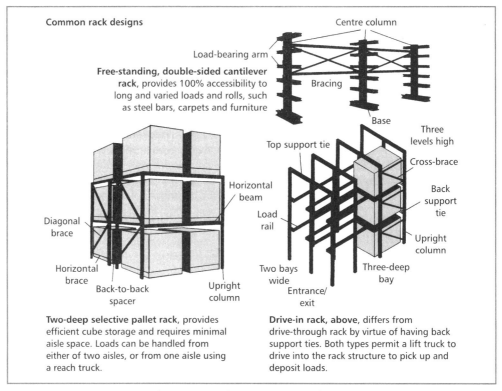

Common rack designs

Centre column

Load-bearing arm

Free-standing, double-sided cantilever rack, provides 100% accessibility to long and varied loads and rolls, such as steel bars, carpets and furniture

Bracing

Base

Top support tie

Three levels high

Cross-brace

Horizontal beam

Back support tie

Load rail

Diagonal brace

Upright column

Horizontal brace

Back-to-back spacer

Upright column

Two bays wide

Three-deep bay

Entrance/ exit

Two-deep selective pallet rack, provides efficient cube storage and requires minimal aisle space. Loads can be handled from either of two aisles, or from one aisle using a reach truck.

Drive-in rack, above, differs from drive-through rack by virtue of having back support ties. Both types permit a lift truck to drive into the rack structure to pick up and deposit loads.

FIGURE 9.1 Non-automated Storage Units – Storage Racks

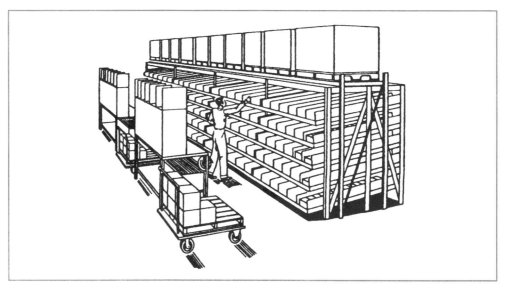

FIGURE 9.2 Gravity Flow Rack

Source: Department of the Navy, Naval Supply Systems Command, Publication 529. From Edward H. Frazelle, *Small Parts Order Picking: Equipment and Strategy*. Oak Brook, IL: Warehousing Education and Research Council, 1988, p. 3. Reprinted with permission.

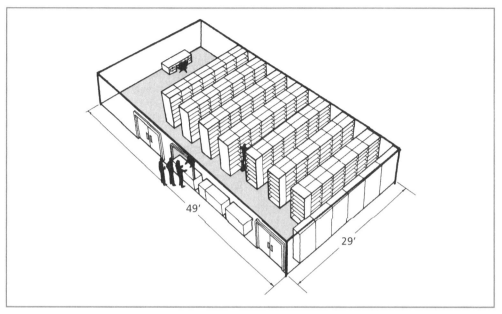

FIGURE 9.3 Bin Shelving Systems

Source: Edward H. Frazelle, *Small Parts Order Picking: Equipment and Strategy*. Oak Brook, IL: Warehousing Education and Research Council, 1988, p. 1. Reprinted with permission.

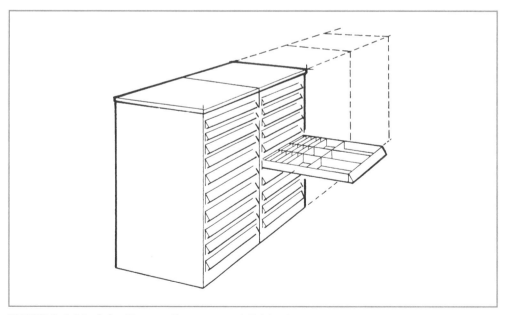

FIGURE 9.4 Modular Storage Drawers and Cabinets

Source: Department of the Navy, Naval Supply Systems Command, Publication 529. From Edward H. Frazelle, *Small Parts Order Picking: Equipment and Strategy*. Oak Brook, IL: Warehousing Education and Research Council, 1988, p. 2. Reprinted with permission.

The storage systems described previously are classified as 'fixed' systems because they are stationary. Others can be classified as 'movable' because they are not in fixed positions. The bin shelving systems shown in Figure 9.3 can be transformed from a fixed to a movable system (see Figure 9.5). In the bin shelving mezzanine, wheels on the bottom of the bins follow tracks in the floors, allowing the bins to be moved and stacked together when not being accessed. This allows maximum utilization of space, because full-width aisles are not needed between each bin.

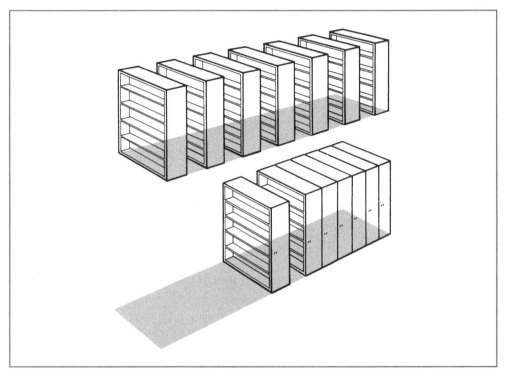

FIGURE 9.5 Bin Shelving Mezzanine

Source: Courtesy of White Storage & Retrieval Systems. From Edward H. Frazelle, *Small Parts Order Picking: Equipment and Strategy*. Oak Brook, IL: Warehousing Education and Research Council, 1988, p. 8. Reprinted with permission.

Products are picked from the various storage systems, using some order-picking approach. In a manual system, the personnel doing the order picking go to the location of the items, walking with a cart or riding a mechanized cart. In many cases, the order picker retrieves items from a flow-through gravity storage rack (see Figure 9.2).

Transportation and Sorting

The order picker can use a large selection of powered and non-powered equipment for transporting and sorting items located in the racks, shelves and drawers. Examples of this type of apparatus include forklift trucks, platform trucks, hand trucks, cranes and carts.

Manual sorting of items is a very labour-intensive part of warehousing. It involves separating and regrouping picked items into customer orders. Personnel physically examine items and place them onto pallets or slipsheets, or into containers for shipment to customers. This is a time-consuming process subject to human error. As a result, most firms attempt to minimize manual sorting.

Shipping of products to customers involves preparing items for shipment and loading them onto the transport vehicle. The powered and non-powered equipment previously described is used for this purpose. Pallets, palletizers, strapping machines and stretch wrappers are also important.

In addition, the shipping and receiving activity requires equipment for handling outbound and inbound transportation vehicles. Therefore, shipping and receiving docks are important elements of the material handling process. For example, new highway regulations increasing the amount of weight a truck trailer can haul, and regulations allowing wider and longer trailers such as 44-tonne lorries in the UK,[2] have placed new demands on shipping and receiving docks. Some of the changes that have occurred are shown in Figure 9.6, which represents a modern shipping and receiving dock. As stated previously, manual or non-automated equipment is often used in combination with automated equipment.

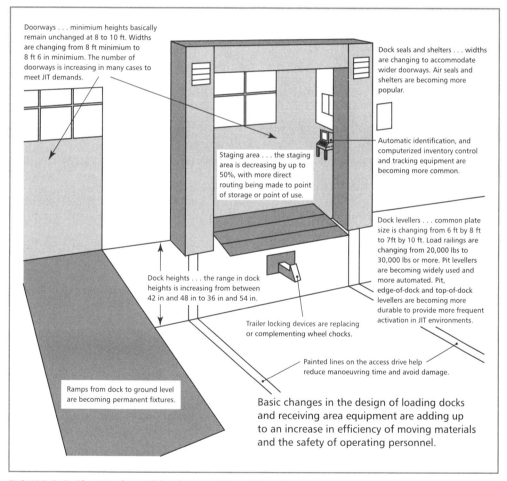

Doorways . . . minimium heights basically remain unchanged at 8 to 10 ft. Widths are changing from 8 ft minimium to 8 ft 6 in minimium. The number of doorways is increasing in many cases to meet JIT demands.

Dock seals and shelters . . . widths are changing to accommodate wider doorways. Air seals and shelters are becoming more popular.

Staging area . . . the staging area is decreasing by up to 50%, with more direct routing being made to point of storage or point of use.

Automatic identification, and computerized inventory control and tracking equipment are becoming more common.

Dock levellers . . . common plate size is changing from 6 ft by 8 ft to 7ft by 10 ft. Load railings are changing from 20,000 lbs to 30,000 lbs or more. Pit levellers are becoming widely used and more automated. Pit, edge-of-dock and top-of-dock levellers are becoming more durable to provide more frequent activation in JIT environments.

Dock heights . . . the range in dock heights is increasing from between 42 in and 48 in to 36 in and 54 in.

Trailer locking devices are replacing or complementing wheel chocks.

Painted lines on the access drive help reduce manoeuvring time and avoid damage.

Ramps from dock to ground level are becoming permanent fixtures.

Basic changes in the design of loading docks and receiving area equipment are adding up to an increase in efficiency of moving materials and the safety of operating personnel.

FIGURE 9.6 The Modern Shipping and Receiving Dock

Source: 'Docks and Receiving – Where it all Begins', *Modern Materials Handling, 1985 Warehousing Guidebook* 40, no. 4 (Spring 1985), p. 36. *Modern Materials Handling*, copyright 1985 by Cahners Publishing Company, Division of Reed Holdings, Inc.

Automated Materials Handling Systems

Automated storage and retrieval systems (AS/RS), carousels, case-picking and item-picking equipment, conveyors, robots and scanning systems have become commonplace in warehouses. As a result, many firms have been able to achieve improvements in materials handling efficiency and productivity.

For example, the Danish supermarket Netto A/S built a new automated distribution centre, supplied by Daifuku Europe Limited, near Copenhagen in 2003/04. The automated systems in the distribution centre include 32 automatic stacker cranes, two tilt tray sorters, three automatic layer depalletizers, 40 sortation transfer vehicles (STV) and a computer-based warehouse management system (WMS). The STV system provides high-speed pallet sorting for up to 700 pallets per hour. The WMS manages the entire facility from the point at which supplier trucks arrive to the dispatch of trucks from the loading docks. The WMS manages picking for a batch of 56 shops every four hours.[3]

Automated equipment can be grouped into the same categories used to describe non-automated equipment: storage and order-picking, transportation and sorting, and shipping. Table 9.1 listed examples of automated storage and order-picking equipment. Bausch & Lomb, Chek Lap Kok Airport in Hong Kong, Compaq, General Electric, Nike, Packard Bell, Posten PaketFrakt in Sweden, Rothmans Tobacco in the Netherlands, Toyota, and many other firms, have employed automated systems with great success. The Creative Solutions box describes Packard Bell's integrated solution to materials handling.

CREATIVE SOLUTIONS

Providing Integrated Solutions through Computers and Automation

At Packard Bell's facility in Angers, France, computers are assembled on a last-minute basis before delivery to the final customer. This strategy requires a materials handling system with the responsiveness to operate in a JIT environment.

The system consists of five kitting (assembly) stations, 32 modular and interchangeable conveyor workstations, where the computers are assembled and tested, and an in-line burn-in testing system.

At the kitting stations, a paperless, pick-to-light system speeds picking and ensures that the kits are error-free by guiding workers through the process by means of a series of light displays mounted to flow racks.

To minimize work-in-process inventory and streamline the workflow, the assembly area operates on a pull strategy. While the system automatically determines the workflow (through a bar code on the transport trays), product is not released from a workstation until the destination location is free to accept it.

Question: What other JIT issues could affect this operation?

Source: 'Total Solutions: Problem Solving with Materials Handling Systems', *Modern Materials Handling* 50, no. 7 (June 1995), p. 22.

The Toyota Marketing Company's Parts Distribution Centre in South Africa was partially automated between 1984 and 1991 at a cost of about 33 million Rand, or US$5.6 million. Additional storage and handling facilities were added; new receiving, binning and order-processing systems were introduced; and a high-rise bulk warehouse was constructed. The benefits were significant:

- order-processing productivity increased 300 per cent
- product damage rates declined by 50 per cent
- stock accuracy and service rates improved by 65 per cent
- the work of three clerks was eliminated and an additional three clerks were reassigned to more essential tasks.[4]

Among the most important items of storage and order-picking equipment are automated storage and retrieval systems (AS/RS). In comparison with manual systems, an AS/RS provides reduced labour cost and floor space, while increasing inventory accuracy. An AS/RS is applicable to virtually all types of product and many warehouse configurations.

Advantages of Automated Systems

Automated systems can provide several benefits for warehouse operations. Table 9.2 lists some of the most important benefits of automated systems as identified by users. Generally, the benefits can be categorized into operating cost savings, improved service levels, and increased control through more and better information. The Technology box provides an example of how an automated system can also reduce manual handling for a more efficient warehouse system.

Benefit	Percentage of respondents that 'Agree' or 'Strongly agree'
Labour cost reduction	98.8
Ability to increase output rate	95.2
Improvement in consistency of service	92.1
Reduction in materials handling	92.1
Increased accuracy level	89.5
Service availability	87.0
Improvement of speed of service	81.0

TABLE 9.2 Benefits of Automated Materials Handling Scheme

Source: Kofi Q. Dadzie and Wesley J. Johnston, 'Innovative Automation Technology in Corporate Warehousing Logistics', *Journal of Business Logistics* 2, no. 1 (1991), p. 76.

TECHNOLOGY

Quick Read Through

The Libri book distribution centre in Bad Hersfeld, Germany, receives pallets of books from publishers, stores them in its warehouses, and fulfils orders received daily from book stores and individual Internet customers throughout Europe. The facility, built in 2001, consolidated six warehouses into one. It holds about 300,000 titles, yet it takes only about 100 employees to pick up to 40,000 books per hour due to advances in technology during the latter part of the 1990s to handle such volumes.

The dynamic picking system provided by the facility designer, Witron, eliminates all labour-intensive and non-value-adding handling steps through automation. When pallets arrive from publishers, employees break them down, repacking smaller quantities of books into plastic storage bins. Repacking compresses storage needs while making products accessible to multiple picking workstations at the same time.

All bins have a fixed bar code label on them and are conveyed to the bin storage warehouse for storage in one of 425,000 locations. These locations are serviced by 156 miniload AS/RS cranes and every hour an estimated 2300 bins move through the picking area. Each picking area is configured so that nearly 80 per cent of the picks are within one step of the workstation. Thus, the average pick rate per picker is up to 450 picks per hour.

While operators pick books, printers generate product labels that the operators place on the back of each book prior to placing the books in the order bin. The pick workstation is equipped with a scanner that verifies the accuracy of the pick. A scale built into the workstation automatically weighs the order bin to ensure the correct order quantity. Then the bin is conveyed to another pick station or to the shipping area.

As order bins move towards the shipping area, an invoice is automatically printed and dropped into the bins. The bins are then automatically sealed, and conveyed to the order consolidation area, where they are staged for loading onto trucks. Most customer orders that arrive at Libri have fewer than nine books per order; in fact the majority of orders are for a single book. From the time a customer order arrives, it takes about 2.5 hours to fulfil the order and place it on a delivery truck.

Question: **What other types of fast moving consumer goods (fmcg) could benefit from this type of automatic materials handling technology?**

Source: Leslie Langnau, 'Come on a European Picking Tour', *Materials Handling Management Online* (June 2001), http://www.mhmonline.com/viewStory.asp?nID=1024&pNum=1&CaH=1.

Disadvantages of Automated Systems

However, automated systems are not without their disadvantages. Typical problems faced by firms choosing to automate materials handling operations include the following:

- initial capital cost
- downtime or unreliability of equipment/maintenance interruptions
- software-related problems (e.g. poor documentation, incompatibility, failure)
- capacity problems
- lack of flexibility to respond to changing environment
- maintenance costs
- user interface and training
- worker acceptance
- obsolescence.[5]

Types of Equipment

The initial capital outlay is usually the most significant obstacle. For example, a miniload AS/RS (see Figures 9.7 and 9.8), where a storage/retrieval (S/R) machine travels horizontally and vertically simultaneously in a storage aisle, transporting containers to and from an order-picking station at one end of the system, generally costs between €120,000 and €350,000 per aisle.[6]

> When the unit to pick is a full pallet or similar large load, the AS/RS offers complete automation from storage to retrieval in minimal space. Unit-load AS/RSs are installed up to 100 feet high with aisles only inches wider than the load to be stored. The S/R machines operate at speeds much faster than industrial trucks and travel simultaneously in horizontal and vertical directions. They are used when inventories, throughput, and space costs are high. In totally automated systems, AS/RSs are supplied by conveyors, automated guided vehicles, or electrified monorail systems.[7]

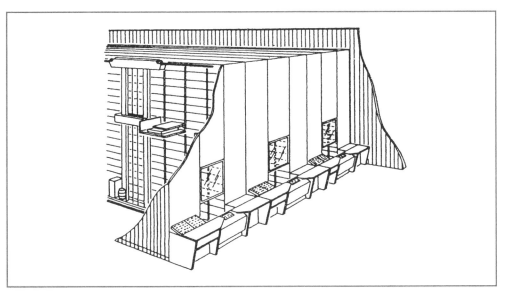

FIGURE 9.7 Miniload AS/RS

Source: Department of the Navy, Naval Supply Systems Command, Publication 529. From Edward H. Frazelle, *Small Parts Order Picking: Equipment and Strategy*. Oak Brook, IL: Warehousing Education and Research Council, 1988, p. 6. Reprinted with permission.

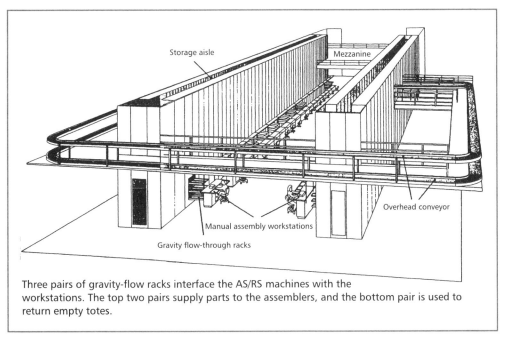

Three pairs of gravity-flow racks interface the AS/RS machines with the workstations. The top two pairs supply parts to the assemblers, and the bottom pair is used to return empty totes.

FIGURE 9.8 Minimizing Inventory at Apple Computer with a Flexible Miniload AS/RS

Source: 'Mini-Load AS/RS Trims Inventory, Speeds Assembly', *Modern Materials Handling* 39, no. 13 (21 September 1984), pp. 48–9. *Modern Materials Handling*, copyright 1984 by Cahners Publishing Company, Division of Reed Holdings, Inc.

Carousels

A form of AS/RS is the carousel. Carousels are mechanical devices that house and rotate items for order picking. The most frequently utilized carousel configurations are the horizontal and vertical systems.

A horizontal carousel (see Figure 9.9) is a linked series of rotating bins of adjustable shelves driven on the top or bottom by a drive motor. Rotation takes place on an axis perpendicular to the floor at approximately 80 feet a minute. Costs for horizontal carousels begin at €30,000 a unit.[8]

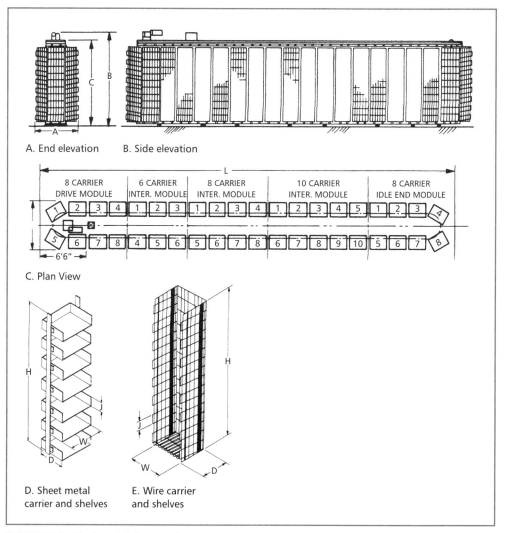

A. End elevation   B. Side elevation

C. Plan View

D. Sheet metal carrier and shelves   E. Wire carrier and shelves

FIGURE 9.9 Horizontal Carousels

Source: courtesy of SPS Technologies, Inc. From Edward H. Frazelle, *Small Parts Order Picking: Equipment and Strategy*. Oak Brook, IL: Warehousing Education and Research Council, 1988, p. 4. Reprinted with permission.

A vertical carousel is a horizontal carousel turned on its end and enclosed in sheet metal (see Figure 9.10). Like horizontal carousels, an order picker operates one or multiple carousels. The carousels are indexed either automatically by way of computer control or manually by the order picker operating a keypad on the carousel's work surface. The cost of a typical vertical carousel is around €77,000.

Transportation and sorting activities are typically performed in combination with storage and order picking. The three pieces of transportation equipment most frequently used are conveyors, automatic guided vehicle systems (AGVS) and operator-controlled trucks or tractors.

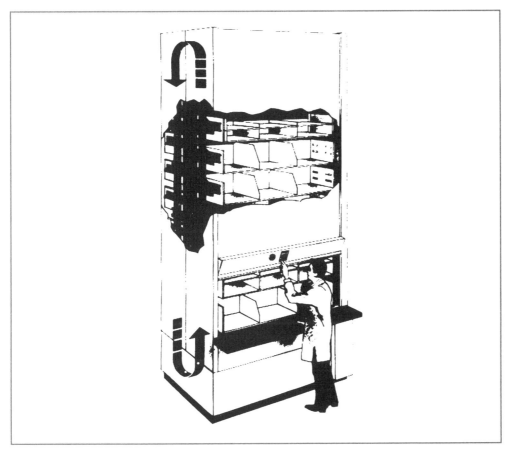

FIGURE 9.10 Vertical Carousel

Source: courtesy of Kardex Systems, Inc. From Edward H. Frazelle, *Small Parts Order Picking: Equipment and Strategy*. Oak Brook, IL: Warehousing Education and Research Council, 1988, p. 5. Reprinted with permission.

Conveyors

Sorting equipment for case dispensing can be assembled from other components such as conveyors and diverters, or it can be specialized such as a tilt-tray sorter with built-in diverting mechanisms.

Carte Blanche, the UK greetings card and gift manufacturer, installed an overhead monorail conveyor system for order picking in its new 3255-square-metre warehouse and office facility in Chichester, England. The system uses specially designed carriers containing bins with picking lists and circulates past 2500 pick locations over three floors. The carriers are fitted with transponders for monitoring their movement such that they enter only the necessary pick lanes and can swivel through 360 degrees and be locked in place, ensuring that pick operators always have the bins in the best position for order fulfilment with the minimum effort. Once an order is completed the carrier is routed to a packing zone before beginning the process again. The over-head conveyor was designed using modular components such as powered incline/decline conveyors that simplify carrier flow and increasing design flexibility by eliminating permanent physical barriers at floor level.[9]

Automatic Guided Vehicle Systems (AGVSs)

Automatic guided vehicle systems (AGVSs) are 'battery-powered driverless vehicles that are controlled by computers for task assignment, path selection, and positioning'.[10] AGVSs are often used in automated warehouse operations involving AS/RSs. The benefits of AGVSs include 'lower handling costs, reduced handling-related product damage, improved safety, the ability to interface with other automated systems, and reliability'.[11] Automated guided vehicles (AGVSs) cost about €23,000 for a single model at the low end of the scale and about €54,000 for a more advanced model.[12]

Robots

The robot is another type of equipment used in many phases of materials handling. Robots have been used in the manufacturing process for some time, but advances in robotics technology have expanded their use to a larger number of applications. It is likely that materials handling robots will have steady growth in many application areas (see Figure 9.11).

Automation in the shipping area has also occurred. The two aspects of shipping activity that have been most affected by automation are packaging and optical scanning. We have previously described pass-through and rotary stretchwrapping machines, and will further address packaging later in this chapter.

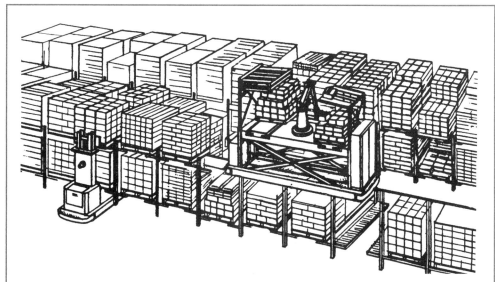

Artist's impression of guided vehicles and robotics in the warehouse. An automatically guided and programmed lift truck is shown stacking a load into a pallet rack. The robot is shown on a floor-supported vehicle capable of raising and lowering; the robot is picking from gravity flow racks and building pallet loads.

FIGURE 9.11 Robots in the Warehouse

Source: 'Warehousing Flexibility Aided by Robots'. *Material Handling Engineering* 40, no. 9 (September 1985), p. 103. Reproduced by permission of the St Onge Company, York, PA.

Computerized Tracking and Information Systems

Another aspect of shipping automation is documentation. As other components of the warehouse become automated, firms need to computerize their tracking and information systems with new technologies, as discussed in Chapter 3. French liqueur producer Grand Marnier invested in a new WMS with bar code scanning and the ability to also use radio frequency identification (RFID), as discussed in Chapter 3, voice and web-based technologies to manage traceability of its food and non-food components and finished goods to meet new EU requirements in 2005.[13] Items entering the warehouse are bar code scanned and the data collected become part of the warehouse information system, which is used for a variety of purposes, including the preparation of business-related documentation for customers.

Many companies are utilizing various computerized documentation procedures. Whether such technology is being used by Avia Presto of Holland (air freight cargo handling), the Royal Marines of Britain or the Barrow-upon-Soar site of British Gypsum (plaster production), firms are recognizing the benefits of automating the materials handling process, including increased productivity, better space utilization, higher customer service levels, reduced operating expenses and improved flow of materials.

The type and scope of benefits a company receives will vary according to product characteristics, labour intensity of the operation, existing customer service levels and present level of company expertise.

Warehousing in a Just-in-time Environment

As manufacturing and merchandising firms adopt and implement just-in-time (JIT) programmes, logistics components such as warehousing will be directly affected. Because JIT stresses reduced inventory levels and more responsive logistics systems, greater demands are placed on warehousing to maximize both efficiency and effectiveness. Examples of these demands include the following.

- *Total commitment to quality*: warehouse employees must perform their tasks (inbound and outbound) at levels specified by customers.
- *Reduced production lot sizes*: items are packaged in smaller lots, and warehouse deliveries are smaller and in mixed pallet quantities.
- *Elimination of non-value-added activities*: non-essential and inefficient physical movement and handling activities are identified and eliminated, resulting in improved facilities layout and warehouse operating efficiencies.
- *Rapid flow-through of materials*: because JIT stresses low or even zero inventory, emphasis is placed on the mixing function of the warehouse rather than storage.

Rio Bravo Electricos, a firm in Juarez, Mexico, that assembles electric wiring harnesses for General Motors vans, supplies its 36 subassembly and assembly lines with parts every two hours or so. A wiring harness can be assembled in one hour within Rio Bravo's JIT environment.[14]

Computer Technology, Information and Warehouse Management

We saw in Chapter 8 that the basic functions of warehousing were movement, storage and information transfer. In each of these areas, the use of computer technology has become widespread, as discussed in Chapter 3. Warehousing is moving towards greater computer utilization. The fully computerized warehouse will likely have a structure similar to that shown in Figure 9.12, where all activities of the warehouse interface with the system, including receiving, quality control, storage, order picking, error control, packing and shipping. Significant advantages will result, including improved customer service, lower costs, and more efficient and effective operations.

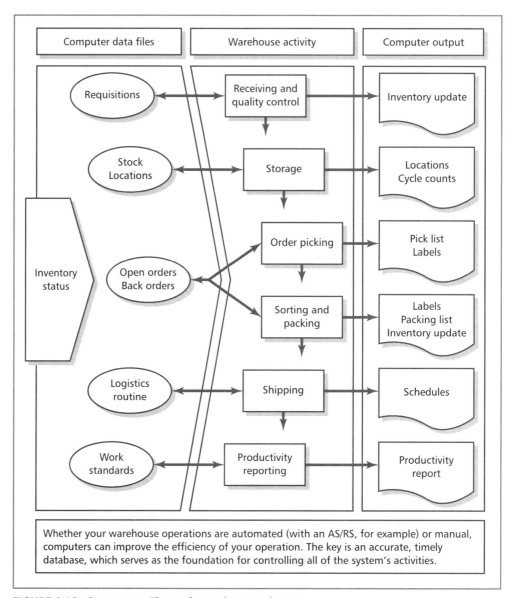

| Computer data files | Warehouse activity | Computer output |

FIGURE 9.12 Computers Throughout the Warehouse

Source: 'Increase Productivity with Computers and Software', *Modern Materials Handling, 1986 Warehousing Guidebook* 41, no. 4 (Spring 1986), p. 68. *Modern Materials Handling*, copyright 1986 by Cahners Publishing Company, Division of Reed Holdings, Inc.

Malaysia Airlines, Southeast Asia's largest airline company, was the first Asia-Pacific air carrier to implement advanced bar code and client-server technology in cargo-tracking and warehouse operations.[15] Other companies in diverse industries (toy retailing, blood-testing laboratories, textbook manufacturing) are utilizing similar technology to improve materials handling efficiency and effectiveness.

Information is the key to successful warehouse management. However, many warehousing operations exhibit symptoms resulting from a lack of information. Not many warehouse managers operate in a total information vacuum, but many information gaps exist in warehousing operations.

The importance of information in warehouse management is significant. Accurate and timely information allows a firm to minimize inventories, improve routing and scheduling of transportation vehicles, and generally improve customer service levels. A typical warehouse management system achieves these improvements in three ways:

1 reducing direct labour
2 increasing materials handling equipment efficiency
3 increasing warehouse space utilization.

Networks are communications systems that allow transmission of data between a number and variety of devices such as terminals, word processors, bar code readers, robots, conveyors, automatic guided vehicles and AS/RSs. A local area network (LAN), whose devices are located in close proximity to one another, is typically used in warehousing. Figure 9.13 shows an example of a local area network.

Many approaches are possible to setting up a LAN system. No matter which approach a firm uses, the objectives are the same: to provide better control over information flows and to allow the warehouse facility to maximize its effectiveness and efficiency. Due to direct connection and a common database, information feeds and flows directly to the next. This reduces redundant data entry, excessive paperwork and the potential for error.

Packaging

Packaging is an important warehousing and materials management concern, one that is closely tied to warehouse efficiency and effectiveness. The best package increases service, decreases cost and improves handling. Good packaging can have a positive impact on layout, design and overall warehouse productivity.

Functions of Packaging

Packaging serves two basic functions: marketing and logistics. In its marketing function, the package provides customers with information about the product and promotes the product through the use of colour and shape. It is the consumer's 'encounter with the product and should not be deceptive or misleading'.[16]

> The [package] is the 'silent sales [person],' and it is the final interface between the company and its consumers . . . Consumers generally choose to buy from the image they perceive that a product has, and what they perceive is heavily influenced by the cues given on the product's packaging: brand name, color and display.[17]

From a logistics perspective, the function of packaging is to organize, protect and identify products and materials. In performing this function, packaging takes up space and adds weight. Industrial users of packaging strive to gain the advantages packaging offers while minimizing the disadvantages, such as added space and weight. We are getting closer to that ideal in several types of packaging, including corrugated containers, foam-in-place packaging, stretch-wrapping and strapping.

More specifically, packaging performs six functions, as described below.

1 *Containment*: products must be contained before they can be moved from one place to another. If the package breaks open, the item can be damaged or lost, or may contribute to environmental pollution if it is a hazardous material.
2 *Protection*: to protect the contents of the package from damage or loss from outside environmental effects (e.g. moisture, dust, insects, contamination).

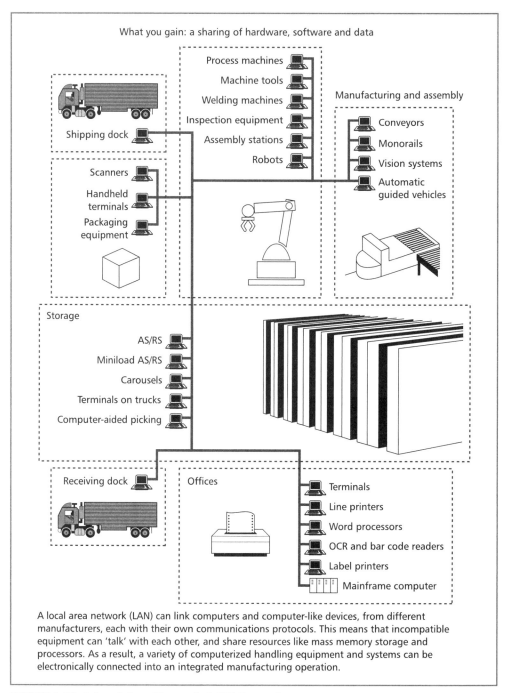

What you gain: a sharing of hardware, software and data

Process machines
Machine tools
Welding machines
Inspection equipment
Assembly stations
Robots

Manufacturing and assembly

Conveyors
Monorails
Vision systems
Automatic guided vehicles

Shipping dock

Scanners
Handheld terminals
Packaging equipment

Storage

AS/RS
Miniload AS/RS
Carousels
Terminals on trucks
Computer-aided picking

Receiving dock

Offices

Terminals
Line printers
Word processors
OCR and bar code readers
Label printers
Mainframe computer

A local area network (LAN) can link computers and computer-like devices, from different manufacturers, each with their own communications protocols. This means that incompatible equipment can 'talk' with each other, and share resources like mass memory storage and processors. As a result, a variety of computerized handling equipment and systems can be electronically connected into an integrated manufacturing operation.

FIGURE 9.13 A Local Area Network (LAN) Example

Source: 'Local Area Networks – The Crucial Element in Factory Automation', *Modern Materials Handling* 39, no. 7 (7 May 1984), p. 51. *Modern Materials Handling*, copyright 1986 by Cahners Publishing Company, Division of Reed Holdings, Inc.

3 *Apportionment*: to reduce the output from industrial production to a manageable, desirable 'consumer' size – that is, translating the large output of manufacturing into smaller quantities of greater use to customers.

4 *Unitization*: to permit primary packages to be unitized into secondary packages (e.g. placed inside a corrugated case); the secondary packages are unitized into a stretchwrapped pallet, and ultimately into a container that is loaded with several pallets. This reduces the number of times a product must be handled.

5 *Convenience*: to allow products to be used conveniently – that is, with little wasted effort by customers (e.g. blister packs, dispensers).

6 *Communication*: the use of unambiguous, readily understood symbols such as a UPC (universal product code).[18]

The package should be designed to provide the most efficient storage. Good packaging interfaces well with the organization's materials handling equipment and allows efficient utilization of storage space, as well as transportation cube and weight constraints.

Effects of Packaging on Costs and Customer Service

In the past, packaging trade-offs were frequently ignored or downplayed in logistics decision-making. Like all logistics decisions, packaging has an impact on both costs and customer service levels. From a cost perspective, suppose a company uses a carton that is 25 centimetres × 25 centimetres × 15 centimetres instead of a carton that measures 25 centimetres × 25 centimetres × 30 centimetres. Assume the smaller carton costs €0.40 less and requires less loose fill, which can save half a cubic foot of fill material costing €0.65. In this example, that is a saving of €1.05 per carton. Multiplied by hundreds, thousands or millions of packages distributed during a year, the savings quickly add up .

At the same time as costs are reduced, service levels are improved because customers are able to obtain more of the same amount of product in less space, enabling them to achieve cost savings. The customer is likely to realize fewer partial or split shipments from suppliers because more products can be placed on the transport vehicle that makes the delivery.

Saving Money through Efficient and Effective Packaging

Packaging is becoming a more visible issue with current environmental concerns about recycling and the reuse of packaging. The environmental aspects of packaging are important because of reverse logistics, which is discussed in the next section of this chapter. Investing in efficient and effective packaging can save a company money in the following ways:

- lighter packaging may save transportation costs
- careful planning of packaging size/cube may allow better space utilization of warehousing and transportation
- more protective packaging may reduce damage and requirements for special handling
- more environmentally conscious packaging may save disposal costs and improve the company's image
- use of returnable containers provides cost savings as well as environmental benefits through the reduction of waste products.

Factors Governing Good Package Design

Good package design is influenced by (1) standardization, (2) pricing (cost), (3) product or package adaptability, (4) protective level, (5) handling ability, (6) product packability, and (7) reusability and recyclability. With the growth in automation and computerization of warehousing, the ability to utilize 'high' storage space and convey information are key. The importance a firm places on each factor, as well as the cost–service trade-offs it makes, varies by company, industry and geographic location.

Due to differences in the cost and physical characteristics of products, a food processor, for example, is more concerned than a computer manufacturer with having a package that minimizes shipping and storage costs. A computer manufacturer emphasizes the protective aspects of packaging because of the fragile, expensive nature of computer systems.

Another illustration would be a company that completed construction of a fully automated warehouse. Managers of such a facility would be very concerned with handling ability, cube utilization and the ability to convey information so that it could be 'read' by the equipment.

On the other hand, a company doing business in Germany would be concerned with the reusability and recyclability aspects of packaging because of Germany's strict environmental laws. The packaging decision is truly one that requires the use of a systems approach in order to understand the true 'total cost' picture.

Examples of the packaging and logistics cost trade-offs are shown in Table 9.3. There are many important interfaces between packaging and activities such as transportation, inventory, warehousing and information systems.

Logistic activities	Trade-offs
Transportation	
Increased package information	Decreases shipment delays; increased package information decreases tracking of lost shipments
Increased package protection	Decreases damage and theft in transit, but increases package weight and transport costs
Increased standardization	Decreases handling costs, vehicle waiting time for loading and unloading; increased standardization; increases modal choices for shipper and decreases need for specialized transport equipment
Inventory	
Increased product protection	Decreases theft, damage, insurance; increase product availability (sales); increases product value and carrying costs
Warehousing	
Increased package information	Decreases order filling time, labour cost
Increased product protection	Increases cube utilization (stacking), but decreases cube utilization by increasing the size of the product dimensions
Increased standardization	Decreases material handling equipment costs
Communications	
Increased package information	Decreases other communications about the product such as telephone calls to track down lost shipments

TABLE 9.3 Packaging Cost Trade-offs with Other Logistics Activities

Source: Professor Robert L. Cook, Department of Marketing and Hospitality Services Administration, Central Michigan University, Mt Pleasant, MI, 1991.

Of course, other factors influence the product package, such as the channel of distribution and institutional requirements. This is often true in retail channels. For example, when compact discs (CDs) first came out, retailers were concerned whether they could utilize the racks they were using for albums. There was also concern about potential pilferage because of CDs' small size. A larger, environmentally unfriendly package was chosen as a way to address these concerns.[19]

Procter & Gamble (P&G) has examined the full implications of the packaging decision. The company developed a programme called Direct Product Profitability, which identified product costs through the entire channel of distribution, including those associated with packaging. Some of the results achieved by P&G included the following.

- A shampoo bottle was redesigned in a squarer configuration that took up less space, saving distributors' storage costs.
- A washing powder detergent was reformulated so that P&G was able to shrink the size of the box without reducing the number of washes per box. P&G was able to pack 14 boxes in a case instead of 12, thus reducing handling and storage costs. Later the firm introduced an even smaller package that allowed for a comparable number of wash loads but reduced the case size.[20]

The next section on reverse logistics, and Box 9.1, describe some of the trade-offs between packaging materials and other logistics aspects as a result of green marketing and environmental concerns.

Box 9.1 Green Manufacturing Has Major Implications for Logistics

Design for disassembly is a hot new trend in manufacturing. The goal is to design, develop and produce product with the goal of reducing the waste created when the product reaches the end of its useful life. That could involve recycling, refurbishing or safely disposing of a product and its components. It has major implications for how a company designs its logistics and purchasing systems.

Germany has been the leader in the green movement by requiring manufacturers to 'take back' their product's packaging. To address this requirement, manufacturers banded together to form a private company that collects, recycles and disposes of packaging material. In the first two years of implementation, this has reduced that amount of waste due to packaging materials by 4 per cent.

This has major implications for materials handling equipment and packaging design. For example, companies have been designing product to use less packaging. Colgate and several other manufacturers are now using a design for a toothpaste tube that stands on the cap. Thus no box is needed.

Hewlett-Packard has designed workstations in a green way, which has many implications for logistics. Instead of using internal metal 'frames' to hold the parts in place, HP uses a polypropylene foam chassis with cut-outs for each component and connection. This is so effectively protective that external packaging can be reduced – by as much as 30 per cent. The product is lighter, which reduces transportation cost. Disassembly time has been reduced by 90 per cent.

As components are reused, new ways of transporting, storing and handling the unusable materials and inventory need to be found. Logistics will play a key role in this process.

Source: Gene Bylinsky, 'Manufacturing for Reuse', *Fortune*, 6 February 1995, pp. 102–12.

Reverse Logistics

Reverse logistics is a growing area in logistics management. Concern about environmental issues such as pollution, traffic congestion, global warming, disposal and clean-up of hazardous materials has led to a number of environmental laws and EU directives that affect logistics systems design and strategies.[21]

Reverse logistics encompasses all of the activities in the CSCMP definition of logistics provided in Chapter 1. However, the difference is that reverse logistics encompasses all of these activities as they occur in reverse. Thus reverse logistics can be defined as:

> The process of planning, implementing, and controlling the efficient, cost effective flow of raw materials, in-process inventory, finished goods and related information from the point of consumption to the point of origin for the purpose of recapturing value or proper disposal.[22]

Essentially, reverse logistics is the process of moving goods from their 'point of consumption' to the appropriate link in the supply chain for the purpose of capturing any residual value through remanufacturing or refurbishing, or for proper disposal. Thus, reverse logistics management represents a 'systematic business model . . . to profitably close the loop on the supply chain'.[23]

But, reverse logistics is not only about reusing containers, recycling packaging materials, redesigning packaging to use less material, or reducing the energy use and pollution that results from transportation. These activities, while important, are more in the realm of green logistics. If goods or materials are not sent 'backwards' through the supply chain then an activity is probably not a reverse logistics activity. Reverse logistics also includes processing returned merchandise due to damage, seasonal inventory, restock, salvage, recalls and excess inventory.

Key management elements in reverse logistics include the following.

- *Gatekeeping* to screen defective and unwarranted returned merchandise at the entry point into the reverse logistics process.
- *Short disposition cycle times* related to return product decisions, movement and processing to avert a lengthy ageing process on returns.
- *Reverse logistics information systems* to properly track returns, and measure disposition cycle times and vendor performance.
- *Central return centres* or processing facilities dedicated to handling returns quickly and efficiently.
- *Zero returns* policies that avoid accepting any physical returns and instead set maximum values of returned products that are payable to customers.
- *Remanufacture and refurbishment* of products, which has the following categories: repair, refurbishing and remanufacturing to recondition or upgrade products, and cannibalization and recycling to use or dispose of products.
- *Asset recovery* classifies and disposes returned goods, surplus, obsolete, scrap, waste and excess material products, and other assets, to maximize returns to the owner, and minimize the costs and liabilities associated with disposition.
- *Negotiation* of the value of returned material without any pricing guidelines. This task is often performed by specialist third parties, who advise the primary participants in the supply chain who are working to transfer ownership of the material back to the original source.
- *Financial management* policies to properly handle accounting and reconciliation issues related to returned products.
- *Outsourcing* reverse product flow, to reverse logistics outsource suppliers who can be used as a benchmark to help determine what and how reverse activities should be performed, and how much those activities should cost.

Benefits to firms from practising reverse logistics management include cost reductions, added value for customers and proper compliance with legislative regulations. Critical success factors for reverse logistics programmes to capture the key elements above include the following.

- *Management and control* by mapping or flowcharting the reverse logistics process through the firm, developing an environmental management system, educating customers, employees, suppliers and others supply chain members, and developing partnerships to achieve reverse logistics goals and economies of scale.

■ *Measurement* by adopting full product life cycle and end of product life costing as they relate to reverse logistics activities and the product supply chain.

■ *Finance* to properly allocate sufficient resources for reverse logistics activities and environmental initiatives.[24]

Reverse and green logistics come together as a result of regulation and legislation, such as the EU Waste Electrical and Electronic Equipment (WEEE) Directive introduced in 2005 and adopted into EU law. The directive was designed to reduce the amount of WEEE, of which there was some 915,000 tonnes across Europe in 1998, and increase levels of refurbishment and recycling by requiring EU member states to force producers and retailers to take responsibility for the return and reverse logistics of such products.[25]

The need to consider reverse and green logistics has also seen the growth of third-party reverse logistics providers who assist companies to meet new guidelines and enhance their business opportunities. For example, Caledonian Alloys in the United Kingdom provides a global revert or scrap management and processing outsourced service for several industries, particularly aerospace. Caledonian Alloys collects from various customers used high-value nickel and titanium alloy solid products, such as jet aircraft engine vanes and components, and process turnings, and then sorts and processes these materials for onward sale to alloy-melting processors or disposal.

The importance and growth of reverse and green logistics issues and legislation will require firms to incorporate appropriate strategies that transcend national boundaries. The Global box discusses the efforts of a Norwegian cooperative food retailer in addressing reverse and green logistics with its multinational partners.

GLOBAL

'Green' Logistics and 'Green' Packaging at NKL

Coop NKL BA is a Norwegian food cooperative and since 2002 has been a 20 per cent co-owner of the Nordic cooperative Coop Norden AB, together with Kooperativa Förbundet, Sweden (42 per cent), and FDB, Denmark (38 per cent). Norden's vision is for a better, more secure everyday life through profitable shops owned by members. It sees its future as being the Scandinavian retail business with the greatest geographical coverage and thus the best route for suppliers to reach Scandinavian customers. This strategy enables Norden to quickly and responsively implement joint activities, product ranges and agreements across national borders.

NKL, together with its Norden partners, believes reverse and green logistics provides it with a competitive advantage in the difficult Nordic marketplace. NKL's initiatives in being a green retailer provided a 220 per cent increase of environmental product sales in the late 1990s. It used 1.5 million reusable containers for fruits and vegetables, and increased rail transport use from 50 to 60 per cent, which led to reduced costs, improved order cycle times and reduced pollutants and energy wastage. Overall, the Norden group in 2003 sold 93,800 tonnes of eco-labelled/organic foods and used some 2350 tonnes of recycled plastic and some 17,600 tonnes of recycled cardboard in its terminals and warehouses.

Packaging design has also played an important role at NKL, which has sought to improve under-utilized transportation and warehouse capacity, handling and damage problems, and reduce extra work due to returns and co-distribution problems. For example, it developed a new consumer and retail package for a potato flakes snack food in conjunction with the manufacturer Nestlé and NKL's retailers. The packaging supply chain project realized total savings of 800,000 Norwegian Kroner (NOK) or about €97,000 per year, of which retailers saved NOK 375,000, and NKL saved the balance on warehouse and transportation costs.

Question: **Are you aware of your local retailers' reverse or green logistics initiatives and, if so, what are they?**

Sources: James R. Stock, *Development and Implementation of Reverse Logistics Programs*. Oak Brook, IL: Council of Logistics Management 1999, pp. 41, 207–9; and *Coop NKL BA website* (http://www.coop.no, accessed February 2005).

SUMMARY

In this chapter, we described warehousing materials handling, automation, JIT warehousing, computerization, packaging issues and reverse logistics. All these factors are closely related and interact in creating efficient, effective warehousing and materials handling operations. They support the decisions related to warehouse facilities presented in Chapter 8.

Within a warehouse, manual (non-automated) or automated materials handling equipment can be employed. Standard equipment can be categorized by the function it performs: storage and order picking, transportation and sorting or shipping. Automated equipment includes items such as automated storage and retrieval systems (AS/RS), carousels, conveyors, robots and scanning systems.

Computer technology has had a significant impact on warehousing by improving the speed and accuracy of movement storage and information transfer.

Packaging decisions affect warehousing, promotion, transportation, information transfer and a number of other key logistics decisions. Environmental concerns about packaging, recyclability and reusability are becoming more important, and the implementation and management of reverse logistics will also affect logistics decision-making in the future.

KEY TERMS

Page
Automatic guided vehicle systems (AGVSs)   **274**
A full Glossary can be found at the back of the book.

QUESTIONS AND PROBLEMS

1 Compare the advantages and benefits of automated materials handling systems with those of manual systems.
2 What are some potential pitfalls of automated materials handling? How do they affect the growth of automation in warehousing?
3 Discuss the relationship between warehousing and JIT.

4 Packaging serves two basic functions: marketing and logistics. Identify the role of packaging in each of these.

5 What marketing and logistics conflicts might occur in consumer goods packaging decisions? Use trade-off analysis to show how those conflicts might be analysed and resolved.

6 What are some of the key trends and their implications in warehousing automation and computerization?

7 Do firms need to adopt reverse logistics policies? If so, what are the key reverse logistics criteria they will need to consider in the twenty-first century?

THE LOGISTICS CHALLENGE!

PROBLEM: DELIVERING THE NEWS ON TIME

Audax, based in Gilze, Netherlands, provides distribution for newspaper and book publishers in northern Europe to airlines, bookstores and retailers' news shelves on time, every day. Audax also delivers stationery and other paper products to various business customers.

Audax has enjoyed increasing customer volume and product diversity in the last few years. Audax management knew they needed to upgrade their product picking and distribution systems from manual to automatic to accommodate these increased volume levels.

Retailers place orders daily, usually before noon, although a few push the order deadline as late as possible. One result is that orders are usually small case loads. Customers prefer to order smaller amounts to reduce overstock issues. However, the size demand means that Audax needs a flexible system.

Leo Nolen, managing director of Media Logistics, said, 'Our logistics processes have a nearly 30 million turnover. Thus, it's an important part of our operations.'

Audax requires a cost-effective automated distribution system to meet its operational and customer needs. What would you recommend it does?

What Is Your Solution?

Source: Leslie Langnau 'It's a Small World After All', *Material Handling Management Online* (August 2004), http://www.mhmonline.com/viewStory.asp?sID={2547C8B7-FCAF-4528-BF4E-629AC0F4E4DE}&S=1.

SUGGESTED READING

BOOKS

Ackerman, Kenneth B., *Warehousing Profitably: A Manager's Guide*. Columbus, OH: Ackerman Publications, 1994.

Frazelle, Edward H., *World-Class Warehousing and Materials Handling*. New York: McGraw-Hill, 2002.

Hensher, David A. and Kenneth J. Button (eds), *Handbook of Transport and the Environment*. Amsterdam: Elsevier, 2003.

Nofsinger, John B., 'Storage Equipment', in *The Distribution Management Handbook*. James A. Tompkins and Dale Harmelink (eds). New York: McGraw-Hill, 1994.

Olson, David R., 'Material Handling Equipment', in *The Distribution Management Handbook*. James A. Tompkins and Dale Harmelink (eds). New York: McGraw-Hill, 1994.

Owens, Greg and Robert Mann, 'Materials Handling System Design', in *The Logistics Handbook*. James F. Robeson and William C. Copacino (eds). New York: Free Press, 1994.

Rogers, Dale S. and Ronald S. Tibben-Lembke, *Going Backwards: Reverse Logistics Trends and Practices*. University of Nevada-Reno: Reverse Logistics Executive Council, 1998.

Stock, James R., *Development and Implementation of Reverse Logistics Programs*. Oak Brook, IL: Council of Logistics Management, 1998.

JOURNALS

Ackerman, Kenneth B., 'The Changing Role of Warehousing', *Warehousing Forum* 8, no. 12 (November 1993), pp. 1–4.

Andel, Tom, 'The Environment's Right for a Packaging Plan', *Transportation & Distribution* 34, no. 11 (November 1993), pp. 66–74.

Andel, Tom, 'Pallets Take New Directions', *Transportation & Distribution* 33, no. 1 (January 1992), pp. 32–4.

Dadzie, Kofi Q. and Wesley J. Johnston, 'Innovative Automation Technology in Corporate Warehousing Logistics', *Journal of Business Logistics* 12, no. 1 (1991), pp. 76–90.

Forger, Gary, 'We Turned an AS/RS into a Materials Control Center', *Modern Materials Handling* 50, no. 7 (June 1995), pp. 54–5.

Fraedrich, John and John Cherry, 'New Technology: Its Effects on International Distribution Systems of LDCs', *International Journal of Physical Distribution and Logistics Management* 23, no. 2 (1993), pp. 15–24.

Gray, Victor and John Guthrie, 'Ethical Issues of Environmental Friendly Packaging', *International Journal of Physical Distribution and Logistics Management* 20, no. 8 (1990), pp. 31–6.

Harrington, Lisa, 'Cross-docking Takes Costs out of the Pipeline', *Distribution Logistics* 92, no. 9 (September 1993), pp. 64–6.

Lancioni, Richard A. and Rajan Chandran, 'The Role of Packaging in International Logistics', *International Journal of Physical Distribution and Logistics Management* 20, no. 8 (1990), pp. 41–3.

Stock, James R., 'Managing Computer, Communication and Information Technology Strategically: Opportunities and Challenges for Warehousing', *Logistics and Transportation Review* 26, no. 2 (June 1990), pp. 32–54.

Tibben-Lembke, Ronald S. and Dale S. Rogers, 'Differences Between Forward and Reverse Logistics in a Retail Environment', *Supply Chain Management: An International Journal* 7, no. 5 (2002), pp. 271–82.

REFERENCES

[1] Edward H. Frazelle and James M. Apple, Jr, 'Materials Handling Technologies', in *The Logistics Handbook*, ed. James F. Robeson and William C. Copacino. New York: Free Press, 1994, p. 547.

[2] Alan C. McKinnon, 'The Economic and Environmental Benefits of Increasing Maximum Truck Weight: The British Experience', *Transportation Research Part D* 10 (2005), pp. 77–95.

[3] 'Daifuku Europe Delivers Award Winning Automated Distribution Centre for Netto A/S Denmark', *Logistics Manager* 12, no. 1 (February 2005), pp. 36–7.

[4] C.M. Baker, 'Case Study: Development of National Parts Distribution Center', *Proceedings of the Conference on the Total Logistics Concept*. Pretoria, South Africa, 1991.

[5] Kofi Q. Dadzie and Wesley J. Johnston, 'Innovative Automation Technology in Corporate Warehousing Logistics', *Journal of Business Logistics* 12, no. 1 (1991), p. 72.

[6] Edward H. Frazelle, *World-class Warehousing and Materials Handling*. New York: McGraw-Hill, 2002, p. 125.

[7] *The Warehouse Manager's Guide to Effective Orderpicking*, Monograph Series no. M0008. Raleigh, NC: Tompkins Associates, n.d., p. 21.

[8] Frazelle, *World-class Warehousing and Materials Handling*, p. 140.

[9] 'Increased Efficiency', *Logistics Manager* 11, no. 2 (March 2004), pp. 34–5.

[10] David R. Olson, 'Material Handling Equipment', in *The Distribution Management Handbook*, ed. James A. Tompkins and Dale Harmelink. New York: McGraw-Hill, 1994, pp. 19, 17.

[11] Les Gould, 'Selecting an AGVS: New Trends, New Designs', *Modern Materials Handling* 50, no. 6 (May 1995), pp. 42–3.

[12] James Aaron Cooke, 'Should You Automate Your Warehouse?', *Traffic Management* 34, no. 11 (November 1993), pp. 6-S–8-S.

[13] 'Grand Marnier in the Race to Trace', *Supply Chain Europe* 13, no. 9 (November/December 2004), p. 16.

[14] Karen A. Auguston, 'Feeding the JIT Pipeline from Across the Border', *Modern Materials Handling* 50, no. 6 (May 1995), pp. 34–5.

[15] 'Airline Adopts Barcoding', *Logistics and Materials Handling* 4, no. 2 (21 April 1995), p. 39.

[16] David Jobber, *Principles and Practice of Marketing*, 4th edn. Maidenhead, UK: McGraw-Hill, 2004, p. 294.

[17] Rod Sara, 'Packaging as a Retail Marketing Tool', *International Journal of Physical Distribution and Logistics Management* 20, no. 8 (1990), p. 30.

[18] Gordon L. Robertson, 'Good and Bad Packaging: Who Decides?', *International Journal of Physical Distribution and Logistics Management* 20, no. 8 (1990), pp. 38–9.

[19] Example provided by Dr William A. Cunningham, Air Force Institute of Technology, Wright-Patterson Air Force Base, Ohio.

[20] Martin Christopher, 'Integrating Logistics Strategy in the Corporate Financial Plan', in *The Logistics Handbook*, ed. James F. Robeson and William C. Copacino. New York: Free Press, 1994, pp. 255–7.

[21] This section draws heavily on Dale S. Rogers and Ronald S. Tibben-Lembke, *Going Backwards: Reverse Logistics Trends and Practices*. University of Nevada-Reno: Reverse Logistics Executive Council, 1998; and James R. Stock, *Development and Implementation of Reverse Logistics Programs*. Oak Brook, IL: Council of Logistics Management, 1998.

[22] Rogers and Tibben-Lembke, *Going Backwards: Reverse Logistics Trends and Practices* (1998), p. 2.

[23] Stock, *Development and Implementation of Reverse Logistics Programs* (1998), p. 20.

[24] Stock, *Development and Implementation of Reverse Logistics Programs* (1998), pp. 9–10.

[25] Gordon Scott, 'WEEE Developments: Who Will be Accountable?', *Logistics & Transport Focus* 7, no. 2 (March 2005), pp. 30–3; and 'Working Out WEEE', *Logistics & Transport Focus* 7, no. 2 (March 2005), p. 34.

CHAPTER 10
ORGANIZING FOR EFFECTIVE LOGISTICS

chapter 10 Organizing for Effective Logistics

CHAPTER OBJECTIVES

- To identify what impacts an effective logistics organization can have on a firm's efficiency and effectiveness
- To describe various types of logistics organizational structures
- To explore the factors that can influence the effectiveness of a logistics organization
- To examine an approach to developing an optimal logistics organization
- To identify attributes that can be used to measure organizational effectiveness

In the 1990s, quality and customer service became the focus of top management. In the twenty-first century, speed and supply chain management are taking hold as key competitive issues. Embracing the concepts is only the first step. A firm must be able to implement the strategies, plans and programmes to deliver acceptable levels of quality and service to its customers. Logistics and the people that are part of the logistics function play vital roles in that process. Professor Hans-Christian Pfohl, head of the Research and Development Committee for the European Logistics Association (ELA) has commented that 'business is people. At the end of the day, the success or failure of a business depends on management's ability to harness the willing participation and creativity of people.'[1]

A study by a consulting firm highlighted the importance of quality programmes in over 200 European and United States businesses. The study identified several barriers to instituting a high-quality programme. Interestingly, the top six barriers identified were related to employees or organizational issues. In order of importance, they included:

1 changing the corporate culture
2 establishing a common vision throughout the organization
3 establishing employee ownership of the quality process
4 gaining senior executive (top-down) commitment
5 changing management processes
6 training and educating employees.

The roles of individual employees and logistics departments are especially important in strategic logistics management, and each will be explored in this chapter.

Logistics executives have seen their discipline develop over the past 30 years, from infancy, when logistics functions were dispersed throughout the organization, to a highly structured, computerized and big-budget activity. The role of the logistics executive is far removed today from what it was 30 years ago, and probably quite different from what it will be 30 years hence. The twenty-first century promises unprecedented challenges.

The logistics executive has been affected by a multitude of factors, including economic uncertainty, inflation, product and energy shortages, environmentalism, green marketing, regulatory constraints, and rising customer demands and expectations (see Global box). Logistics activity is becoming increasingly difficult to manage. In this chapter, we will examine the issues of how to organize logistics within the firm and how to measure its effectiveness. We will see how important an effective logistics organization is to firms, and the types of organizational structure that exist. Although no single 'ideal' structure is appropriate for all companies, we will see how to evaluate various organizational structures and describe the approaches that can be used to develop an effective logistics organization.

GLOBAL

EU Working Time Directive Affects Freight Transport

In early 2005 the European Union's (EU) Working Time Directive (WTD) came into effect, which reduced allowable hours of work per week in EU countries from 55 to 48 hours. Truck drivers can work up to 60 hours a week providing the average does not exceed 48 over a four-month period. Self-employed drivers are exempt until 2009. However, the WTD has serious implications for long-haul freight transport across Europe and into Asia and Africa.

The United Kingdom's Road Haulage Association report estimates the extra cost of WTD in 2005 will be more than £3.8 billion, while members of the UK Freight Transport Association believe it will require them to recruit another 44,000 drivers and leave customers facing larger transport bills. A Lex Transfleet *Report on Freight Transport 2003* determined the WTD could add £18 on every UK family's monthly shopping bill, based on 4000 tonne-kilometres of freight carried each year for every man, woman and child, and could increase a typical specialist commercial vehicle operator's costs by around £350,000 a year within the next five years.

Lex Transfleet surveyed 250 fleet managers of specialist haulier and own account capacities. The survey revealed a typical specialist commercial vehicle operator with a turnover of £4.5 million employing 30 people and running 25 trucks would have to employ on average seven more drivers and buy on average two more trucks to meet the 12 per cent of additional journeys they calculate the EU directive will add to their workload.

Michael Nuttall, contract hire director of Lex Transfleet, explained that 'the UK is already finding it hard to compete with its European rivals on the back of higher fuel costs, rapidly growing insurance premiums and higher labour costs. The EU directive will be the final nail in the coffin for many operators unless they review their vehicle management and staffing practices and make significant changes in readiness for the new legislation.'

Question: **How do you think workers affected by the WTD will react to this new legislation?**

Sources: adapted from Jim Rowley, 'Working Time Directive: ILT Freight Transport Forum – Seminar', *Logistics & Transport Focus* 5, no. 6 (July/August, 2003), pp. 15–17; and Lex Transfleet, press release, 1 April 2003, www. lextransfleet.co.uk/article.cfm/id/173.html.

The Importance of an Effective Logistics Organization

An effective and efficient logistics organization is a vital part of a firm's strategic management process. The problems and challenges that organizations face do not lie primarily with strategic decision-making, but how their systems, networks, formal procedures and processes, and people are integrated and coordinated;[2] the way these interact to create a synergistic system is critical. For this reason, many organizations have engaged in reengineering – that is, they are essentially 'recreating' their organizations and systems rather than making minor changes.

Many firms have not employed their strategic resources properly. Some argue that their lack of success may have been due to the lack of a competitive organizational design or structure. An example is the information technology-enabled process reengineering implemented for the Holland Supply Bank (HSB), a Dutch flower sales intermediary. The Netherlands leads the world in producing and distributing cut flowers and potted plants, with world market share of over 50 per cent, and annual sales from the seven Dutch flower auctions exceeding €2.2 billion. The auctions are large events with about 2000 buyers conducting 50,000 transactions daily with growers at auction sites almost 750,000 square metres in size. The auctions use the traditional 'Dutch auction' method for determining price, where a clock hand starting at a high price drops until a buyer 'stops the clock' to bid for a lot at the price determined by the clock hand. Logistically, an auction site acts as a central hub for transferring products from sellers to buyers. The weaknesses of this approach are multiple handling and repackaging of products that incur higher costs and damage. Two flower auctions created HSB to mediate purchases by large customers. Prices, lot sizes, and product and delivery specifications are set by contract, and electronic data interchange (EDI) is used to communicate orders and coordinate settlements and delivery. The logistical benefits include fewer logistics activities at the auction site as growers can directly and efficiently transfer products in bulk to buyers with no repackaging costs.[3]

Other companies, as diverse as Eastman Kodak, Karolinska Hospital of Sweden (see the Creative Solutions box), and Rohm and Haas have undergone similar changes and achieved much the same results as the Holland Supply Bank.

CREATIVE SOLUTIONS

Organizing for the Twenty-first Century

The Karolinska Hospital in Stockholm, Sweden, underwent a significant reorganization in which it redesigned its traditional structure around patient flow. In other words, the patient stay in the hospital was viewed as a process, with admitting, surgery, and so on, being individual steps in the process.

The hospital had 47 separate departments and was highly decentralized. Customer service levels for patient care were unacceptable. For example, some patients spent 255 days between their first contact with the hospital and the time they received treatment. And treatment represented only 2 per cent of the entire process.

The departments in the hospital were reduced in number and redesigned. Eleven departments were formed, and two new positions were created: nurse coordinator and medical chief. Nurse coordinators are responsible for ensuring that all operations within and between departments occur smoothly. Medical chiefs are responsible for maintaining high levels of medical expertise within each department. One of the unique results of the organizational restructuring is that doctors report to nurses on administrative matters.

The results have been significant. 'Waiting times for surgery have been cut from six or eight months to three weeks. Three of 15 operating rooms have been closed, yet 3000 more operations are performed annually – a 25 per cent increase.'

Question: **What other public services in your country would benefit from better logistics organization?**

Source: adapted from Rahul Jacob, 'The Struggle to Create an Organization for the 21st Century', *Fortune*, 3 April 1995, pp. 90–9.

In traditional organizations, logistics functions were scattered throughout the firm, with no single executive, department or division responsible for managing the entire distribution process. This type of situation is depicted in Figure 10.1.

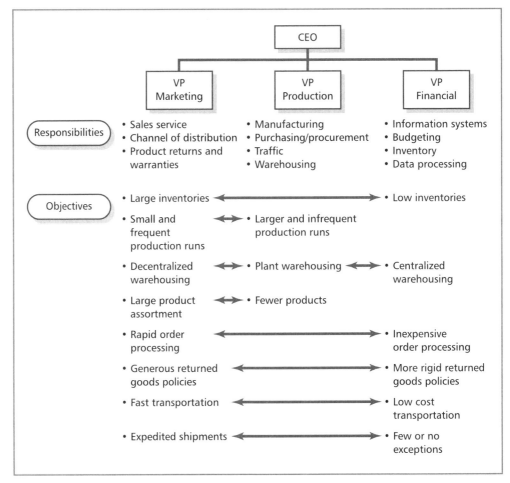

FIGURE 10.1 Traditional Approach to Logistics Management

The lack of an organizational structure that combines the activities of logistics under a single, high-level executive indicates a failure to adopt and implement the integrated logistics management concept (discussed in Chapter 1).

Since the 1960s, the trend has been towards the integration of many logistics functions under one top-ranking corporate executive. Table 10.1 lists the range of activities over which the logistics executive has had authority. In general, there has been an expansion of the logistics executive's span of control to include transportation, warehousing, inventories, order processing, packaging, materials handling, forecasting and planning, and purchasing. Coordination of the various logistics activities is crucial to the well-being of a firm.

In the next section, we will examine the major organizational types of logistics found in business firms.

Activities	Percentage of reporting companies					
	1966[a] (%)	1976[b] (%)	1985[c] (%)	1990[d] (%)	1995[e] (%)	2004[f] (%)
Transportation	89	94	97	98	95	87
Warehousing	70	93	95	97	94	92
Inventory control	55	83	81	79	80	67
Order processing	43	76	67	61	59	54
Packing	8	70	37	48	41	39
Purchasing and procurement	15	58	44	51	47	34
Number of reporting companies	47	180	161	216	208	63

TABLE 10.1 Control Exercised by the Logistics Executive over Selected Logistics Activities

Sources: [a]John F. Spencer, 'Physical Distribution Management Finds its Level', *Handling and Shipping* 7 no. 11 (November 1966), pp. 67–9.

[b]Bernard J. La Londe and James F. Robeson, 'Profile of the Physical Distribution Executive', in *Proceedings of the Fourteenth Annual Conference of the National Council of Physical Distribution Management*, 1976, pp. 1–23.

[c]Data reported are for directors of logistics. From Bernard J. La Londe and Larry W. Emmelhainz, 'Where Do You Fit In?', *Distribution* 8, no. 11 (November 1985), p. 34

[d]James M. Masters and Bernard J. La Londe, 'The 1990 Ohio State University Survey of Career Patterns in Logistics', *Proceedings of the Annual Conference of the Council of Logistics Management* 1 (1990), pp. 33–52.

[e]James M. Masters and Bernard J. La Londe, 'The 1995 Ohio State University Survey of Career Patterns in Logistics', *Proceedings of the Annual Conference of the Council of Logistics Management* Survey (1995), pp. 195–215.

[f]Bernard J. La Londe and James L. Ginter, 'The 2004 Ohio State University Survey of Career Patterns in Logistics', http://www.cscmp.org (2005), p. 9.

Types of Logistics Organizational Structure

To understand how various departments within a firm interact with one another, it is helpful to understand how business organizations have developed over the past 100 years. This will provide a background for understanding major types of interaction pattern, including functional silos, teams and committees.[4]

Development of Business Structures

Over 150 years ago companies tended to be 'one-person' operations. Only a few people at most were needed to run the entire operation. The companies were generally small and specialized, and served a local area.

Towards the middle of the nineteenth century, as companies began to grow, it was no longer possible for one person or a few people to manage all of an organization's operations. Companies began hiring people to specialize in working with or managing various functions, such as manufacturing, sales, distribution and accounting. It was believed that this created efficiency and expertise.

Divisions

By the beginning of the twentieth century, as companies continued to grow and diversify their product or service offerings, functional specialization by itself was no longer sufficient for effective management. Large organizations began to set up divisions organized vertically around similar product or service offerings. Employees became specialized in terms of function and product.

In some organizations, functions that did not directly affect the organization's product or service offering, and that cut across divisional boundaries, were left at a 'corporate' level, supporting various divisions. This was common for human resources, accounts payable, purchasing and treasury. There was no reporting relationship between 'line' divisional employees and 'staff' or corporate employees.

Matrix Organizations

By the 1950s, some large organizations realized that the divisionalized structure was not working well. It did not provide linkages between line people in various divisional and corporate positions, so the synergies of being part of a large corporation were lost. To combat this problem, many organizations began to implement **matrix organizations**.

Instead of replacing the divisional structure, a matrix structure overlays the divisional structure. In addition to divisional reporting relationships, managers in matrix organizations have reporting responsibility to another person in their function outside of the division, often at a corporate level. This structure may also be used to create reporting relationships for special projects that straddle two or more divisions.

Hollow Corporations

As we enter the twenty-first century, there has been much speculation about the prevailing form of organizations. As organizations have increased their outsourcing, contracting for many activities that had been done internally, some speculate that a **hollow corporation** will develop.

This hollow corporation, also called a network, will exist as a small organization of managers and 'ideas' people who hire or outsource external companies to perform all types of activity, including manufacturing, logistics, distribution, billing, and even sales and marketing. There are just as many experts who say that this will not happen because of loss of control, coordination issues and a host of other concerns.[5]

The rationale for the hollow organization is that organizations should specialize in and focus on what they do best, and hire specialists to perform other activities. A variation on this concept is the **virtual corporation**, where a number of companies come together to develop, produce and distribute or sell a product or service of limited scope. These organizations establish a very close working relationship, which exists only as long as the product or service is viable.

Interorganizational or Interfunctional Teams

An organization may be engaged simultaneously in a number of such relationships across a variety of products and services. These organizations focus on the product or service to be delivered to the customer, relying heavily on **interorganizational and interfunctional teams**. This type of organization is apparent in strategic alliances, such as the relationship between Apple Computer, IBM and Motorola to develop a comprehensive microprocessor and operating system for future generations of computers.[6] With the evolution of various organizational forms in mind, the discussion now turns to how this affects relationships within the firm.

Functional Silos

The term **functional silos** signifies the type of organization in which each individual functional area, such as purchasing, finance, marketing and accounting, focuses primarily on its internal operations, rather than on its obligations to the success of the corporation as a whole. While the function may have defined its goals with the overall corporate perspective in mind, it may not be sensitive to how its activities interact with the efforts of other functions within the firm in supporting overall corporate objectives.

To use a simple illustration, the purchasing function may have the goal of providing the organization with the lowest-priced inputs that meet specifications. The manufacturing group may be rewarded for providing high-quality products at the lowest cost per unit. The distribution/logistics group may have the goal of getting the product to the customer in a timely, cost-effective manner.

In this example, by focusing primarily on the goal of low price, purchasing chooses a supplier with a varying delivery lead-time. That supplier's delivery may frequently arrive late or early. So, while purchasing is meeting its goal of a low price and at the same time contributing to manufacturing's cost concerns by providing low-cost materials, it is ignoring the impact of additional production set-up costs on total product costs and may inadvertently decrease customer service by creating stockouts of key products.

This may in turn create significant costs for the company as line stoppages occur, orders are expedited and orders must be shipped to customers using expensive 'overnight' methods to meet the customers' due dates. In this way, the savings accrued from using a low-price supplier may increase the company's total costs.

Therefore, while purchasing may think that the purchasing function is meeting its goals, purchasing is not really supporting the organization as a whole. By not examining the broader picture of how the purchasing function's decisions affect other functional areas and the organization's overall efficiency, higher costs may be incurred. However, some research has shown that organizations that 'perform well in one functional area tend to be good performers in other functional areas' and that, thus, 'customers can and should expect overall firm excellence rather than compensating tradeoffs between functional areas'.[7]

A functional silo mentality or culture is extremely difficult to change. Each member of a functional area tends to develop a loyalty and commitment to his or her function, with the needs of the total organization coming second. Interactions with other functions become a zero-sum proposition. Adversarial relationships may develop among functions within the organization, as they vie for scarce resources and strive to achieve goals that may be in conflict. The advantages of specialized functions, such as focus, expertise and scale, may be overwhelmed by myopia and poor communications. However, process integration across functional areas can be a source of competitive advantage for better-performing organizations and logistics can act as a bridge for such cross-functional integration.

Generic Logistics Strategies

Michael Porter[8] proposed three competitive corporate strategies for organizations to pursue to meet their overall corporate objectives. Each of these strategies is briefly described below.

1 *Cost leadership*, where an organization focuses on those factors that will help it achieve and maintain a low-cost position in its industry.
2 *Differentiation*, where an organization creates a unique image or value for a product or service.
3 *Focus*, where an organization takes actions to compete in a particular industry segment or 'niche' that can be based on a variety of criteria including customers, products and geography.

However, to implement any or a combination of these overall corporate strategies, organizations must also develop complementary specific functional area strategies.[9] Manufacturers, wholesalers and retailers all perform logistics activities, but they are often organized differently. Manufacturers may use one of the three following organizational strategies: process-based, market-based or channel-based strategy.

1 *Process-based strategy* is concerned with managing a broad group of logistics activities as a value-added chain. The emphasis of a process strategy is on achieving efficiency from managing purchasing, manufacturing scheduling and physical distribution as an integrated system.

2 *Market-based strategy* is concerned with managing a limited group of logistics activities across a multidivision business or across multiple business units. The logistics organization that follows a market-based strategy seeks (1) to make joint product shipments to customers on behalf of different business units or product groups, and (2) to facilitate sales and logistical coordination by a single order-invoice. Often the senior sales and logistics executives report to the same senior manager.

3 *Channel-based strategy* focuses on managing logistics activities performed jointly in combination with dealers and distributors. The channel orientation focuses a great deal of attention on external control.

Wholesalers are structured differently from manufacturers because of their position in the channel of distribution and the nature of the activities they perform. In addition to the traditional wholesaling functions such as transport and storage, wholesalers offer a number of value-added services, including light manufacturing and assembly, pricing, order processing, inventory management, logistics system design and development of promotional materials.

Retailers, because of their direct contact with final customers and the high level of competition they face, often place more emphasis on inventory, warehousing and customer service activities than manufacturers. They tend to be more centralized than manufacturers and wholesalers. Rapid response is critical due to low margins, product substitutability at competitive retailers and first-level contact with customers. Many retailers are requesting deliveries direct to the store from manufacturers and are purchasing many logistics services from third parties rather than performing those activities themselves.

Logistics Coordination

Coordination of the various logistics activities can be achieved in several ways. The basic systems are generally structured through a combination of the following:

- strategic versus operational structure
- centralized versus decentralized structure
- line versus staff structure.

Strategic vs Operational

Strategic versus operational refers to the level at which logistics activities are positioned within the firm. Strategically, it is important to determine the position of logistics in the corporate hierarchy relative to other activities, such as marketing, manufacturing and finance/accounting. Equally important is the operational structure of the various logistics activities – warehousing, inventory control, order processing, transportation and others – under the senior logistics executive.

Centralized vs Decentralized

The term *centralized distribution* can reflect a system in which logistics activities are administered at a central location, typically a corporate headquarters, or a system in which operating authority is controlled under a single department or individual. Central programming of activities, such as order processing, traffic, or inventory control, can result in significant cost savings due to economies of scale.

On the other hand, decentralization of logistics activities can be effective for some firms. Some argue, with justification, that decentralizing logistics activities can lead to higher levels of customer service. Developments in computer technology and information systems, however, make it possible to deliver high levels of customer service with a centralized logistics activity.

Line vs Staff

Within the three basic types of organizational structure, logistics activities can be line, staff, or some combination of both. Logistics as a line activity is comparable to sales or production, in that employees are 'doing things' – that is, performing various tasks. When this is done, one individual is made responsible for doing the distribution job.

In the staff organization, the line activities, such as order processing, traffic and warehousing, may be housed under a logistics vice president, or under production, marketing or finance/accounting. The various staff activities assist and coordinate the line functions. The combination of line and staff activities joins these two organizational types, thus eliminating the shortcomings inherent in systems where line and staff activities are not coordinated.

In the typical staff approach to organization, logistics finds itself primarily in an advisory role. In line organizations, logistics responsibilities are operational – that is, they deal with the management of day-to-day activities. Combinations of line and staff organizations are possible, and most companies are structured in this fashion.

Other organizational approaches are possible. Examples include logistics as a function, logistics as a programme, logistics as a process and the matrix organization approach. Figure 10.2 shows the organizational design for logistics as a *function*.

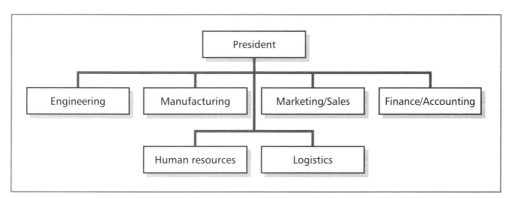

FIGURE 10.2 Organization Design for Logistics as a Function

It has been argued by some logistics experts that if a firm treats logistics as a functional area, without regard to other activities, the results will be less than optimal. Logistics is cross-functional and therefore requires a different organizational structure, not the 'functional silo' approach.

When logistics is organized as a *programme* (see Figure 10.3), the distribution activity assumes the role of a programme in which the total company participates. Individual functional areas are subordinate to the programme.

It can be argued that the optimal logistics organization lies between the two extremes represented by the functional and programme approaches. One approach has been termed the *matrix organization* and was explained earlier in the chapter (see Figure 10.4). Many firms utilize a matrix management approach, including the ABB Group (power and automation technologies), Caterpillar, Inc. (earthmoving and construction equipment), and Royal Dutch/Shell Group (petroleum, gas and chemicals).

The matrix management approach requires the coordination of activities across unit lines in the organization. Therefore, it is essential that top-level management wholeheartedly support the logistics executive. Even with high-level support, the complexities of coordination are difficult to master.

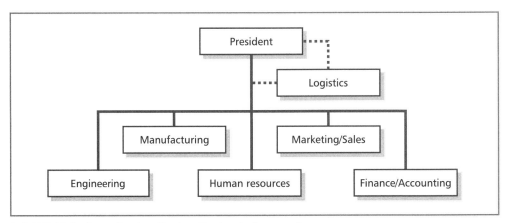

FIGURE 10.3 Organization Design for Logistics as a Programme

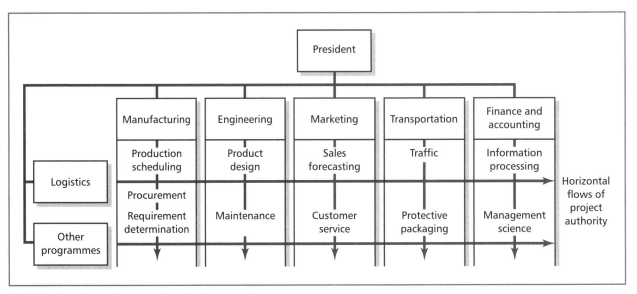

FIGURE 10.4 Logistics in a Matrix Organization

Source: adapted from Daniel W. DeHayes, Jr and Robert L. Taylor, 'Making "Logistics" Work in a Firm', *Business Horizons* 15, no. 3 (June 1972), p. 44.

For example, any time there are multiple reporting responsibilities – common in matrix organizations – problems may arise from reporting to multiple managers who may have different goals. As a result, many organizations have adopted a team structure.

For some industries, team organizations can be very effective. High-technology firms are especially suited to organizing in a team structure because of the high incidence of task or project-orientated activities that overlap several functional areas. A team structure also supports the 'flattening' of organizational layers many firms are experiencing today.[10]

A team structure involves a small group of people with complementary skills, a common goal, mutual accountability, and the resources and empowerment to achieve that goal. This differs from a work group, which is more like the traditional matrix organization, because people on the team hold themselves mutually accountable for results, rather than only individually accountable.[11] Because of the number of people involved, it is often difficult to make decisions in matrix organizations. For this reason, team structures are becoming increasingly popular.

Committees

To counteract some of the problems of poor interfunctional communications, organizations are increasingly using work groups, of which there are a wide variety in practice. The goal of a work group is to combine the skills and expertise of a number of people, generally from different functional areas, to develop a better plan, decision or execution of some action than if each worked on the problem individually. These work groups can take on a number of forms: ad hoc committees, standing committees, task teams and work teams. The level of time commitment and ongoing responsibility varies; ad hoc committees generally represent the lowest level, while work teams represent the highest level. The differences among these various types of work group are shown in Figure 10.5.

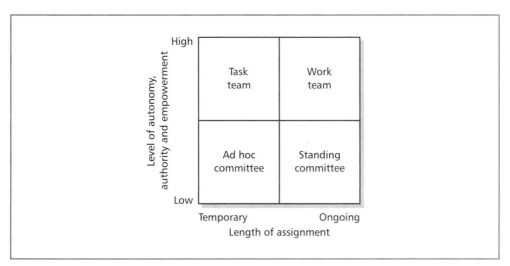

FIGURE 10.5 Types of Work Group

Source: Lisa M. Ellram and Laura M. Birou, *Purchasing for Bottom Line Impact*. Burr Ridge, IL: Irwin Professional Publishing, 1995, p. 87.

Ad Hoc Committees

Generally **ad hoc committees** are formed to resolve a certain, focused issue within a specified time frame. For example, a committee may be instituted to recommend improvements in the way office supplies are ordered and managed. Once the committee has made an analysis and recommendation, it will likely disband, never to exist in that form again.

Ad hoc committees vary significantly in intensity and duration, depending on the type of decision to be made and the urgency of the decision. Committee members must be dedicated to the goal they are trying to achieve or to the decision they are trying to make. They may not need to be dedicated to the other team members or the functioning of the team, because they know it is a temporary assignment. They may not, then, want to expend much effort on issues of group dynamics. By definition, committees are generally not all-encompassing; the employee still has other duties to maintain while serving as a member of a committee.

Standing Committees

Standing committees are distinguished from ad hoc committees because their duration is indefinite. For example, there may be a committee in charge of carrier certification. As long as the organization continues to certify carriers, that committee will exist. Because these are ongoing committees, their structure and functioning tends to take on more importance than that of ad hoc committees.

Teams

A team is defined as 'a small number of people with complementary skills who are committed to a common purpose, set of performance goals and approach for which they hold themselves mutually accountable'.[12] A key difference between teams and committees relates to their accountability and the manner in which they actually perform work, as expanded on below. A distinction is made between task teams and work teams. Task teams exist for a specific, identifiable purpose with a clear end, whereas **work teams** are ongoing in nature, much like a divisional structure, with specific, continuing goals.

Task Teams

While both task teams and ad hoc committees focus on a discrete task, they function differently. The key distinction between a task team and an ad hoc committee is that the members of an ad hoc committee still function primarily as individuals rather than as a team. Ad hoc committees may meet to decide what has to be done, but each person individually completes his or her 'piece' and is accountable for the results. In a task team, there is mutual accountability and much of the work may even be performed with other teams' members.

Another important distinction is that the task team 'owns' the project it is working on. For example, a task team to implement carrier certification would be fully empowered to design and implement the programme without further approvals. An ad hoc committee would require approval from parties outside the team, generally top management, before proceeding with each step. Thus, a task team has a much higher level of task and group responsibility than an ad hoc committee.

Work Teams

Returning to the definition of teams above, work teams, also known as **self-directed teams**, are distinguished from other types of organization by their commitment to a common purpose and goals, but more importantly by their ongoing mutual accountability to achieve those goals. These cross-functional groups are often organized around a product or a service, and may be responsible for all aspects of that product or service, from design and development to customer support.

The work team framework is unique in that it is not a temporary structure that overlays another organizational form. In practice, the work team is relatively rare. Its use represents a major change for most organizations. A key question is whether joint performance and extensive cross-functional interaction and participation will yield significantly better results than a more traditional approach. Only if the answer is yes should an organization implement work teams. Otherwise, committees may suffice.

One firm that has implemented a work team approach (empowered people on empowered teams) is DaimlerChrysler in its model platform teams. Each team is responsible for the design, development, logistics, engineering and purchasing aspects of a new model, such as the Neon or PT Cruiser.

Empowered teams should not be confused with consensus management. Team members are empowered to commit for the team on many business issues. Consensus is used for systemic issues and problem resolution. Japanese quality circles are based more on consensus decision-making.

Using Teams and Committees

Organizations are not limited to the use of one type of team or committee at a time. The organizational structure used should fit both the activity to be performed and the culture of the organization. Because the team structure is a new idea, organizations may evolve towards it slowly, using committees first, then task teams, before evolving to work teams. Conversely, an organization may decide that work teams would be of little benefit to its performance.

Because most people are educated and employed in environments where individual results and accountabilities are paramount, use of teams and committees can be threatening and even dysfunctional. To combat potential problems, most organizations train employees to function as a contributing group or team member. Training may include topics such as appreciating diversity and individual differences, team interaction and team accountability. It could involve team-building exercises or activities outside the workplace. Other issues that must be addressed in developing teams and committees are accountability, degree of responsibility and decision-making authority, and the impact of team/committee performance on individual performance appraisals.

Regardless of the formal structures chosen, interfunctional relationships remain a key component of the logistics job. Good working relationships with other functions are critical to a logistics manager's effectiveness. Companies that revamp their logistics systems by decentralizing authority and implementing self-directed teams may have difficulty. The reason is that the nature of logistics demands centralized control. Thus logistics is often a 'matrixed' team member, retaining dual reporting to both the team and a centralized logistics function, where efficient and effective strategies can be leveraged on a company-wide scale.

A review of the multitude of organizational types found in companies reveals a variety of structural forms. Firms can be successful utilizing one or more organizational structures. However, the form that is best for any given company is a difficult question to answer. Instead of examining organizational structures of several companies and speculating about 'ideal' organizations, we need to employ some empirical measures to correlate organizational structure and efficiency-productivity. The logistics executive must not only determine the firm's organizational structure, but evaluate its performance or effectiveness.

Decision-making Strategies in Organizing for Logistics

In the face of higher costs of operation and increasing pressures from customers for better service, the logistics organization must adapt to meet the challenge. An understanding of the factors that make organizations effective, and a knowledge of how these factors interrelate, are the first steps towards developing the optimal system for a firm's customers.

Mission Statements

Like individuals, an organization can and should establish a mission statement to define its overall purpose for existence.

Personal Mission Statement

A personal mission statement sets a person's overall guidelines for living. An organization is simply a group of individuals. It cannot be more focused or effective than the individuals that make up the organization. As a result of this realization, many individuals are trying to discover their true 'missions' – what they want out of work. Many are being supported in their efforts by their employers. Organizations are encouraging self-mastery and personal fulfilment. The reasons are not entirely selfless. How can an organization successfully empower employees who feel powerless and not in command of their own lives? Some of the key messages stressed are the need for continuous learning, a growth in self-confidence and personal responsibility, and a tolerance for ambiguity and change. These beliefs should work together to create a balance in life – as expressed in a personal mission statement.

The Logistics Mission Statement[13]

On an organizational level, mission statements provide a foundation upon which a company develops strategies, plans and tactics. The mission statement defines the basic purpose of an organization and identifies the parameters under which the firm will operate.

As corporate mission statements serve to provide the starting point for development corporate goals and objectives, so too will **logistics mission statements** provide direction for developing business strategies.

> Logistics has the potential to become the next governing element of strategy as an inventive way of creating value for customers, an immediate source of savings, an important discipline on marketing, and a critical extension of product flexibility.[14]

The components of a corporate mission statement or a logistics mission statement will be similar. They will vary in their specific content because the logistics mission statement is only one element of a firm's total corporate mission, but both will contain similar components. Typically, mission statements will contain eight key components, as follows.[15]

1 *Targeted customers and markets.* Who are the firm's customers? The selection of target markets and the development of marketing strategies to research those segments are vital components of a firm's activities. The following is illustrative of mission statements including such material:

> To deliver quality products to our customers and manufacturing groups in the fastest and most cost-effective mode possible. [Hospital supply company]

2 *Principal products and services.* What products or services does the firm produce? Most firms mention these in the logistics mission statement. An example is:

> [To] provide timely and effective services for the storage and commercial movement of all company finished products and of materials and supplies necessary for company operations. [Tobacco company]

3 *Geographic domain.* Where are the firm's markets located. Few firms refer to competitive markets in their logistics mission statements.

4 *Core technologies.* What technologies does the firm utilize? Few firms include core technologies in their logistics mission statements, which can perhaps be attributed to the distinction between manufacturing technologies and logistics technologies.

5 *Survival, growth and profitability.* Corporate mission statements almost universally include the issues of survival, growth and profitability. An example is:

> [To] provide timely, cost-effective shipment and delivery; to enhance our position . . . and to provide career growth for our employees. [Metal products manufacturer]

6 *Company philosophy.* What are the basic priorities to the firm? Logistics mission statements seldom include overall company or logistics philosophy. When logistics mission statements do mention philosophy, they include the notion that logistics can create competitive advantage for the firm:

> Logistics will be active in the integration and differentiation strategies that produce competitive advantage. [Lighting manufacturer]

7 *Company self-concept.* What are the firm's strengths and weaknesses relative to its competitors? When firms include such statements, they are often stated as follows:

[To] provide our customers the quality product they need when they desire it so that our service is better than our competitors. [Optical manufacturer]

8 *Desired public image*. What is the firm's social responsibility and what image does it wish to project? Most logistics mission statements include the firm's desired public image, if only generally stated. For example:

To promote the firm's image, physical distribution is committed to excellence. [Pharmaceutical firm]

Firms need a clear statement of purpose in order to develop the best combination of activities that must be performed in the day-to-day operations of the enterprise. In sum, the logistics mission statement is an important document to guide the planning, implementation and control of a firm's logistics activities.

Components of an Optimal Logistics Organization

Many factors can influence the effectiveness of a logistics organization. In general, the factors contributing to organizational effectiveness can be summarized as (1) organizational characteristics, (2) environment characteristics, (3) employee characteristics, and (4) managerial policies and practices.

Organizational Characteristics

Structure and technology are the major components of a firm's organizational characteristics. *Structure* refers to the relationships that exist between various functional areas: interfunctional (marketing, finance, operations, manufacturing, logistics) or intrafunctional (warehousing, traffic, purchasing, customer service). The relationships are frequently represented by a company's organization chart. Examples of structural variables are decentralization, specialization, formalization, span of control, organization size and work-unit size.

Technology 'refers to the mechanisms used by an organization to transform raw inputs into finished outputs. Technology can take several forms, including variations in the materials used, and variations in the technical knowledge brought to bear on goal-directed activities.'[16]

Environmental Characteristics

The effectiveness of the organization is influenced by factors internal and external to the firm. Internal factors, which are more or less controllable by the logistics executive, are known as **organizational climate**. Sometimes, this is referred to as **corporate culture**.

External factors, sometimes referred to as uncontrollable elements, include the political and legal, economic, cultural and social, and competitive environments.

Employee Characteristics

The keys to effective organizations are the employees who 'tick the boxes' on the organization chart. The ability of individuals to carry out their respective job responsibilities ultimately determines the overall effectiveness of any organization.

All employees possess different outlooks, goals, needs and abilities. These human variations often cause people to behave differently, even when placed in the same work environment. Moreover, individual differences can have a direct bearing on two important organizational processes that can have a marked impact on effectiveness. These are **organizational attachment**, or the extent to which employees identify with their employer, and individual **job performance**. Without attachment and performance, organizational effectiveness becomes all but impossible.

Managerial Policies and Practices

Policies at the macro (entire company) level determine the overall goal structure of the firm. Policies at the micro (departmental) level influence the individual goals of the various corporate functions, such as warehousing, traffic, order processing and customer service. Macro and micro policies in turn affect the procedures and practices of the organization. The planning, coordinating and facilitating of goal-directed activities – which determine organizational effectiveness – depend on the policies and practices adopted by the firm at the macro and micro levels.

A number of factors can aid the logistics executive in improving the effectiveness of the organization. Six of the most important factors that have been identified are:

1 strategic goal setting
2 resource acquisition and utilization
3 performance environment
4 communication process
5 leadership and decision-making
6 organizational adaptation and innovation.[17]

Strategic Goal Setting

Strategic goal setting involves the establishment of two clearly defined sets of goals: the overall organization goal or goals, and individual employee goals. Both sets must be compatible and aimed at maximizing company–employee effectiveness. For example, the company may have an overall goal to reduce order cycle time by 10 per cent, but the actions of each employee attempting to improve his or her component of the order cycle are what bring about achievement of that goal.

Resource Acquisition and Utilization

Resource acquisition and utilization includes the use of human and financial resources, as well as technology, to maximize the achievement of corporate goals and objectives. For example, this involves having properly trained and experienced persons operating the firm's private truck fleet, using the proper storage and retrieval systems for the company's warehouses, and having the capital necessary to take advantage of forward buying opportunities, massing of inventories and other capital projects.

Performance Environment

The performance environment is concerned with having the proper organizational climate to motivate employees to maximize their effectiveness and, subsequently, the effectiveness of the overall logistics function. Strategies that can be utilized to develop a goal-directed performance environment include (1) proper employee selection and placement, (2) training and development programmes, (3) task design, and (4) performance evaluation, combined with a reward structure that promotes goal-orientated behaviour.

Communication Process

One of the most important factors influencing logistic effectiveness in any organization is the communication process. Without good communications, logistics policies and procedures cannot be effectively transmitted throughout the firm, and the feedback of information concerning the success or failure of those policies and procedures cannot take place. Communication flows within the logistics area can be downward (boss–employee), upward (employee–boss) or horizontal (boss–boss or employee–employee).

Leadership and Decision-making

Comparable to the importance of effective communication in an organization is the quality of leadership and decision-making expertise exercised by the senior logistics executive. In many companies, the logistics department or division is a mirror image of the top logistics executive. If the top executive is a highly capable and respected individual who makes thoughtful, logical and consistent decisions, then the logistics organization that reports to him or her is also likely be highly effective. Conversely, a logistics organization led by an executive who lacks the necessary leadership and decision-making skills will not usually be as efficient.

Organizational Adaptation and Innovation

Finally, organizational adaptation and innovation is an important attribute of effective organizations. The environment that surrounds the logistics activity requires constant monitoring. As conditions change, logistics must adapt and innovate to continue to provide an optimal cost–service mix to the firm and its markets. Examples of fluctuating environmental conditions include changes in transportation regulations, service requirements of customers, degree of competition in the firm's target markets, economic or financial shifts in the marketplace and technological advances in the distribution sector. It is important that adaptation and innovation not be haphazard and unplanned.

An effective organization must exhibit stability and continuity; it must find a unique offering it can deliver to the market and stick with it to provide customer value.

An Approach to Developing an Optimal Logistics Organization[18]

Logistics organizations evolve and change – that is, there is probably a variety of good organizational designs for a firm and, over time, a company may have to modify its design to reflect environmental or corporate changes. As an executive attempts to structure a new logistics organizational unit or perhaps restructure an existing one, he or she should proceed through the following steps or stages:

1 research corporate strategy and objectives
2 organize functions in a manner compatible with the corporate structure
3 define the functions for which the logistics executive is accountable
4 know his or her management style
5 organize for flexibility
6 know the available support systems
7 understand and plan for human resource allocation so that it complements the objectives of both the individual and organization

Corporate Objectives

Overall corporate strategy and objectives provide the logistics activity with long-term direction. They provide the underlying foundation and guiding light for each functional component of the firm – finance, marketing, production and logistics. The logistics structure must support the overall corporate strategy and objectives. It is imperative that logistics executives completely understand the role their activity will play in carrying out corporate strategy. Furthermore, the logistics organizational structure must be compatible with the primary objectives of the firm.

Corporate Structure

While the specific organizational structure of the logistics activity is affected by the overall corporate structure, logistics is increasingly being centralized. In reporting relationships, logistics will typically report to the marketing group if the firm is a consumer goods company and to manufacturing/operations/administration if the firm is primarily an industrial goods producer.

Logistics is often a separate organizational activity reporting directly to the CEO in firms that have a combination of consumer and industrial goods customers. This practice is growing as the strategic importance of logistics is more widely recognized.[19]

Functional Responsibilities

Identifying a clear definition of the function of the logistics organization can be difficult, especially if the function has been restructured from an organizational structure having a traditional responsibility.[20] It is important to most of the logistics subfunctions housed under a single division or department. Full functional responsibility in one department allows the firm to implement the concepts of integrated logistics management and total cost trade-offs. Many of the functional responsibilities of the logistics organization are shown in Table 10.1.

Management Style

Almost as important as the formal structure of the organization is the **management style** of the senior logistics executive. Many firms have undergone significant changes in personnel, employee morale and productivity as a result of a change in top management. Organizational restructuring does not necessarily have to occur. The style or personality of the senior logistics executive, and to a lesser degree his or her lower-level managers, influences the attitudes, motivation, work ethic and productivity of employees at all levels of the organization.

Management style is one of those intangibles that can make two companies with identical organizational structures perform at significantly different levels of efficiency, productivity and profitability. Management style is a vital ingredient to the success of a firm's logistics mission and is one of the primary reasons that many different organizational structures can be equally effective.

Flexibility

Any logistics organization must be able to adapt to the changes that will inevitably occur. Unresponsive and unadaptable organizations typically lose their effectiveness after a period of time. While it may be difficult to anticipate future changes in the marketplace or the firm, the logistics organization must be receptive to those changes and respond to them in ways that are beneficial to the firm.

Support Systems

The nature of the logistics activity makes support systems essential. The logistics organization cannot exist on its own. There must be a variety of support services as well as support specialists available to aid the logistics department or division. As we saw in Chapter 3, a good management information system (MIS) system, manual or automated, is an important facet of an effective logistics network. Other support services or systems include legal services, computer systems, administrative services and financial/accounting services. The Technology box discusses how technology affects the way we work.

TECHNOLOGY

New Technologies and Processes are Changing the Way We Work

Voice technology has revolutionized the use of labour in warehouse and distribution centres (DCs). DC workers can dramatically affect an organization's profitability; they can operate very efficiently or waste time, make mistakes and make wrong decisions.

Voice technology is part of a process termed dynamic workforce optimization (DWO), which is a 'people-centric solution to the challenges of labour use' in the DC to improve floor logistics and maximize the 'contribution of the front-line workforce'.

Voice technology in order picking works as follows. Pickers in a DC receive computerized voice instructions through headsets, based on the picking list, that direct them to the right storage bins and to pick the right quantities for an order. The pickers wear fingertip scanners to enter products picked, and this information is then fed back to the computer to update the order on a real-time basis. Picking errors can be corrected immediately and the voice technology software program can provide instructions in different languages as required.

But DWO goes further than that – it focuses on everything a DC worker does regarding floor logistics and 'leverages the fact that the worker is always in constant, real-time communication'. For example, a worker can advise the system of blocked aisles and coordination issues between workers who finish tasks at different times, and identify why certain exception situations occur.

So while voice technology emphasizes picking and data capture, the benefits come when it is strategically used, thus taking DC workers to new levels of effectiveness.

Question: **What benefits besides effectiveness would warehouse workers receive from DWO practices?**

Source: adapted from Steven Gerrard 'Optimising Floor Logistics', *Logistics Manager* (November 2003), pp. 36–9.

Human Resource Considerations

Perhaps the most important component of an effective logistics organization is people. It is the people who ultimately determine how well the company operates. Therefore, employees' skills and abilities, pay scales, training programmes, selection and retention procedures, and other employee-related policies, are vital to the structuring or restructuring of a logistics organization.

Logistics managers are essential to a successful organization. Productive and efficient employees must be effectively led. Managers must possess certain important qualities or characteristics:

- personal integrity and an awareness of business ethics
- the ability to motivate
- planning skills
- organization skills
- self-motivation
- managerial control
- effective oral communication
- supervisory ability
- problem-solving ability
- self-confidence.[21]

Successful organizations are those that blend the optimal combination of organizational structure, planning process, people and style.

Organizational Structures of Successful Companies

While there is no single best organizational form for a firm's logistics activity, benefits can be obtained by examining the organizational structures of successful companies. First, as a purely graphical representation, an organization chart allows a person to view how the functional areas of the firm relate and how the logistics activities are coordinated. Second, viewing several organization charts of companies in a variety of industries shows that there is no single ideal structure. Third, the commonality of the logistics activities across industry types leads to marked similarities in the various organization charts. Companies have found through experience that certain logistics functions should be structured or organized in certain ways.

As an example, the organizational chart of the chemical manufacturer Rohm and Haas is shown in Figure 10.6. Rohm and Haas manufactures speciality chemicals worldwide and generated sales of €5.6 billion in fiscal 2004. Figure 10.6 shows the firm's logistics activity, which was reorganized in 1996 to incorporate all logistics functions and 400 logistics employees under a director of supply chain and logistics with a annual logistics budget of €92 million. The company integrated its 16 producing locations and 30 offsite warehouses, and in 2003 began using SAP's R/3 software for its enterprise resources planning (ERP). Rohm and Haas utilizes all transport modes to ship its chemical products to locations throughout the world.

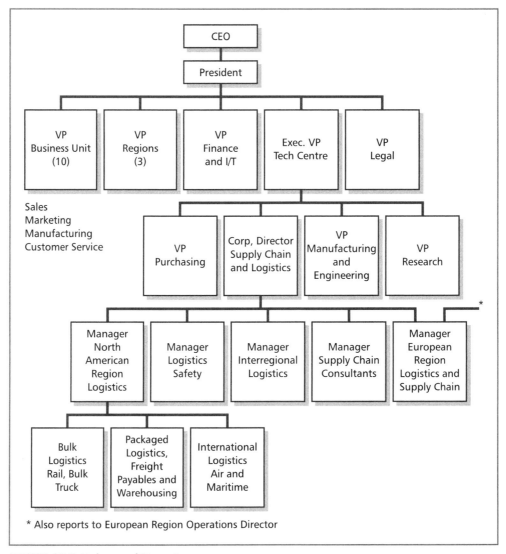

FIGURE 10.6 Rohm and Haas Company

Source: Rohm and Haas Company. Used with permission.

Measuring the Effectiveness of the Logistics Organization

Organizational performance can be measured against many criteria. Of course, it is not enough to merely identify the dimensions of organizational effectiveness, although this is a necessary first step.

The second step is to prioritize the various categories of effectiveness and to develop specific measuring devices to evaluate the level of effectiveness achieved by the logistics organization. It is vital that management identifies the measures of organizational effectiveness it wishes to utilize and puts them in order of priority. It is impractical in most instances to employ every effectiveness measure in the evaluation process. Time and monetary constraints impede the collection and monitoring of all the data needed for this type of evaluation.

Usually, it is sufficient to examine only a portion of the available measures, because patterns or trends are often exhibited very early in the evaluation process. The selection of particular measures of logistics organizational effectiveness depends on a firm's characteristics and needs. Perhaps the most difficult process is developing the techniques or procedures needed to measure the criteria of effectiveness. In this regard, there are a number of alternatives.

Cost-to-sales Ratios

Businesses use **cost-to-sales ratios** extensively to evaluate organizational effectiveness. However, in isolation, these ratios are poor measures. For example, in retailing, transportation costs are often measured as a percentage of sales. When buyers purchase the wrong product and big markdowns are taken, transportation is expected to reduce costs. Also, problems exist with regard to which costs to include under the logistics activity. For example: 'Were net or gross sales used? What logistics functions were included in the cost total? Were management salaries included? Was inventory carrying cost included? Has there been a change in order mix or service levels?'[22] There are no simple answers to these questions. All costs that are rightfully logistics costs should be included when computing measures of cost-effectiveness. If management has implemented the integrated logistics management concept, there is greater likelihood that all relevant costs will be included. Caution should be exercised when comparing these ratios across companies.

Predetermined Standards

Every measure must be evaluated against some predetermined standard. Financial measures are presented more fully in Chapter 13. Many managers believe that the firm's standards should be based on those of other firms within the same industry or of the leading firms in other industries with similar characteristics. There are many arguments in favour of this approach, but the major one is that a firm should be most concerned with its position in relation to its competition; therefore, the competition should influence the way management evaluates the firm's effectiveness. After all, customers are evaluating a firm's performance through their buying decisions.

A limitation of this approach is that each competitor has a different marketing mix and perhaps slightly different target markets. Thus the competitive benchmark that is most valuable is 'How are we doing in terms of satisfying the customers, compared with our competitors?'

Logistics Management Personnel

One area in which performance measurement is critical is logistics management personnel. Typically, managers are evaluated on the following three attributes.

1 *Line management ability*. This criterion considers the manager's ability to manage the department's day-to-day operations and to meet the established goals for productivity, utilization and all aspects of performance, including budget.
2 *Problem-solving ability*. This criterion covers the ability to diagnose problems within the operation and to identify opportunities for savings, service improvement or increased return on investment.
3 *Project management ability*. This refers to the ability to structure and manage projects designed to correct problems, improve productivity and achieve improvement benefits.

Other Measures

Firms may employ other measures (e.g. the ability to motivate and train employees) but they are not as easily measurable.

Many public- and private-sector firms have used an approach known as a **360-degree evaluation** to assess their managers. Decision-making usually involves anonymous inputs from the boss, workers/peers and subordinates. While the results generated are more qualitative than quantitative, the 360-degree evaluation generates a clear picture of how the employee is perceived at all levels and identifies areas of ambiguity and conflict between participants. Once these problems are overcome, the manager becomes much more efficient and effective. Performed effectively, the bottom line is better managers.

If management is to measure the firm's organizational effectiveness, it must employ a variety of factors that are measurable, and standards of performance need to be established. Finally, management should compare the firm with others in its industry. There is no single ideal organizational structure that every company should adopt. The most logical approach to organizing a firm's logistics activities is to understand the factors that contribute to effective organizational performance, and to include them in the planning, implementation and control of the organization.

SUMMARY

What is it that makes some organizations more effective than others? Where there is no vision, the people will perish.[23] Logistics organizations with clear statements of purpose, specific and measurable objectives, strategies and plans for achieving those objectives, and a committed workforce will undoubtedly achieve higher levels of efficiency.

Logistics organizations must of necessity become more cost- and service-efficient. An understanding of the factors that affect a firm's organizational effectiveness, along with strategies to reveal weaknesses or deficiencies, can help create more efficient logistics systems. Organizational changes form the basis for procedural modifications that can reduce costs or improve service.

In this chapter, we described the importance of an effective logistics organization to a firm. Many firms have shown significant improvements in their logistics cost–service mix as a result of organizational changes. The most important ingredient in successful management is the integration of all of the logistics activities under a single individual, department or division.

Logistics organizations are generally structured along the following lines: strategic versus operational, centralized versus decentralized and line versus staff, in various combinations. There is no single ideal organizational structure, but there are important elements that comprise an effective organization. In general, the factors contributing to organizational effectiveness can be categorized as organizational characteristics, environmental characteristics, employee characteristics, and managerial policies and practices.

Logistics managers can use a number of approaches to measure the effectiveness of their organizations. Each approach requires management to identify the elements that impact effectiveness, and to evaluate their relative importance. Next, the elements must be measured and performance evaluated. Evaluation requires that standards of performance be established.

With this and the preceding chapters as background, the concepts and principles already learned can be applied to logistics financial management issues – the subject of Chapter 11.

chapter 10 Organizing for Effective Logistics

KEY TERMS

A full Glossary can be found at the back of the book

QUESTIONS AND PROBLEMS

1   Describe the relationship between a firm's organizational structure and the integrated logistics management concept.
2   Coordination of the various logistics activities can be achieved in a variety of ways. Explain each of the following within the context of logistics organizational structure:
    a) process-based versus market-based versus channel-based strategies
    b) strategic versus operational
    c) centralized versus decentralized
    d) line versus staff.
3   'There is no single ideal or optimal logistics organizational structure.' Do you believe this statement is accurate? Briefly present the arguments for and against such a statement.
4   How do personnel affect the degree of organizational effectiveness or productivity of a firm's logistics activity?
5   What is the major value of a personal mission statement? A logistics mission statement? A corporate mission statement?
6   What role does the communication process have in influencing logistics effectiveness? Describe several strategies that can be followed to improve communication within a firm.
7   Identify how a firm's logistics management can be evaluated on each of the following factors:
    a) total logistics cost
    b) cost-specific logistics functions
    c) performance.

THE LOGISTICS CHALLENGE!

PROBLEM: WORKFORCE MOTIVATION

MFI Furniture Group is the UK's major retailer and manufacturer of kitchen and bedroom furniture. MFI has a UK retail presence of over 200 MFI stores and 200 Howden's trade stores, international operations in Europe and the Far East, and ambitions of penetrating other international markets. MFI is a labour-intensive business and faces intense global competition and pressure on profit margins. It considers its workforce is key to its success.

In 1999, the new chief executive John Hancock undertook a fundamental review to stabilize the company financially and to turn the core business around focusing on brand repositioning, a customer-service orientated approach and better efficiencies.

The logistics group was one of the key levers in delivering Hancock's new corporate strategy. It was under great pressure to meet current customer demands, support the growth of the business and make operational improvements within the new distribution network.

Vital to this was the issue of motivating a 1450-employee workforce to deliver a first-class service to MFI customers as well as reducing the logistics cost as a percentage of sales. To meet these growing demands, MFI logistics decided to invest in its most valuable asset: its people.

The logistics management team developed three primary objectives: increase productivity to gain additional capacity; reduce costs to meet an objective of 10 per cent logistics cost as a percentage of sales; and develop performance management tools to allow management to monitor and account for performance at all operational levels. How do you think it could meet these objectives?

What Is Your Solution?

Source: adapted from European Logistics Association and Kurt Salmon Associates, *Success Factor PEOPLE in Distribution Centres*. Brussels: European Logistics Association/Kurt Salmon Associates, 2004.

SUGGESTED READING

BOOKS

Bowersox, Donald J., Patricia J. Daugherty, Cornelia L. Dröge, Richard N. Germain and Dale S. Rogers, *Logistical Excellence: It's Not Business as Usual*. Burlington, MA: Digital Press, 1992.

Gunn, Thomas G., *In the Age of the Real-time Enterprise*. Essex Junction, VT: Oliver Wight, 1994.

Jackson, Thomas L. and Constance E. Dyer, *Corporate Diagnosis: Setting the Global Standard for Excellence*. Portland, OR: Productivity Press, 1996.

Kanter, Rosabeth M., Barry A. Stein and Todd D. Jick, *The Challenge of Organizational Change – How Companies Experience it and Leaders Guide it*. New York: Free Press, 1992.

Mintzberg, Henry, John P. Kotter and Abraham Zaleznik, *Leadership: The Definitive Resource for Professionals*. Harvard, MA: Harvard Business School Press, 1998.

Porter, Michael E., *Competitive Strategy*. New York: Free Press, 1980.

JOURNALS

Bartlett, Christopher A. and Sumantra Ghoshal, 'Changing the Role of Top Management: Beyond Strategy to Purpose', *Harvard Business Review* 72, no. 6 (November–December 1994), pp. 79–88.

Bowersox, Donald J. and Cornelia Dröge, 'Similarities in the Organization and Practice of Logistics Management among Manufacturers, Wholesales and Retailers', *Journal of Business Logistics* 10, no. 2 (1989), pp. 61–72.

Carlsson, Jan and Hans Sarv, 'Mastering Logistics Change', *The International Journal of Logistics Management* 8, no. 1 (1997), pp. 45–54.

Chow, Garland, Trevor D. Heaver and Lennart E. Henriksson, 'Strategy, Structure and Performance: A Framework for Logistics Research', *The Logistics and Transportation Review* 31, no. 4 (1995), pp. 285-308.

Cooke, James Aaron, 'CEOs Seize Logistics Opportunities', *Traffic Management* 34, no. 3 (March 1995), pp. 29–35.

Dröge, Cornelia and Richard Germain, 'The Design of Logistics Organizations', *Logistics and Transportation Review* 34, no. 1 (1998), pp. 25–37.

Gooley, Toby B., 'Logistics in the Boardroom', *Logistics Management* 35, no. 5 (May 1996), pp. 51–2.

Henkoff, Ronald, 'Delivering the Goods', *Fortune*, 28 November 1994, pp. 64–78.

Mitroff, Ian I., Richard O. Mason and Christine M. Pearson, 'Radical Surgery: What Will Tomorrow's Organizations Look Like?', *Academy of Management Executive* 8, no. 2 (1994), pp. 11–21.

Sharman, Graham, 'The Rediscovery of Logistics', *Harvard Business Review* 62, no. 5 (September–October 1984), pp. 71–9.

Stuart, Ian, 'Purchasing in an R&D Environment: Effective Teamwork in Business', *International Journal of Purchasing and Materials Management* 27, no. 4 (1991), pp. 29–34.

Totoki, Akira, 'Management Style for Tomorrow's Needs', *Journal of Business Logistics* 11, no. 2 (1990), pp. 1–4.

REFERENCES

[1] European Logistics Association and Kurt Salmon Associates, *Success Factor PEOPLE in Distribution Quality in Logistics*. Brussels: European Logistics Association, 2003, Preface.

[2] Cornelia Dröge and Richard Germain, 'The Design of Logistics Organizations', *The Logistics and Transportation Review* 34, no. 1 (1998), pp. 25–37.

[3] Ajit Kambil and Eric van Heck, 'Reengineering the Dutch Flower Auctions: A Framework for Analyzing Exchange Organizations', *Information Systems Research* 9, no. 1 (1998), pp. 1–12.

[4] This section draws heavily on Lisa M. Ellram and Laura M. Birou, *Purchasing for Bottom Line Impact*. Burr Ridge, IL: Irwin Professional Publishing, 1995, pp. 84–7.

[5] Ronan McIvor and Marie McHigh, 'The Organizational Change Implications of Outsourcing', *Journal of General Management* 27, no. 4 (2002), pp. 41–62.

[6] William H. Davidow and Michael S. Malone, *The Virtual Corporation*. New York: HarperCollins, 1992.

[7] Edward A. Morash, Cornelia Dröge and Shawnee Vickery, 'Boundary Spanning Interfaces Between Logistics, Production, Marketing and New Product Development', *International Journal of Physical Distribution & Logistics Management* 26, no. 8 (1996), p. 54.

[8] Michael E. Porter, *Competitive Strategy*. New York: Free Press, 1980.

[9] Garland Chow, Trevor D. Heaver and Lennart E. Henriksson, 'Strategy, Structure and Performance: A Framework for Logistics Research', *The Logistics and Transportation Review* 31, no. 4 (1995), pp. 285–308.

[10] Donald J. Bowersox, Patricia J. Daugherty, Cornelia Dröge, Richard N. Germain and Dale S. Rogers, *Logistical Excellence: It's Not Business as Usual*. Burlington, MA: Digital Equipment Corp., 1992, pp. 29–32.

[11] Jon R. Katzenback and David K. Smith, 'The Discipline of Teams', *Harvard Business Review* 71, no. 2 (March–April 1993), pp. 111–23.

[12] Ibid., pp. 111–20.

[13] This material is based on James R. Stock and Cornelia Dröge, 'Logistics Mission Statements: An Appraisal', *Proceedings of the Nineteenth Annual Transportation and Logistics Educators Conference*, ed. James M. Masters and Cynthia L. Coykendale. Columbus: Ohio State University, 1990, pp. 79–91.

[14] Joseph B. Fuller, James O'Conor and Richard Rawlinson, 'Tailored Logistics: The Next Advantage', *Harvard Business Review* 72, no. 3 (May–June 1993), pp. 87–98.

[15] Stock and Dröge, 'Logistics Mission Statements', pp. 82–4.

[16] Richard M. Steers, *Organizational Effectiveness: A Behavioral View*. Santa Monica, CA: Goodyear, 1977, pp. 7–8.

[17] Ibid., p. 136.

[18] Much of the material in this section has been developed and adapted from James P. Falk, 'Organizing for Effective Distribution', in *Proceedings of the Eighteenth Annual Conference of the National Council of Physical Distribution Management*. Chicago: National Council of Physical Distribution Management, 1980, pp. 181–99.

[19] Helen L. Richardson, 'Get the CEO on Your Side', *Transportation and Distribution* 36, no. 9 (September 1995), pp. 36–8.

[20] Falk, 'Organizing for Effective Distribution', p. 188.

[21] Paul R. Murphy and Richard F. Poist, 'Skill Requirements of Senior-level Logistics Executives: An Empirical Assessment', *Journal of Business Logistics* 12, no. 2 (1991), pp. 83–7.

[22] A.T. Kearney, Inc., *Measuring and Improving Productivity in Physical Distribution*. Oak Brook, IL: National Council of Physical Distribution Management, 1984, pp. 307–8.

[23] Proverbs 29:18, King James Bible.

CHAPTER 11
LOGISTICS FINANCIAL PERFORMANCE

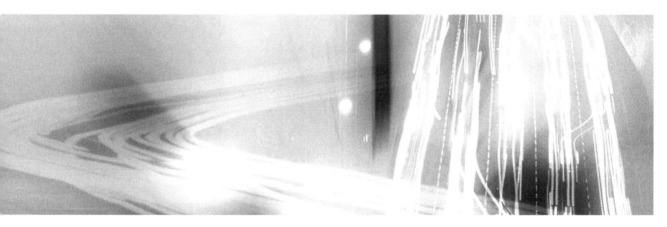

OBJECTIVES

CHAPTER OBJECTIVES

■ To demonstrate how to use logistics costs for decision-making

■ To explain how to measure and control performance of the logistics function

■ To show how to cost-justify changes in logistics structure

INTRODUCTION

Logistics costs may exceed 25 per cent of the cost of doing business at the manufacturing level. For this reason, better management of the logistics function offers the potential for large savings, which can contribute to improved corporate profitability. In mature markets – in which large percentage sales increases are difficult to achieve and corporate profitability is continuously being eroded by increasing costs and competition – it is necessary to look for ways to improve productivity.

In many firms, logistics has not been managed as an integrated system. Even in firms that have accepted the concept of integrated logistics management, evidence suggests that the cost data required for successful implementation are not available. The accurate measurement and control of logistics costs offer significant potential for improving cash flow and return on assets. In this chapter, we will concentrate on the financial control of logistics performance.

The Importance of Accurate Cost Data

Prior to 1960, logistics was viewed as a fragmented and often uncoordinated set of activities spread throughout various organizational functions. However, many major corporations have since accepted the notion that a firm's total logistics costs can be reduced, customer service improved and interdepartmental conflicts substantially reduced by the coordination of logistics activities. Computers, operations research techniques and the systems approach brought high-speed processing and the logic of mathematics to the field of logistics, and led not only to changes in transportation strategy, inventory control techniques, warehousing location policy, order-processing systems and logistics communication, but also to the desire to manage the costs associated with these functions in an integrated format.

Most of the early obstacles confronting full implementation of the concept of integrated logistics management have been overcome. The lack of adequate cost data, however, has prevented logistics management from reaching its full potential. In general, accounting has not kept pace with developments in logistics. In an attempt to solve this problem, many organizations are using **activity-based costing/activity-based management (ABC/ABM)** to analyse and manage costs.[1]

316

Accurate cost data are required for successful implementation of the integrated logistics management concept using total cost analysis. They are also required for the management and control of logistics operations.

Total Cost Analysis

The key to managing the logistics function is **total cost analysis**.[2] That is, at a given level of customer service, management should minimize total logistics costs, rather than attempt to minimize the cost of individual activities. The major shortcoming of a non-integrative approach to logistics cost analysis is that attempts to reduce specific costs within the logistics function may be less than optimal for the system as a whole, leading to greater total costs.

Total logistics costs do not respond to cost-cutting techniques individually geared to warehouse, transportation or inventory costs. Reductions in one cost invariably result in increases in one or more of the others. For example, aggregating all finished goods inventory into fewer distribution centres may minimize warehousing costs and increase inventory turnover, but it also may lead to increased transportation expense. Table 11.1 shows an analysis of this type of situation. Similarly, savings resulting from favourable purchase prices on large orders may be entirely offset by greater inventory carrying costs. Thus, to minimize total cost, management must understand the effect of trade-offs within the distribution function, and how various cost factors interact.

Annual savings from reduction in number of distribution centres (operating costs)	€350,000
Savings inventory carrying costs associated with higher turnover	€550,000
Gross savings	€900,000
Less increased expenses in transportation	
Increased distance	450,000
Premium transportation to maintain same customer service level/lead-time	200,000
Total cost increase	€650,000
Net savings from DC reduction	€250,000

TABLE 11.1 Potential Savings Due to Reduction in Number of Distribution Locations

Note: this would still be a good decision because it results in a *net* saving of €250,000. Cost increases that result from a system change should be offset against the savings to give a true picture of the impact of the change.

Cost trade-offs among the various components of the logistics system are essential. Profit can be enhanced, for example, if the reduction in inventory carrying cost is more than the increase in the other functional costs (see Figure 11.1), or if improved customer service yields greater overall revenue. If knowledgeable trade-offs are to be made, management must be able to account for the costs of each component and to explain how changes in each cost contribute to total costs. Too often, managers are concerned only with the impact on their own functional costs or revenues.

Logistics costs across Europe decreased during the 1980s and 1990s largely due to supply chain and network restructuring and the effective development of European logistics operations through more efficient planning and better logistics procurement.[3] Total logistics costs have been cut in half since 1987 and are currently just over 6 per cent of revenue or turnover. However, a recent survey by the European Logistics Association and A.T. Kearney Management

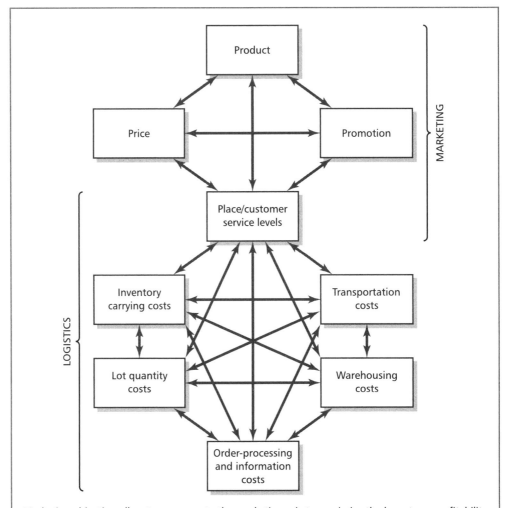

Marketing objective: allocate resources to the marketing mix to maximize the long-term profitability of the firm.

Logistics objectives: minimize total costs given the customer service objective where total costs = transportation costs + warehousing costs + order-processing and information costs + lot quantity costs + inventory carrying costs.

FIGURE 11.1 Cost Trade-offs Required in Marketing and Logistics

Source: adapted from Douglas M. Lambert, *The Development of an Inventory Costing Methodology: A Study of the Cost Associated with Holding Inventory*. Chicago: National Council of Physical Distribution Management, 1976, p. 7.

Consultants notes that decreases in logistics costs have plateaued, and will likely increase to 7 or 8 per cent of revenue by 2008. This increase will be due to increased product complexity with shorter product life cycles and higher unit costs, a global network of suppliers and customers, and continuously rising quality requirements from demanding customers.[4]

As the cost of logistics increases, the need for accurate accounting for the costs becomes increasingly more critical. Because the logistics function is relatively more asset- and labour-intensive than most other areas of the firm, its ratio of costs to total company costs has been increasing in many companies, although in some firms, because of TQM, JIT and other pro-grammes, logistics costs as a portion of total costs may have stabilized or actually gone down. Management cannot realize the full potential of logistics cost trade-off analysis until it can fully determine the costs related to separate functional areas and their interaction.

Further, as firms in a supply chain look to add value, end users or consumers begin to consider total supply chain costs that are reflected in the price of the goods on offer in the market. For example, while it took 20 years for a video cassette recorder to fall in price by 90 per cent from £400 to £40, it only took four years for a DVD player to fall by the same amount. Thus firms need to adopt a proper view of costs from 'end to end' in their supply chain 'since all costs will ultimately be reflected in the price of the finished product in the final marketplace'.[5]

Types of Cost Data Needed

The quality of the accounting data influences management's ability to enter new markets, take advantage of innovative transportation systems, choose between common carriers and private trucking, increase deliveries or inventories, make changes in distribution centre configuration, restructure the levels of inventories, make changes in packaging and determine the extent of automation in the order-processing system. The accounting system must be capable of providing information to answer questions such as those listed below.

- How do logistics costs affect contribution by product, territory, customer and salesperson?
- What are the costs of providing additional levels of customer service? What trade-offs are necessary, and what are the incremental benefits or losses?
- What is the optimal amount of inventory? How sensitive is the inventory level to changes in warehousing patterns or customer service levels? How much does it cost to hold inventory?
- What mix of transport modes or carriers should be used?
- How many field warehouses should be used and where should they be located?
- How many production set-ups are required? Which plants will be used to produce each product? What are the optimum manufacturing plant capacities based on alternative product mixes and volumes?
- What product packaging alternatives should be used?
- To what extent should the order-processing system be automated?
- What distribution channels should be used?

To answer these and other questions, management must know what costs and revenues will change if the logistics system changes. That is, the determination of a product's contribution should be based on how corporate revenues, expenses and hence profitability would change if the product line were dropped. Any costs or revenues that are unaffected by this decision are irrelevant to the problem. For example, a relevant cost is the public warehouse handling charges associated with a product's sales. An irrelevant cost is the overhead associated with the firm's private trucking fleet.

Implementation of this approach to decision-making is severely hampered by the non-availability of accounting data or the inability to use the right data when they are available. The best and most sophisticated models are only as good as the accounting input. A number of studies attest to the inadequacies of logistics cost data.[6]

Controlling Logistics Activities

A major reason for improving the availability of logistics cost data is to control and monitor logistics performance. Without accurate cost data, performance analysis is next to impossible. How can a firm expect to control the cost of shipping a product to a customer if it does not know what the cost should be? How can management determine if distribution centre costs are high or low in the absence of performance measurements? What is 'good' performance for the order-processing function? Are inventory levels satisfactory, too high or too low? These and similar questions illustrate the need for accurate cost data.

As the cost of logistics continues to rise, the need for management to account for the costs associated with each component becomes increasingly critical.[7] It also is necessary to know how changes in the costs of each component affect total costs and profits. Estimates of logistics costs ranging from 10 to 30 per cent of total sales are not uncommon, depending on the nature of the company. At best, these are only educated guesses because they are usually based on costs incorrectly computed by management.

From a corporate standpoint, the inability to measure and manage logistics costs leads to missed opportunities and expensive mistakes. Boxes 11.1 to 11.3 (presented later in this chapter) provide actual case studies that highlight the problems associated with most logistics accounting systems. These are interspersed throughout the chapter where they are relevant.

Limitations of Current Profitability Reports

Gross inadequacies exist in the segment profitability reports used by managers in the majority of corporations. Most segment profitability reports are based on average cost allocations rather than on the direct assignment of costs at the time a transaction occurs. Period costs such as fixed plant overhead and general/administrative costs are allocated to customers and products based on direct labour hours, sales revenue, cost of sales and other similarly arbitrary measures. Opportunity costs, to cover investments in inventories and accounts receivable, are not included.

Insufficient accounting information is a significant problem in most firms. Many of the problems encountered by manufacturers result from the use of a 'full costing' philosophy that allocates all indirect costs (e.g. overhead and general administrative expenses) to each product or customer group on some arbitrary basis. As a result, companies use management controls that focus on the wrong targets: direct manufacturing labour or sales volume.

Figure 11.2 shows how costs of a warehousing operation might be allocated using 'full costing'. Reward systems based on such controls drive behaviour towards either simplistic goals

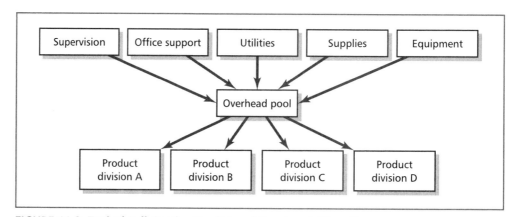

FIGURE 11.2 Typical Full Costing/Traditional Overhead Allocation

that represent only a small fraction of total cost (labour) or single-minded sales efforts (volume). They ignore more effective ways to compete, such as emphasizing product quality, on-time delivery, short lead-times, rapid product innovations, flexible manufacturing and distribution, and efficient deployment of scarce capital.

Unfortunately, most managers do not know the true cost of their company's products or services, how to reduce expenses most effectively, or how to allocate resources to the most profitable business segments, because of the following factors.

■ Accounting systems are designed to report the aggregate effects of a firm's operations to its stockholders, creditors and governmental agencies.

■ Accounting costs are computed to provide a historical record of the company's operations. All of the firm's costs are allocated to the various business segments. Because costs common to multiple segments are allocated, the process is necessarily subjective and arbitrary.

■ Accounting systems typically record marketing and logistics costs in aggregated natural accounts, and seldom attempt to attach the costs to functional responsibilities and to individual products or customers.

■ Profitability reports do not show a segment's contribution to profitability, but include fixed costs, joint product or service costs and corporate overhead costs. Top management often encourages this approach because it fears that knowledge of variable costs will lead to unrealistically low prices. In most cases, however, prices are set by the marketplace, not on the basis of costs.

■ In most standard cost systems, fixed costs are often treated the same as variable costs, which masks the true behaviour of the fixed costs.

Solving the Problem of Insufficient Cost Data

One of the difficulties in obtaining logistics costs is that they may be grouped under a series of natural accounts instead of by function, in order to satisfy generally accepted accounting principles (GAAP). **Natural accounts** are used to group costs for financial reporting on the firm's income statement and balance sheet. For example, all payments for salaries might be grouped into a salaries account. Whether they apply to production, marketing, logistics or finance, they usually are lumped together and the total shown on the financial statements at the end of the reporting period. Other examples of natural accounts include rent, depreciation, selling expenses, general and administrative expenses, and interest expenses. These costs may be lumped into such diverse catchalls as overhead, selling or general expense.

In addition, freight bills are often charged directly to an expense account as they are paid, regardless of when the associated orders are recognized as revenue. These conditions make it difficult to determine logistics expenditures, to control costs or to perform trade-off analyses.

The challenge is not so much to create new data, since much of it already exists in one form or another, but to tailor the existing data in the accounting system to meet the needs of the logistics function. By improving the availability of logistics cost data, management is in a better position to make both operational and strategic decisions. Abnormal levels of costs can be detected and controlled only if management knows what they ought to be for various levels of activity.

As Figure 11.3 shows, logistics performance can be monitored by using standard costs, budgets, productivity standards, statistical process control and activity-based management.[8] When using a third party, open-book contracts can help improve the availability and understanding of logistics cost data. An example of this is shown in the Creative Solutions box.

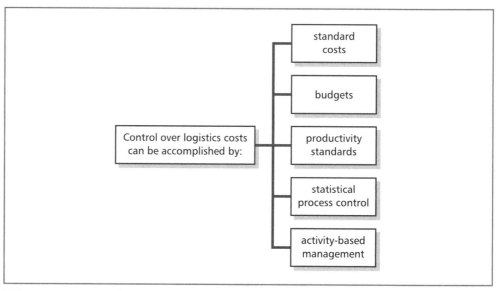

FIGURE 11.3 Controlling Logistics Activities

CREATIVE SOLUTIONS

Open Book Accounting Adds Flexibility and New Information

The global market for communication equipment has shown a growing demand for measurement systems that can handle sound and picture signals through digital technology and thus increase transmission capacity. For the last 15 years, LeanTech in Denmark has been an actor in this market, where it manufactures high-quality products for telecoms companies, and radio and TV stations all over the world.

In recent years, the company has expanded heavily and turnover has multiplied. Traditionally, LeanTech handled a large part of its production work in-house, but due to a steep increase in activities it started to use suppliers to ensure production capacity relative to demand. Recently, the firm outsourced the whole production department to suppliers. Outsourcing the production meant that, suddenly, interorganizational relations became 'highly significant', the logistics manager said. He elaborated the new challenges of the interorganizational relations: 'The subcontractors were handling the main part of our production, which meant that flexibility primarily had to come from that side. The whole thing depended on whether we were able to cooperate more closely with the suppliers in an effort to reduce costs and deliver faster.'

An intense debate among managers as to how to regain control of production processes was taking place. When production was inside, managers were said to have a direct sense of what was going on in production. After outsourcing, this confidence was lost. As an answer to the frustrations of putting production processes at a distance, the logistics management introduced open-book accounting as a medium to control and facilitate production flexibility. Sales forecasts were open to the subcontractors, and suppliers' production accounting system information was opened to LeanTech.

The logistics manager stressed that, 'By opening the books they learn more about us and we learn more about them, and thereby we could ensure flexibility.' Open-book accounting provided the logistics management with access to time and cost information about production processes. All material flows from delivery of the individual types of components to the final shipment of the product to the customer had been bar coded. In addition, information about suppliers' cost structures was also provided. It included adjustment times for assembling machines, the size of the intermediate product inventory and rate of turnover.

The firm was said to have lost touch with production after it had been outsourced, but this was being restored by the new approach. The logistics manager commented on the open access to time and cost information: 'Before I was quite convinced that we were doing everything we could in order to optimize production. We were quite close. Everyone knew each other and their responsibility. I think that the system was transparent because we were so close. This transparency was lost when we outsourced the production system. But it was regained with open-book accounting. Today, open access to time and cost information at assembly suppliers is the key element in our efforts to improve production flexibility.'

Open-book accounting made it possible to benchmark suppliers and to redesign suppliers' production and distribution processes. Furthermore, the logistics manager pointed out, production planning improved, as the open books gave him an overview of the capacity utilization at the different suppliers.

Open-book accounting made LeanTech an organizer of interorganizational relations setting up a form of virtual organization that used information to coordinate other companies. Open-book accounting inscribed a network of companies in terms of productivity, capacity, financial resources, competencies, and so on, and it let the logistics manager play a new role as interorganizational coordinator between firms.

The insights offered by the open-book arrangement into the production processes, which previously were unusual in LeanTech due to a rather informal control system, had other effects than just improving production flexibility. Open-book accounting's information about the production process gave the logistics management an unprecedented opportunity to discuss competitive advantage in terms of faster delivery time and competitive prices.

The production manager commented, 'Suddenly, we had a whole arsenal of information that disclosed the consequences and problems of decisions made by development and sales engineers in relation to the new market demands. Today, our open-book arrangement gives us this opportunity.'

Through open-book accounting with suppliers, logistics management gained access to financial information that disclosed time and cost in production processes. Logistics management used this information to increase production flexibility, but this work took a direction that it had not originally planned. Efforts became concentrated on development and sales activities as the two action areas.

In all, open-book accounting appeared to be a resource in new translations of competitive strategy, which effected development and sales activities, and the interpretation of what a technological edge and customization meant to the firm. The search for flexibility through open-book accounting showed up to be a resource for advancing productivity matters in the firm.

Question: **Can you think of circumstances where open-book accounting would not be appropriate?**

Source: adapted from J. Mouritsen, A. Hansen and C.Ø. Hansen, 'Inter-organizational Controls and Organizational Competencies: Episodes Around Target Cost Management/Functional Analysis and Open Book Accounting', *Management Accounting Research*, 12 (2001), pp. 221–44.

Standard Costs and Flexible Budgets

Control of costs through predetermined standards and flexible budgets is the most comprehensive type of control system available. A *standard* can be defined as a benchmark or 'norm' for measuring performance. **Standard costs** are what the costs should be if the firm is operating as planned. Using management by exception, managers direct their attention to variances from standard. A flexible budget is geared to a range of activity. Given the level of activity that occurs, managers can determine what the costs should have been; the use of standard costs with a flexible budget represents a direct, effective approach to the logistics costing problem. No longer are future cost predictions based simply on past cost behaviour.

Managers who use standard costs must systematically review logistics operations to arrive at the most effective means of achieving the desired output. Obviously, this will not work if logistics costs are pooled, and there is one 'standard' number for freight regardless of weight, cube or destination. This is how 'standard costing' for logistics is performed at most companies.

Accounting, logistics and engineering personnel must work together, using techniques such as regression analysis, time and motion studies, and efficiency studies, so that a series of flexible budgets can be drawn up for various operating levels in different logistics costs centres. Standards can and have been set for virtually all warehouse operations, in order processing, transportation and clerical functions. However, the use of standard costs in logistics has not been widespread.

Only recently has the importance of logistics cost control been recognized. However, management accountants and industrial engineers have a wealth of experience in establishing standard costs in the production area, which, with some effort, could be expanded into logistics.

Standard costs for logistics may be more complex to develop because the output measures can be considerably more diverse than in the case of production. For example, in developing a standard for the picking function, it is possible that the eventual control measure could be stated as a standard cost per order, per order line, per unit shipped or per shipment, or even a combination of these factors. Despite the added complexities, work measurement does appear to be increasing in logistics activities.

For example, one firm used a computerized system with standard charges and routes for 25,000 routes and eight different methods of transportation.[9] Up to 300,000 combinations were possible, and the system was updated regularly. Clerks at any location could obtain from the computer the optimum method of shipment. A monthly computer printout listed the following information by customer:

- destination
- standard freight cost to customer
- actual freight charges paid for shipments to customer
- standard freight to warehouse cost
- total freight cost
- origin of shipment
- sales district office
- method of shipment
- container used
- weight of shipment
- variance in excess of a given amount per hundredweight.

Another monthly report listed the deviation from standard freight cost for each customer, and the amount of the variance. This system provided the firm with a measure of freight performance, but, equally important, the standards provided the means for determining individual customer profitability and identifying opportunities for logistics cost trade-offs. Because this firm used standards as an integral part of its management information system, it could determine the impact of a system change (e.g. an improved, automated order-processing system) on transportation costs fairly easily.

The use of standards as a management control system is depicted in Figure 11.4. As the figure indicates, standards may result from either formal investigation, philosophy/intuition, or both.

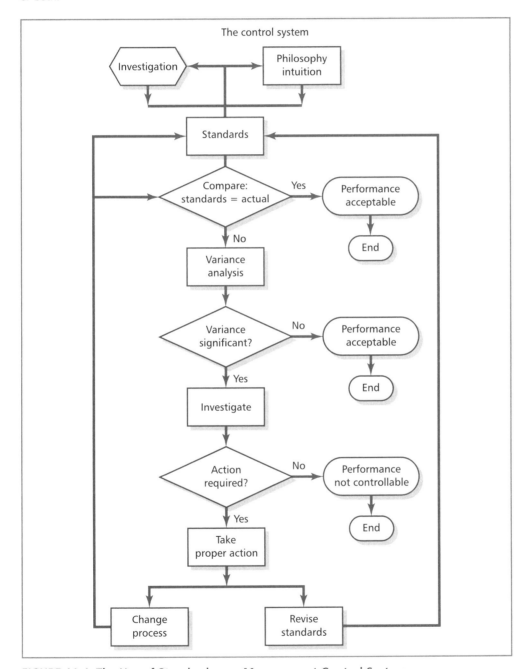

FIGURE 11.4 The Use of Standards as a Management Control System

Source: Richard J. Lewis and Leo G. Erickson, 'Distribution System Costing: An Overview', in *Distribution System Costing: Concepts and Procedures*, ed. John R. Grabner and William S. Sargent. Columbus, OH: Transportation and Logistics Research Foundation, 1972, p. 17A.

Variance Management

Once standards have been set, the actual performance is compared to the standard to see if it is acceptable. If so, the system is under control and that is the end of the control process. Where performance differs from the standard, investigation may be warranted. It is most meaningful to judge variances in terms of their *practical* significance. How significant is the variance in its effects on bottom-line performance – that is, net profit? If significant, the next question to ask is whether some action is required.

If the variance is significant but uncontrollable, no action may be indicated and the control process terminated. If action is indicated, it may be that the standard is considered wrong and must be changed, or the process itself is creating the problem and thus should be changed. The new process or standard will be judged following the same cycle.

A standard tells management the expected cost of performing selected activities and allows management to make comparisons that point out any operating inefficiencies. For example, Table 11.2 illustrates a report that is useful at the operating level. It shows why the warehouse labour for the picking activity was €320 over budget. The costs of logistics activities can be (1) reported by department, division, function, product group or total, (2) compared with their standard, and (3) included as part of regular weekly or monthly performance reports.

Items picked during week	14,500
Hours accumulated on picking activities	330
Standard hours allowed for picks performed based on 50 items per hour	290
Variation in hours	40
Standard cost per labour hour	€8
Variation in cost due to inefficiencies	€320*

TABLE 11.2 Summary of Warehouse Picking Operations for Week X

* The cost was €320 over budget because there were 40 picking hours in excess of the standard number of hours allowed for efficient operation, at a standard cost per labour hour of €8.

Table 11.3 shows a level of aggregation that would interest the firm's president. This report allows the president to see at a glance why targeted net income has not been reached. On the one hand, there is a £3 million difference due to ineffectiveness, which indicates the net income the company has forgone because of its inability to meet its budgeted level of sales.

On the other hand, there is an inefficiency factor of £1.4 million, which indicates that at the actual level of sales the segment-controllable margin should have been £18.0 million. The difference between £18.0 million and the actual outcome of £16.6 million is a £1.4 million variation due to inefficiencies within the marketing and logistics functions.

Budgetary Practices

Conceptually, standard costs are generally superior to budgetary practices for control purposes. Sometimes, however, the use of standards is inappropriate. This is particularly true in situations with essentially non-repetitive tasks and for which work-unit measurements are difficult to establish. In these situations, control can be achieved through budgetary practices. However, the extent to which the budget is successful depends on whether individual cost-behaviour patterns can be predicted and whether the budget can be flexed to reflect changes in operating conditions.

| | Budget | Explanation of variation from budget | | | Actual results |
		Variance due to ineffectiveness	Standard allowed for output level achieved	Variance due to inefficiency	
Net sites	£90,000	£10,000	£80,000	–	£80,000
Cost of goods sold (variable manufacturing cost)	40,500	4,500	36,000	–	36,000
Manufacturing contribution	£49,500	£5,500	£44,000	–	£44,000
Variable marketing and logistics costs (out-of-pocket costs that vary directly with sales to the segment)*	22,000	2,500	20,000	£1,400	21,400
Segment contribution margin	£27,000	£3,000	£24,000	£1,400	£22,600
Assignable non-variable costs (costs incurred specifically for the segment during the period)†	6,000	–	6,000	–	6,000
Segment controllable margin	£21,000	£3,000	£18,000	£1,400	£16,600

TABLE 11.3 Segmental Analysis Using a Contribution Approach (€000)

Notes: this analysis can be performed for segments such as products, customers, geographic areas and divisions.

Assumption: actual sales revenue decreased, a result of a lower volume. The average price paid per unit sold remained the same. (If the average price per unit changes then an additional variance – the marketing variance – can be computed.)

Difference in income of £4400 (£21,000–16,600) between budgeted and actual results can be explained by the following variances:

Ineffectiveness – inability to reach target sales objective   £3000
Inefficiency at operating level achieved of £80,000   1400
     £4400

*These costs might include: sales commissions, transportation costs, warehouse handling costs, order-processing costs and a charge for accounts receivable.

†These costs might include: salaries, segment-related advertising, bad debts and inventory carrying costs. The fixed costs associated with corporate-owned and operated facilities would be included if, and only if, the warehouse was solely for this segment of the business.

Most logistics budgets are static – that is, they are a plan developed for a budgeted level of output. If actual activity happens to be the same as budgeted, management can make a realistic comparison of costs and establish effective control. However, this is seldom the case. Seasonality or internal factors invariably lead to different levels of activity, the efficiency of which can be determined only if the reporting system can compare the actual costs with what costs should have been at the operating level achieved. Box 11.1 illustrates this problem in more detail.

For instance, a firm's warehousing unit may have an estimated or budgeted level of activity of 10,000 line items per week. The actual level of activity may be only 7500. Comparing the budgeted costs at 10,000 line items with the actual costs at 7500 leads to the erroneous conclusion that the operation has been efficient, since items such as overtime, temporary help, packing, postage and order processing are less than budget. A flexible budget, on the other hand, indicates what the costs should have been at the level of 7500 line items of activity, resulting in a true euro measure of efficiency.

Box 11.1 Control Deficiencies

Control of costs and motivation of key personnel is critical in every business activity. Logistics is no exception. However, the control concepts successfully utilized by other functional areas have not been widely adopted for logistics activities. It might be argued that logistics is different from other disciplines and cannot be evaluated with the same tools. In most cases, however, the application has never been attempted. A particular case in point is the application of the flexible budgeting concept.

Company B maintained an annual budget for its branch warehousing costs. These costs consisted of variable and fixed expenses. Each month, the annual budget was divided by 12 and compared with the actual costs for that month. Differences from the budget were recorded as variances, and management took action on these. However, company B's sales were seasonal, with some months far more active than others. The variances were always unfavourable

during peak months and favourable during slow months. Productivity ratios, on the other hand, gave different results. They were high during peak periods and dropped during slower periods.

In this type of situation, neither cost control nor employee motivation is adequately addressed. Dividing the annual budget by 12 and comparing it with actual monthly costs means that management is trying to compare costs at two different activity levels. The costs should be the same only if actual monthly activity is equal to $1/12$ of the planned annual activity. A far more acceptable approach is to recognize that a portion of the costs are variable and will rise or fall with the level of output. Flexing the budget to reflect what the costs should have been at the operating activity level experienced permits a true measure of efficiency and productivity, and provides more meaningful evaluations of performance.

Source: this material is adapted from Douglas M. Lambert and Howard M. Armitage, 'Management Distribution Cost for Better Profit Performance', *Business* 30, no. 5 (September–October 1980), pp. 46–52. Reprinted with permission from *Business* magazine.

Analysis of Cost-behaviour Patterns

The key to successful implementation of a flexible budget lies in the analysis of cost-behaviour patterns. In most firms, little of this analysis has been carried out in the logistics function. The expertise of the cost accountant and industrial engineer can be invaluable in applying tools such as scatter-diagram techniques and regression analysis to determine the fixed and variable components of costs. These techniques use previous cost data to determine a variable rate per unit of activity and a total fixed cost component. Once this is accomplished, the flexible budget for control becomes a reality.

Unlike engineered standards, the techniques are based on past cost-behaviour patterns, which undoubtedly contain inefficiencies. The predicted measure of cost, therefore, may not be a measure of what the activity *should* cost, but an estimate of what it *will* cost, based on the results of previous periods.

Capital Budgets

Capital budgets are used to control capital expenditures, such as long-term investments in property, facilities and equipment. In logistics, this would include purchases of new trucks, computer equipment, materials handling and warehouse equipment, the building of new distribution centres and similar long-term investments. These items are controlled by limiting the euro amount to be spent in a given year (annual capital expenditure budget), and requiring a minimum net present value or payback on productive investments. A discussion of how these items are calculated is included in any basic finance textbook.

Productivity Standards

Logistics costs can be controlled by the use of productivity ratios. These ratios take the form of:

$$\text{Productivity} = \frac{\text{Measure of output}}{\text{Measure of input}}$$

For example, a warehouse operation might make use of such productivity ratios as:

$$\frac{\text{Number of orders shipped this period}}{\text{Number of orders received this period}}$$

$$\frac{\text{Number of orders shipped this period}}{\text{Average number of orders shipped per period}}$$

$$\frac{\text{Number of orders shipped this period}}{\text{Number of direct labour hours worked this period}}$$

Productivity ratios for transportation might include:[10]

$$\frac{\text{Tonne-kilometres transported}}{\text{Total actual transportation cost}}$$

$$\frac{\text{Stops served}}{\text{Total actual transportation cost}}$$

$$\frac{\text{Shipments transported to destination}}{\text{Total actual transportation cost}}$$

Productivity ratios can be generated for the following transportation resource inputs: labour, equipment, energy and cost. Table 11.4 illustrates the specific relationships between these inputs and transportation activities. An X in a cell of the matrix denotes an activity–input combination that can be measured. Table 11.5 illustrates an activity–input matrix for warehousing. Table 11.6 illustrates how warehouse productivity ratios can be calculated, and interprets the results.

Activities	Labour	Facilities	Equipment	Energy	Overall (cost)
Transportation strategy development	–	–	–	–	X
Private fleet over-the-road trucking					
Loading	X	–	–	–	X
Line-haul	X	–	–	X	X
Unloading	X	–	–	–	X
Overall	X	–	–	X	X
Private fleet pick-up/delivery trucking					
Pre-trip	X	–	–	–	X
Stem driving	X	–	–	X	X
One-route driving	X	–	–	X	X
At-stop	X	–	–	–	X
End-of-trip	X	–	–	–	X
Overall	X	–	X	X	X
Purchased transportation operations					
Loading	–	–	–	–	X
Line-haul	–	–	–	–	X
Unloading	–	–	–	–	X
Rail/barge fleet management	–	–	–	–	X
Transportation/traffic management	–	–	–	–	X

TABLE 11.4 Transportation Activity–Input Matrix

Source: A.T. Kearney, Inc., *Measuring and Improving Productivity in Physical Distribution*. Oak Brook, IL: National Council of Physical Distribution Management, 1984, p. 144.

Activities	Labour	Facilities	Equipment	Energy	Overall (cost)
Company-operated warehousing					
Receiving	X	X	X	–	X
Put-away	X	–	X	–	X
Storage	–	X	–	–	X
Replenishment	X	–	X	–	X
Order selection	X	–	X	–	X
Checking	X	–	X	–	X
Packing and marking	X	X	X	–	X
Staging and order consideration	X	X	X	–	X
Shipping	X	–	X	–	X
Clerical and administration	X	X	X	–	X
Overall	X	–	X	–	X
Public warehousing					
Storage	–	–	–	–	X
Handling	–	–	–	–	X
Consolidation	–	–	–	–	X
Administration	–	–	–	–	X
Overall	–	–	–	–	X

TABLE 11.5 Warehouse Activity–Input Matrix

Source: A.T. Kearney, Inc., *Measuring and Improving Productivity in Physical Distribution*. Oak Brook, IL: National Council of Physical Distribution Management, 1984, p. 195.

Data:

Number of orders shipped this period	2750
Number of orders received this period	2800
Average numbers of order shipped per period	2500
Number of direct labour hours worked this period in shipping	200

Calculation of ratios:

1. $\dfrac{\text{Numbers of orders shipped this period}}{\text{Number of orders received this period}} = \dfrac{2750}{2800} = 98.2\%$

A ratio of less than 100% means the firm is building up unshipped orders, over 100% means the firm is reducing backlog.

2. $\dfrac{\text{Numbers of orders shipped this period}}{\text{Average number of orders shipped per period}} = \dfrac{2750}{2500} = 110\%$

The firm had a busier than usual month (ratio greater than 100%); this combined with ratio 1 shows that the firm is building up a backlog.

3. $\dfrac{\text{Numbers of orders shipped this period}}{\text{Number of direct labour hours worked this period in shipping}} = \dfrac{2750}{200} = 13.75$ orders per hour

This number is only meaningful in comparison with a standard or benchmark. It tells us the number of orders shipped per person hour.

TABLE 11.6 The Calculation of Warehouse Productivity Ratios

Productivity measures of this type have been developed for most logistics activities. In the absence of a standard costing system, they are particularly useful with budgetary practices, because they provide guidelines on operating efficiencies. Furthermore, such measures are easily understood by management and employees. However, productivity measures are not without their shortcomings.

- Productivity measures are expressed in terms of physical units and actual euro losses caused by inefficiencies, and predictions of future logistics costs cannot be made. This makes it difficult to cost-justify any system changes that will result in improved productivity.
- The actual productivity measure calculated is seldom compared to a productivity standard. For example, a productivity measure may compare the number of orders shipped this period to the number of direct labour-hours worked in the same period, but it does not indicate what the relationship *ought* to be. Without work measurement or some form of cost estimation, it is impossible to know what the productivity standard should be in efficient operations.
- Changes in output level may in some cases distort measures of productivity. This distortion occurs because the fixed and variable elements are seldom delineated. Consequently, the productivity measure computes utilization, not efficiency. For example, if 100 orders shipped represents full labour utilization and 100 orders were received this period, then productivity as measured by:

$$\frac{\text{Number of orders shipped this period}}{\text{Number of orders received this period}} \times 100\%$$

The result would be 100 per cent. However, if 150 orders had been received and 100 orders were shipped, productivity would have been 66.67 per cent, even though there was no real drop in either efficiency or productivity.

Statistical Process Control

The Japanese demonstrate a high level of sophistication in the use of statistical methods to enhance the quality of products and services.[11] The use of such approaches is increasing in the automobile industry, in high-tech firms and among consumer products manufacturers. The output of successful logistics is the level of customer service provided. Although many firms measure the proportion of shipments that arrive on time or the average length of the order cycle from a particular supplier, further insight into these areas is seldom obtained through the use of statistical process control techniques.

The use of statistical methods offers an alternative to conventional management control processes. **Statistical process control (SPC)** requires an understanding of the variability of the process itself prior to making management decisions. To analyse delivery times from several suppliers, for example, it is necessary to know the mean, or average, time elapsed from the issuance of a purchase order to the receipt of a shipment, and the likely variation in delivery times.

The SPC Process

Figure 11.5 illustrates the steps of SPC. As with classical approaches to control, the first three steps are:
1 design system
2 establish standards
3 perform process.

Once the first three steps are accomplished, SPC requires that the following questions be raised. First, are the measurements 'in control'? That is, do all measurements fall within rea-

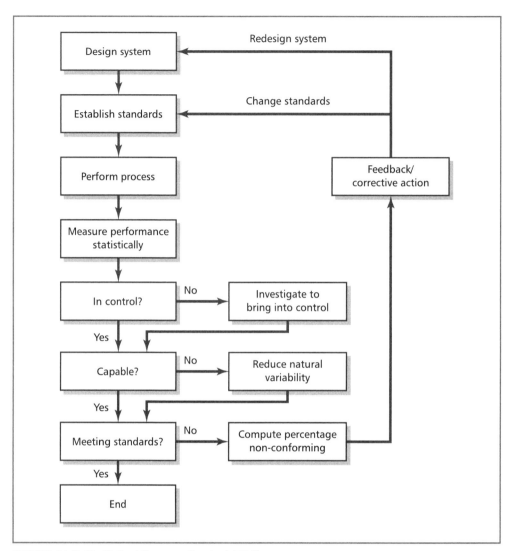

FIGURE 11.5 Statistical Process Control (SPC)

Source: C. John Langley, Jr, 'Information-based Decision Making in Logistics Management', *International Journal of Physical Distribution and Materials Management* 15, no. 7 (1985), p. 50. Copyright MCB University Press Limited.

sonable proximity of the mean? Second, is the process 'capable'? That is, does the observed variability have a lesser magnitude than a pre-specified range? Third, do the measurements meet standards? Only if the answers to these three questions are yes is the process itself said to be in control.

The **control chart** is perhaps the most widely used statistical approach to gain insight into these questions. It permits an examination of process behaviour measurements in relation to both upper and lower control limits. These limits are statistically derived and are used to identify instances where the observed behaviour differs significantly from what was expected and where a problem is likely to exist. The calculation of these limits is beyond the coverage of this text, but will generally be included in a good introductory statistics or quality course. The search for explanations can proceed in an organized efficient manner.

Figure 11.6 shows two control chart applications developed from actual logistics-related data. Part A resulted from an examination of transit time data (in minutes) of shipments travelling between two cities 420 kilometres apart. Once the three points located outside the control limits were identified and isolated, an inquiry determined their causes and assured removal of those causes.

Part B of Figure 11.6 shows the percentages of carrier freight bills that a particular shipper found to contain errors. In this example, the only error percentages of real concern are those

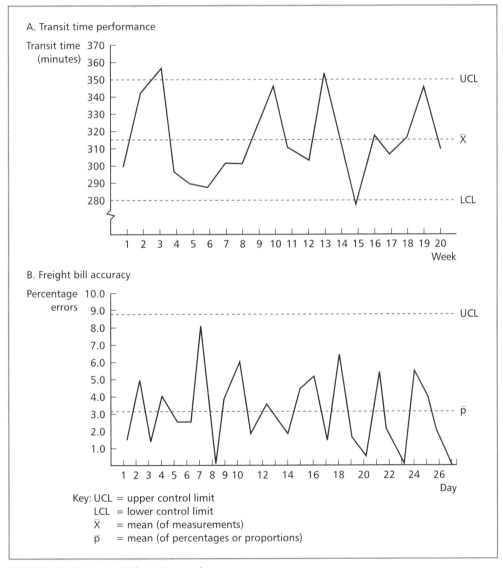

FIGURE 11.6 Control Chart Examples

Source: C. John Langley, Jr, 'Information-based Decision Making in Logistics Management', *International Journal of Physical Distribution and Materials Management* 15, no. 7 (1985), p. 51. Republished with permission, Emerald Group Publishing Limited.

that are excessively high as to be labelled out-of-control in a statistical sense – that is, above the upper control limit (UCL). In this instance, subsequent investigation resulted in the identification and removal of the cause of the problem at hand.

A number of other activities in logistics management could serve equally well for purposes of illustration, including warehousing, materials handling, packaging, inventory control, order processing and customer service. Large cost savings can be captured in these areas through the appropriate use of statistical methods.

Making SPC a Success

There are a number of prerequisites to success with SPC.

1 It is important to recognize that the use of statistical methods is simply a tool that can assist in improving quality. SPC provides valuable insight into the behaviour of the various processes under scrutiny.

2 Top management support is necessary for success. The effective use of SPC approaches may involve a cultural change in a firm. Successful firms have commitment from top management and require a high level of familiarity with the various approaches.

3 Finally, the use of statistical methods should be viewed as a key component of an overall total quality management programme. Other elements of this programme may include the establishment of quality policies, the setting of goals and objectives, supervisory training, quality awareness and programmes for the removal of error causes, and performance management.[12]

The issue of quality in logistics is frequently addressed in conjunction with productivity. However, attention has focused on the identification of specific data and information needs rather than on what to do with the information once it has been acquired. Until more progress is made in this direction, the benefits of improved productivity will continue to represent a largely untapped resource.

Successful implementation of SPC approaches depends on the timely availability of appropriate information. For this reason, SPC provides an opportunity for the logistics and information systems areas to work together to enhance productivity and improve quality.

Logistics Costs and the Corporate Management Information System (MIS)

While substantial savings can be generated when management is able to compare its actual costs with a set of predetermined standards or budgets, even greater opportunities for profit improvement exist in the area of decision-making. If management is to make informed decisions, it must be able to choose between alternatives such as utilizing additional common carrier transportation or enlarging the company's private fleet, increasing the number of deliveries or increasing inventories, expanding or consolidating field warehouses, and automating the order-processing and information system.

The addition or deletion of sales territories, salespeople, products or customers requires a knowledge of how well existing segments are currently performing, and how revenues and costs will change with the alternatives under consideration. For this purpose, management needs a database capable of aggregating data, so that it can obtain routine information on individual segments such as customers, salespeople, products, territories or channels of distribution.

The system must be able to store data by fixed and variable components, so that the incremental revenues and costs linked to alternative strategies can be determined. Some problems connected to the inability to distinguish between variable and fixed costs are presented in Box 11.2.

Box 11.2 Inability to Distinguish between Fixed and Variable Costs

Management of company C used a product reporting statement that deducted manufacturing, logistics and marketing costs from sales to arrive at a net income for each product. It used the profit statement for making decisions about the acceptability of product performance, the assignment of marketing support, and the deletion of products. The allocation of logistics costs to each product was carried out using ABC analysis, in which A products were allocated a certain amount of logistics costs, B products twice as much as A, and C products three times as much as A. These allocations contained costs that varied with activity, such as warehouse labour, supplies and freight expenses. They also included costs that remained fixed regardless of activity levels (e.g. corporate allocations, depreciation and administrative costs of the corporate fleet).

Several of the company's products, including one that was among the 'top 10' in sales performance, were showing losses and were candidates for being discontinued. However, analysis revealed that a large proportion of the total distribution cost, along with approximately 30 per cent of the manufacturing cost, was fixed and would not be saved if the products were eliminated. Indeed, by discontinuing these products, total corporate profitability would decline because all of the revenues related to these products would disappear, but all of the costs would not.

Although the variable costs and the identifiable fixed costs would be saved, the company would continue to incur the majority of fixed costs – which in this case were substantial – regardless of the product deletions under consideration. If the firm discontinued the products, the existing fixed costs would be redistributed to the remaining products, leading to the very real possibility that even more products would appear to be unprofitable.

Source: this material is adapted from Douglas M. Lambert and Howard M. Armitage, 'Management Distribution Cost for Benefit Profit Performance', *Business* 30, no. 5 (September–October 1980), pp. 46–52. Reprinted with permission from *Business* magazine.

Barriers to Effective Information Management

Several types of transaction occur in every business, and each transaction results in the creation of source documents (e.g. customer orders, shipment bill of lading, sales invoices to customers, invoices from suppliers and vendors). In addition, companies perform a variety of internal transactions and activities that are documented (e.g. 'trip reports' for private fleet activities, 'call reports' from salespeople, warehouse labour time cards). Other costs may be recognized by means of standard cost systems, engineering time studies or statistical estimating (e.g. multiple regression techniques).

The key to success is that source documents *must* be computerized. Data need to be linked using the relational database concept (see Chapter 3), so that they can be analysed in a variety of ways that are useful for supporting decision-making.

However, many managers mistakenly feel that the same accounting practices (i.e. the allocation of all costs) used to value inventories and report results to the government for taxation purposes or to the stock exchanges for shareholder information, are required to generate reports for managing the business. Managers also may feel that using only variable and direct fixed costs might encourage suboptimal pricing by salespeople.

Accountants frequently oppose a separate management accounting system, while managers often fail to recognize the differences between fixed and variable costs, the distinction between direct and indirect expenses, or the usefulness and purpose of contribution reports.

Finally, management information systems personnel often discourage the development of such reports by citing the difficulties in creating the databases and operating systems required to assign direct costs to specific product and market segments. With the revolution in information technology, these reasons are no longer valid.

Importance of Good Information

With the information from the relational database, management is in a position to evaluate the profitability of various segments. The database permits the user to simulate trade-off situations and determine the effect of proposed strategic and system changes on total cost. The Technology box provides an example of how one leading software supplier, Germany's SAP AG, can help firms account for costs in operations and logistics within various industrial sectors including the oil industry.

TECHNOLOGY

SAP Software Helps Manage Financial and Logistics Issues

Founded in 1972 in Walldorf, Germany, SAP AG has grown into the leading enterprise software supplier across the globe. It provides collaborative software business solutions for all types of industry in every major market. In 2004 SAP generated sales of €7.5 billion from its 12 million users, 88,700 installations and more than 1500 partners. It is the world's largest inter-enterprise software company and the world's third-largest independent software supplier overall. SAP employs more than 32,000 people in over 50 countries.

Two of its software-product suites are 'mySAP ERP', which contains modules for accounting, financial reporting, performance management and corporate governance, and 'mySAP SCM', which contains modules for planning and execution capabilities for managing enterprise operations, as well as coordination and collaboration technology to extend those operations beyond corporate boundaries.

An example of one product that can benefit a specific industrial sector is its oil and gas accounting suite. When oil companies such as BP, Royal Dutch Shell and Norway's Statoil explore and develop oil reserves in the North Sea, they tend to do so on a joint venture or shared basis regarding costs and resultant revenues. One firm is the managing partner or contractor and is responsible for operations related to production, suppliers, customers, and so on. The joint venture also has a silent partner in government: the national government where the oil-field is located takes a percentage royalty on all oil and gas production revenue. The managing partner also does all the accounting for the joint venture, including allocating costs and revenues, paying the other joint venture partners and billing them if capital expenditures need to be done, such as servicing or reworking the well, and ensuring the government royalties are paid.

SAP has developed SAP PSA and SAP JVA to meet the requirements of upstream oil and gas companies involved in production-sharing contracts (PSCs) and joint venture accounting respectively. SAP PSA enables the managing partner to effectively manage PSCs, plan budgets and project expected financial results. By using SAP PSA and JVA together, the cost data (maintained in SAP JVA accounting ledgers) can be used as the source cost data in SAP PSA, providing consistent accounting.

SAP PSA automates the following processes: recording and classifying costs, allocating production to government royalty and contractor, calculating profit shares and entitlements, and reporting results for use by the joint venture partners and government.

Also, SAP RLM, for remote logistics management, keeps an oil company's offshore production facilities in the North Sea supplied by seamlessly integrating remote logistics with all other logistics processes. The oil company can stock materials onshore, offshore or both. SAP RLM can automatically convert stock transfer requests, created at the remote location, into stock transport orders, or purchase requisitions at an onshore base plant.

Besides this suite of products for upstream oil and gas exploration companies, SAP also provides products to downstream operations for commodity sales, bulk transportation and inventory management of hydrocarbon products, and to service station operations for convenience retailing, fuels management, site and headquarters management, and business analysis and reporting.

Question: **What other logistical functions might benefit from integrated software programs?**

Source: adapted from SAP AG, http://www11.sap.com/, 2005.

The key to measuring logistics performance is an integrated, broad-based computer data file. To track an order and its associated costs from origin to receipt by customer, for example, it is necessary to access a number of files in the logistics information system:

- open orders (for back orders)
- deleted orders (order history file)
- shipping manifest (bills of lading)
- transportation freight bills paid.

With today's information-processing capabilities, it is possible to access the desired information automatically from these and other necessary files (e.g. inventory, customer retail feedback data, damage reports and claims, billing and invoicing files). From these files, management can construct a condensed 'logistics performance' database, which can provide all of the necessary information required to measure overall as well as individual activities on a regular basis. What used to take several 'person-years' of concerted programming effort, and cost large sums of money, can now be developed in a matter of weeks or months, using personal computers and standard statistical packages.

A major consumer products company uses this approach to construct a series of more than 50 reports using a common 'logistic performance' file, and these cover more than just logistics costs. The same data files are used to report financial, customer service or productivity-related reports.

Activity-based Costing

Activity-based costing has received increased attention as a method of solving the problem of insufficient cost data. Traditional accounting systems in manufacturing firms allocate factory/corporate overhead to products based on direct labour. In the past, this method of allocation may have resulted in minor distortions. However, product lines and channels have proliferated and overhead costs have increased dramatically, making traditional allocation methods dangerously inaccurate.

An activity-based system examines the demands made by particular products (or customers) on indirect resources.[13] Three rules should be followed when examining the demands made by individual products on indirect resources:

1 focus on expensive resources
2 emphasize resources whose consumption varies significantly by product and product type
3 focus on resources whose demands are uncorrelated with traditional allocation methods such as direct labour or materials costs.

The process of tracing costs, first from resources to activities that 'drive' resource usage (cost drivers) and then from activities to specific products (or customers), cannot be done with surgical precision.

Limitations of ABC

It is better to be mostly correct with activity-based costing – say, within 5 or 10 per cent of the actual demands a product or customer makes on organizational resources – than to be precisely wrong (perhaps by as much as 200 per cent) using outdated allocation techniques or including indirect common costs.

It should be noted that any time costs are *allocated*, we are admitting that we cannot identify the cause of the cost – on an avoidable cost basis. If this information were available, we could *attach* (assign) the cost to the appropriate segment. Some of the potential problems of cost allocation are detailed in Box 11.3.

Box 11.3 The Pitfalls of Cost Allocation

Most logistics costing systems are in their infancy and rely heavily on allocations to determine the performance of segments such as product, customers, territories, division or functions. Such allocations in company D led to erroneous decision-making and a loss of corporate profits.

Company D was a multidivisional corporation that manufactured and sold high-margin pharmaceutical products and a number of lower-margin packaged goods. The company maintained a number of field warehouse locations managed by corporate staff. These climate controlled facilities were designed for the pharmaceutical business and required security and housekeeping practices far exceeding those necessary for packaging goods.

To fully utilize the facilities, the company encouraged non-pharmaceutical divisions to store their products in these distribution centres. The costs of operating the warehouses were primarily fixed, although overtime or additional warehouse employees were necessary if throughput increased. The corporate policy was to allocate costs to user divisions on the basis of the square footage occupied. Pharmaceutical warehousing requirements made this charge relatively high. Furthermore, the corporate divisions was managed on a decentralized profit centre basis.

The vice president of logistics in a division that marketed relatively bulky and low-value consumer products realized that similar services could be obtained at lower cost to his division by using a public warehouse. He withdrew the division's products from the corporate facilities and began to use public warehouses in these locations. Although the volume of product handled and stored in the corporate distribution centres decreased significantly, the cost savings were minimal in terms of the total costs incurred by these facilities because of the high proportion of fixed costs. Consequently, approximately the same cost was allocated to fewer users, making it even more attractive for the other divisions to change to public warehouses in order to obtain lower rates. The result was higher, not lower, total company warehousing costs.

The corporate warehousing costs were primarily fixed, so whether the space was fully occupied would not significantly alter these costs. When the non-pharmaceutical divisions moved to public warehouses, the company continued to incur approximately the same total expense for the corporate-owned and operated warehouses and in addition incurred the new public warehousing charges. In effect, the costing system motivated the divisional logistics managers to act in a manner that was in the best interest for divisional profitability, but not in the best interest of the total company. Thus costs to the total company escalated, reducing profitability.

Source: this material is adapted from Douglas M. Lambert and Howard M. Armitage, 'Management Distribution Cost for Better Profit Performance', *Business* 30, no. 5 (September–October 1980), pp. 46–52. Reprinted with permission from *Business* magazine.

ABC in a Warehousing Example

Figure 11.7 shows how activity-based costing may be used to apportion activity costs in a warehouse. Compare this with Figure 11.2 to understand the difference between ABC and traditional accounting systems. The major shortcoming of activity-based costing is that it is simply another method of allocation. Caution must be exercised since any method of allocation can result in charges against segment revenue that would not disappear if the revenue stream was lost.

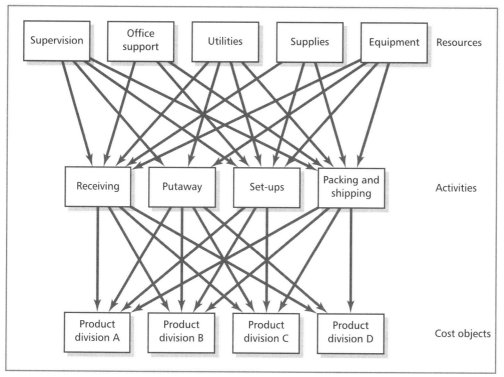

FIGURE 11.7 Assignment of Costs with an Activity-based Cost System

Source: Lisa M. Ellram *et al.*, 'Understanding the Implications of Activity Based Costing for Logistics Management', *Proceedings of the Annual Conference of the Council of Logistics Management*. Oak Brook, IL: Council of Logistics Management, 1994, p. 13.

The potential problems associated with using activity-based costing for logistics were illustrated in a 1991 article in *Management Accounting*. The examples used were simply average costs where selling costs were 5 per cent of sales, advertising costs were 30 cents per unit sold, warehousing costs were 70 cents per pound shipped, packing and shipping costs were 15 cents per unit sold, and general office expenses were allocated at €15 per order.[14] ABC implemented in this manner represents an average cost system and these costs are no longer useful for decision-making.

For example, transportation costs need to be identified by origin and destination postal codes, and by shipment size categories before they can be assigned to customers or products. It is important to identify how costs and revenues will change with a decision that is being made.

Managers responsible for product and customer business segments need to understand the financial implications of their decisions. Executives must be able to talk the language of accountants – to understand the true meaning of data used for decision-making. The support and active participation by top management, including the chief executive, is necessary to improve accounting data because resistance to change is a major barrier facing manufacturers in their quest to become world class.

Segment Contribution Reports

Using segment contribution reports, managers can begin to accurately assess strategic options; for example, which product lines to drop or whether prices can be raised on inelastic products or reduced on high-volume products. They can place added emphasis on those segments that are most profitable, and eliminate unprofitable product lines.

Firms that have developed and implemented segment profitability reports with accurate cost assignments have been able to identify products and customers that were either unprofitable or did not meet corporate financial objectives. Ironically, many of these products or customers were previously thought to be profitable because of the way costs had been arbitrarily allocated.

Direct Product Profitability

Direct Product profitability (DPP) is another method of allocating logistics costs that has become popular in the retail sector.[15] The process behind DPP is to identify all attributable costs accruing to a product, including logistics costs. To make this process practical products will likely need to be grouped together in categories. Table 11.7 shows a generic income statement determination of DPP.

	Sales		Y
−	Direct product costs or cost of goods sold	X	
=	Gross product profit or margin		Y
−	Logistics related costs:		
	Transportation costs	X	
	Warehousing costs	X	
	Inventory costs	X	
	Sourcing or procurement costs	X	
	Operation support costs	X	
	Invoicing and collection costs	X	
	Retail costs	X	
	Sourcing or procurement costs	X	
−	Overheads directly attributable to product	X	
=	Direct product profit		Y

TABLE 11.7 Determination of Direct Product Profitability (DPP)

Sources: adapted from Martin Christopher, *Logistics and Supply Chain Management: Creating Value-adding Networks*, 3rd edn. Harlow, UK: FT Prentice Hall, 2005, pp. 109–11; and Alan Harrison and Remko van Hoek, *Logistics Management and Strategy*. Harlow, UK: FT Prentice Hall, 2002, pp. 60–3.

The DPP process attempts to convert fixed logistics costs into direct costs associated with a product or product category. The result is that DPP should provide better information about which products or product groups contribute most to a firm's profitability.

All supply chain members, including retailers, distributors and suppliers, need to be conscious of and appreciate the effect a product or product group's DPP as it moves their particular logistics system so as to seek to influence it favourably. But, as discussed above, the need for proper cost data is an important factor to undertake DPP analysis, and the question arises again of how and where to source the correct cost data for DPP calculations.

The Role of the Order-processing System

The order-processing system can affect the performance of the logistics function in two major ways. First, the system can improve the quality of the management information system by providing data such as customer names, customer locations, items demanded by customers, sales to customers, sales patterns (when items are ordered), order size, sales by salesperson and sales data for the company's sales forecasting package.

Second, the customer order is the message that sets the logistics function in motion. The speed and quality of the information provided by the order-processing system has a direct impact on the cost and efficiency of the entire logistics process.

Slow and erratic communication can lead to lost customers or excessive transportation, inventory and warehousing costs. It can bring about production inefficiencies because of frequent line changes. Implementation of the latest technology in order-processing and communications systems can lead to significant improvements in logistics performance.

Cost Justification of Logistics System Changes

In Chapter 2 we saw how an integrated approach to the management and control of the logistics function can significantly improve a firm's profitability. However, successful implementation of integrated logistics management depends on total cost analysis. That is, changes in the logistics system structure must be cost-justified by comparing total costs before and after a change. The availability of accurate cost data is critical for the cost-justification of logistics system changes. The Global box illustrates the importance of understanding cost trade-offs in the supply chain.

GLOBAL

When a Global Supply Chain Doesn't Work

A major European manufacturer of industrial components and related products had continual problems with the profitability of its North American subsidiary, which was a major supplier of industrial components to the parent company.

The parent company had significant economies of scale in engineering and manufacturing of industrial components. This should have given the subsidiary a competitive advantage in cost and quality. However, this never materialized. Why?

The North American subsidiary's major supplier (over 60 per cent of finished goods) was the parent, but the parent was the subsidiary's most unreliable source. This created:

- poor supply performance
- long lead-times
- poor supply or fulfilment to its customers
- high logistics costs for transportation and inventory.

This unreliability was compounded by local sourcing on the part of most of the subsidiary's North American competitors, providing them with lower transportation costs and shorter replenishment cycles.

In this case, being the low-cost producer was not enough for the parent company. The company needed to consider the trade-offs between price/cost, delivery cost, inventory and customer service levels. The subsidiary's performance measurements – aggressive inventory turns and high fill rates – were impossible to achieve owing to the parent's poor performance.

While these problems showed up in the poor financial performance of the subsidiary, their roots were a lack of an understanding of the true nature and interrelatedness of supply chain companies, and a dysfunctional performance measurement and reporting system. As a result, the global company has now changed its strategy and realigned its processes to support this new strategy. The strategy now includes:

- integrating supply chain strategic planning
- inventory planning
- performance measurement
- interorganizational information exchange and integration.

Question: **What does this example suggest for interorganizational communication?**

Source: Les B. Artman and Ted Pollock, 'Supply Chain Alignment', *Logistics*! (Fall 1996), pp. 21–6.

In addition to the financial analysis, a number of qualitative – or less easily quantified – benefits can be presented to management. These should not be relied upon to justify the system or process change, but should instead supplement the financial analysis as icing on the cake. For example, the additional benefits associated with an improved order-processing system could (or might) include the following.

- *Customer service improvements*. Customer service may be improved in two ways. First, the improved communication may allow the customer and the customer service representative to arrange for immediate substitution if a stockout occurs, or the representative can provide the customer with a realistic estimated delivery date if it is necessary to wait for the product to be manufactured. The new system may facilitate enquiries about order status after the order has been made. Second, if the improved communication reduces the variability of the order cycle time, this improvement should be documented. Suppose that the current order cycle is 10 days and ranges from 7 to 13 days – a variability of three days. Reducing order cycle variability by two days, to a range of 9 to 11 days, will enable customers to reduce their safety stocks.
- *Improved cash flow*. The advanced order-processing system should result in more accurate and timely invoicing of customers, thereby improving cash flow.
- *Improved information*. Advanced order processing should improve information in two major ways. First, sales data should be captured sooner and more reliably, leading to more timely and better information for sales forecasting and production planning. Second, the improved system can be used as a source of valuable input for the logistics management information system.

Management must have access to good cost data in order to determine the financial impact of purchasing a new forecasting model, an inventory control package or any other logistics system change. With a well-thought-out financial analysis, the logistics executive will be able to determine the probable profit impact of any proposed system.

SUMMARY

Accurate cost data are required to achieve least-cost logistics. Successful implementation of integrated logistics management depends on full knowledge of the costs involved. Cost data are also required to manage logistics operations.

In this chapter, we saw how to use logistics costs for decision-making and how erroneous decisions result from the use of inaccurate costs. We examined the measurement and control of logistics performance, using standard costs and flexible budgets, budgetary practices, activity-based costing and productivity standards. Finally, we described how the total cost concept can be used to cost-justify changes in logistics structure.

chapter 11 Logistics Financial Performance

QUESTIONS AND PROBLEMS

1 Why is it important to have accurate cost data for management of the logistics function?
2 What problems are associated with the use of average cost data for decision-making?
3 How does the inability to distinguish between fixed and variable costs hamper good management practice?
4 What problems are associated with the arbitrary allocation of logistics costs?
5 How do accurate cost data contribute to the motivation of personnel?
6 Why is it difficult to obtain logistics cost data in many firms?
7 Identify and describe the methods that can be used for controlling logistics activities. What are the advantages and disadvantages of each?
8 What is activity-based costing? What are its advantages? What are its potential limitations?
9 How can the order-processing system improve the quality of the logistics information system?
10 What non-financial measures can be used to justify logistics system changes?

THE LOGISTICS CHALLENGE!

PROBLEM: GEARING UP FOR LORRY ROAD USER CHARGING

According to the UK Chancellor of the Exchequer's 2002 Budget statement, the proposed Lorry Road User Charge (LRUC) has a modest and worthy aim: to 'ensure that lorry operators from overseas pay their fair share towards the cost of using UK roads'. Foreign-registered hauliers would have to pay around 15 pence for every kilometre travelled in the UK, assuming they bought their fuel outside the country, as most currently do. This would raise an extra £39 million annually for the UK Treasury.

But this system of road user charging, which the UK government is planning to introduce in 2008, will not only be applied to foreign hauliers. All 430,000 lorries registered in the UK with gross weights of over 3.5 tonnes will also be subject to exactly the same charges.

There will be significant technical and organizational challenges from the proposed system of vehicle tracking and toll collection. This system will involve the integration of a range of telematics and communication technologies, including satellite tracking, cellular telephony, microwave systems and digital tachographs. Registers of vehicles, operators and authorized fitters of tolling equipment will have to be compiled and maintained. Separate revenue streams will have to be created for the inward flow of toll income and return flow of fuel rebates.

Even at the outset this proposed system will be highly complex. To put the complexity of the UK system into context, it is worth comparing it with the German Maut system. In Germany, only vehicles with a gross weight of 12 tonnes or more travelling on autobahns are to be tolled and the level of toll will be fixed. In the UK, all vehicles over 3.5 tonnes are to be charged for their use of all roads at a rate that is likely to be varied by road type and time of day.

It was widely expected that with the introduction of LRUC, vehicle excise duty (VED) would be abolished, but recent reports suggest that it will be retained. Currently, annual taxes on lorries in the UK are £280 million for VED and £3040 million for fuel tax, for a total of £3320 million. In the absence of any offsetting administrative cost savings, the annual operating cost of LRUC could well be £700 million or more – five times as much as the extra revenue from foreign hauliers. Internalizing this total cost within the LRUC scheme would add a further 3 pence per kilometre or more to the charge. Alternatively, if the Chancellor is true to his word and does not increase the total tax burden on the UK haulage industry, the Treasury would have to incur this cost.

The key component in the new charging regime will be the 'on-board unit' (OBU) to be installed in all lorries travelling more than 12,000 kilometres on UK roads each year. This will determine the truck's location using a global positioning system (GPS), match that location against a built-in digital map of the UK road network (to determine the class of road) and communicate with a control centre, relaying information about the distance travelled and toll to be charged. One of the ironies of the current proposal is that the majority of foreign trucks entering the UK travel fewer than 12,000 kilometres annually on UK roads; 59 per cent make fewer than 12 trips to the UK per annum and the average trip length for foreign vehicles within the UK is 644 kilometres – hence total distance travelled is only around 7700 kilometres. These vehicles will be covered by a separate 'occasional user scheme', which will require them to use a 'low-use OBU'. This will be quite different from the standard one and use a microwave system to communicate with roadside beacons located at 'strategic locations' across the UK road network. So, the main LRUC system will not even apply to the majority of foreign vehicles entering the UK.

Some people see this separate arrangement for foreign vehicles as an interim measure until EU countries agree on a standard 'black box' that will permit interoperability between all the national road tolling systems introduced across Europe. This seems a distant prospect, however, if the experience of the new digital tachograph is any guide. Discussion on the EU specification for this tachograph have been under way for over a decade and have still not reached a satisfactory conclusion.

Given all these parameters and uncertainties, what are the cost and accounting factors that a UK or foreign haulier must consider in preparation for the implementation of LRUC?

What Is Your Solution?

Source: Alan C. McKinnon, *Lorry Road User Charging: A Review of the UK Government's Proposals*. Edinburgh, UK: Logistics Research Centre, Heriot-Watt University, 2004.

BOOKS

Christopher, Martin, *Logistics and Supply Chain Management: Creating Value-adding Networks*, 3rd edn. Harlow, UK: FT Prentice Hall, 2005.

Ellram, Lisa M., *Total Cost Modeling in Purchasing*. Tempe, AZ: Centre for Advanced Purchasing Studies, 1994.

Kearney, A.T., Inc., *Measuring and Improving Productivity in Physical Distribution*. Oak Brook, IL: National Council of Physical Distribution Management, 1984.

JOURNALS

Armitage, Howard M., 'The Use of Management Accounting Techniques to Improve Productivity Analysis in Distribution Operations', *International Journal of Physical Distribution and Materials Management* 14, no. 1 (1984), pp. 41–51.

Carter, Joseph R. and Bruce G. Ferrin, 'The Impact of Transportation Costs on Supply Chain Management', *Journal of Business Logistics* 16, no. 1 (1995), pp. 189–212.

Cavinato, Joseph L., 'A Total Cost/Value Model for Supply Chain Competitiveness', *Journal of Business Logistics* 13, no. 2 (1992), pp. 285–301.

Ellram, Lisa M., 'Activity-based Costing and Total Cost of Ownership: A Critical Linkage', *Journal of Cost Management* 8, no. 4 (Winter 1995), pp. 14–21.

'Finding the Hidden Cost of Logistics', *Traffic Management* 34, no. 3 (March 1995), pp. 47–50.

Gustin, Craig, Patricia Daugherty and Theodore P. Stank, 'The Effects of Information Availability on Logistics Integration', *Journal of Business Logistics* 16, no. 1 (1995), pp. 1–22.

Keegan, Daniel P. and Stephen W. Portik, 'Accounting Will Survive the Coming Century, Won't It?', *Management Accounting* 77 (December 1995), pp. 24–30.

La Londe, Bernard J. and Terrance L. Pohlen, 'Issues in Supply Chain Costing', *The International Journal of Logistics Management* 7, no. 1 (1996), pp. 1–12.

Lambert, Douglas M. and Jay U. Sterling, 'What Types of Profitability Reports Do Marketing Managers Receive?', *Industrial Marketing Management* 16, no. 4 (1987), pp. 295–303.

Mentzer, John T. and Brenda Ponsford Konrad, 'An Efficiency/Effectiveness Approach to Logistics Performance Analysis', *Journal of Business Logistics* 12, no. 1 (1991), pp. 33–62.

Novak, Robert A., 'Quality and Control in Logistics: A Process Model', *International Journal of Physical Distribution and Materials Management* 19, no. 11 (1989), pp. 1–44.

Shank, John K. and Vijay Govindarajan, 'Strategic Cost Management and the Value Chain', *Journal of Cost Management* 5, no. 1 (Winter 1992), pp. 5–21.

Shapiro, Jeremy, 'Integrated Logistics Management, Total Cost Analysis and Optimization Modeling', *International Journal of Physical Distribution and Logistics Management* 22, no. 3 (1992), pp. 33–6.

Tyndall, Gene R. and John R. Busher, 'Improving the Management of Distribution with Cost and Financial Information', *Journal of Business Logistics* 6, no. 2 (1985), pp. 1–18.

REFERENCES

[1] Drew Stapleton, Sanghamitra Pati, Erik Beach and Poomipak Julmanichoti, 'Activity-based Costing for Logistics and Marketing', *Business Process Management Journal* 10, no. 5 (2004), pp. 584–97.

[2] This section is adapted from Douglas M. Lambert and Howard M. Armitage, 'Distribution Costs: The Challenge', *Management Accounting* 60, no. 11 (May 1979), p. 33.

[3] Charles Davis, 'Countering the Costs', *Logistics Europe* 12, no. 7 (September 2004), pp. 24–8.

[4] European Logistics Association and A.T. Kearney Management Consultants, *Differentiation for Performance: Results of the Fifth Quinquennial European Logistics Study 'Excellence in Logistics 2003/2004'*. Hamburg: Deutcher Verkehrs-Verlag GmbH, 2004, p. 21; and Davis, 'Countering the Costs', p. 24.

[5] Martin Christopher and John Gattorna, 'Supply Chain Management Cost and Value-based Pricing', *Industrial Marketing Management* 34 (2005), pp. 115–16.

[6] Joseph L. Cavinato, 'A Total Cost Value Model for Supply Chain Competitiveness', *Journal of Business Logistics* 13, no. 2 (1992), pp. 285–302; and Terrance L. Pohlen, *The Effect of Activity Based Costing on Logistics Management*, doctoral dissertation, Ohio State University, 1993.

[7] This material is adapted from Douglas M. Lambert and Howard M. Armitage, 'Management Distribution Cost for Better Profit Performance', *Business* 30, no. 5 (September–October 1980), pp. 46–52. Reprinted with permission from *Business* magazine.

[8] Lambert and Armitage, 'Management Distribution Cost for Better Profit Performance', pp. 50–1.

[9] Michael Schiff, *Accounting and Control in Physical Distribution Management*. Chicago: National Council of Physical Distribution Management, 1972, pp. 4-63–6-70.

[10] A.T. Kearney, Inc., *Measuring and Improving Productivity in Physical Distribution*. Oak Brook, IL: National Council of Physical Distribution Management, 1984, p. 170.

[11] This material is adapted from C. John Langley, Jr, 'Information-based Decision Making in Logistics Management', *International Journal of Physical Distribution and Materials Management* 15, no. 7 (1985), pp. 48–52; and James R. Evans and William M. Lindsay, *The Management and Control of Quality*. Minneapolis, MN: West Publishing Co., 1993, pp. 529–656.

[12] Evans and Lindsay, *The Management and Control of Quality*; and Jari Juga, 'Redesigning Logistics to Improve Performance', *The International Journal of Logistics Management* 6, no. 1 (1995), pp. 75–84.

[13] Martin Christopher, *Logistics and Supply Chain Management: Creating Value-adding Networks*, 3rd edn. Harlow, UK: FT Prentice Hall, 2005, pp. 111–14.

[14] Ronald J. Lewis, 'Activity-based Costing for Marketing', *Management Accounting* 73, no. 5 (November 1991), pp. 33–8.

[15] Christopher, *Logistics and Supply Chain Management*, pp. 109–11; and James H. Bookbinder and Feyrouz H. Zarour, 'Direct Profit Profitability and Retail Shelf-space Allocation Models', *Journal of Business Logistics* 22, no. 2 (2001), pp. 183–208.

CHAPTER 12
GLOBAL LOGISTICS

CHAPTER OBJECTIVES

■ To identify some of the controllable and uncontrollable factors that affect global logistics activities

■ To describe the major international distribution channel strategies – exporting, licensing, joint ventures, ownership and importing

■ To highlight the elements involved in managing export shipments

■ To identify the organizational, financial and managerial issues related to global logistics

One of the most important business developments in recent years has been the expansion of global industry. For an ever-growing number of firms, management is defining the marketplace globally. Table 12.1 identifies the world's largest industrial corporations. The companies may be headquartered in Asia, Europe or North America, but their markets are international in scope.

New markets are opening up and existing markets are expanding worldwide. The economies of the industrialized nations have matured – that is to say, their economic growth rates have slackened, so firms in those countries are seeking market opportunities abroad. A global financial network has developed that allows multinational enterprises to expand their operations. In addition, manufacturers have increased new material and component acquisitions from other countries (i.e. global sourcing). In sum, the world economy is becoming more interdependent.

To support non-domestic (i.e. non-national) markets a company must have a distribution system or network that satisfies the particular requirements of those markets. For example, the distribution systems in the developing countries of Africa, South America or Asia are characterized by large numbers of channel intermediaries supplying an even larger number of small retailers. The systems in these nations are marked by inadequate transportation and storage facilities, a large labour force of mainly unskilled workers, and an absence of distribution support systems. In the developed countries (e.g. most of Europe, Japan, Canada and the United States), the distribution systems are highly sophisticated, have good transportation systems, high-technology warehousing and a skilled labour force.

In a European context there is one important question: 'What is 'international' to a firm in Europe (i.e. is France an international market for Germany and vice versa)?' The 1991 Maastricht Treaty established the European Union (EU) and created concepts of harmonized trade, standards and laws, the abolition of borders and the four freedoms: freedom of movement of persons, goods, service and capital. This indicates that an EU country dealing with another EU country might consider them part of their domestic market.

However, the reality 15 years after the event is somewhat different. The four freedoms are still subject to national protectionist behaviour; not every EU member has adopted the euro common currency; and national corporate tax rates still differ, ranging from 28 per cent in Sweden and Finland to 41.7 per cent in France and 43.6–56.7 per cent in Germany.[1] With the addition of the 10 accession countries in May 2004 the hoped-for integration and harmonization of the EU project may be delayed further still.

Rank	Company	Country	Industry	2003 sales ($ millions)
1	Wal-Mart Stores	USA	General merchandisers	263,009
2	BP	UK	Petroleum refining	232,571
3	Exxon Mobil	USA	Petroleum refining	222,883
4	Royal Dutch/Shell Group	UK/Netherlands	Petroleum refining	201,728
5	General Motors	USA	Motor vehicles	195,324
6	Ford Motor	USA	Motor vehicles	164,505
7	DaimlerChrysler	Germany	Motor vehicles	156,602
8	Toyota	Japan	Motor vehicles	153,111
9	General Electric	USA	Diversified financials	134,187
10	Total	France	Petroleum refining	118,441
11	Allianz	Germany	Insurance	114,950
12	ChevronTexaco	USA	Petroleum refining	112,937
13	AXA	France	Insurance	111,912
14	ConocoPhillips	USA	Petroleum refining	99,468
15	Volkswagen	Germany	Motor vehicles	98,636
16	Nippon Telegraph	Japan	Telecommunications	98,229
17	ING Group	Netherlands	Insurance	95,893
18	CitiGroup	USA	Banks	94,713
19	IBM	USA	Computers	89,131
20	American International	USA	Insurance	81,303
21	Siemens	Germany	Electronics	80,501
22	Carrefour	France	Food and drug stores	79,734
23	Hitachi	Japan	Electronics	76,423
24	Hewlett-Packard	USA	Computers	73,061
25	Honda Motor	Japan	Motor vehicles	72,264

TABLE 12.1 The World's Largest Corporations, 2003

Source: 'The *Fortune* Global 500, The Largest 500 Companies in the World', *Fortune* 150, no. 2, 26 July 2004, pp. F-1, F-16–F-22.

We will consider that trade between individual EU member states is international in this text such that following discussions on international strategies and issues will apply to individual EU member states. However, we also recognize that the EU situation may be different in some circumstances and contexts, and that associated trade might be considered domestic in nature.

In this chapter, we will describe some of the similarities and differences in the management of logistics in domestic and international environments. We will see how to assess the global logistics environment and how to develop meaningful logistics strategies in that environment.

International Distribution Channel Strategies

Many factors can influence a company's decision to enter international markets. They include:

- market potential
- geographic diversification
- excess production capacity and the advantage of a low-cost position due to experience-curve economies and economies of scale
- products near the end of their life cycle in the domestic market could generate growth in the international market
- source of new products and ideas
- foreign competition in the domestic market.[2]

Raw materials, component parts or assemblies are additional reasons for a firm to enter international markets. For example, some raw materials, such as petroleum, bauxite, uranium and certain foodstuffs, are found only in certain geographic regions. A firm may locate a facility overseas or import an item for domestic use, and thereby become international in scope.

Companies that become involved in the international marketplace have many options available to them:

- exporting
- licensing
- joint ventures
- ownership
- importing
- counter-trade.

Several options are available within each channel strategy. Figure 12.1 identifies some of the major participants in an international logistics transaction, including product and information flows.

Successful completion of the various logistics activities in the international distribution channel can contribute to the development of global markets in many ways, including door-to-door freight services, which offer speed and reliability of delivery, and allow order lead-times to be quoted accurately; reduced delivery costs through consolidation; expansion into new world markets that were previously out of reach; ability to offer a reasonable after-sales service or replacement policy to international markets; and, once captured, an overseas market may be held and expanded despite intense competition, because of high levels of customer service offered through distribution services.

Exporting

The most common form of distribution for firms entering international markets is exporting. **Exporting** refers to selling products in another country. Companies can hire independent marketing intermediaries (indirect exporting) or market their products themselves (direct exporting). Exporting requires the least amount of knowledge about foreign markets because domestic firms allow an international freight forwarder, distributor, trading company or some other organization to carry out the logistics and marketing functions.

Advantages of Exporting

Many advantages are associated with exporting, such as greater flexibility and less risk than other international distribution strategies. For example, no additional production facilities or logistics asset investment is needed in the foreign market because firms produce the product domestically and allow the exporting intermediary to handle distribution of the product abroad.

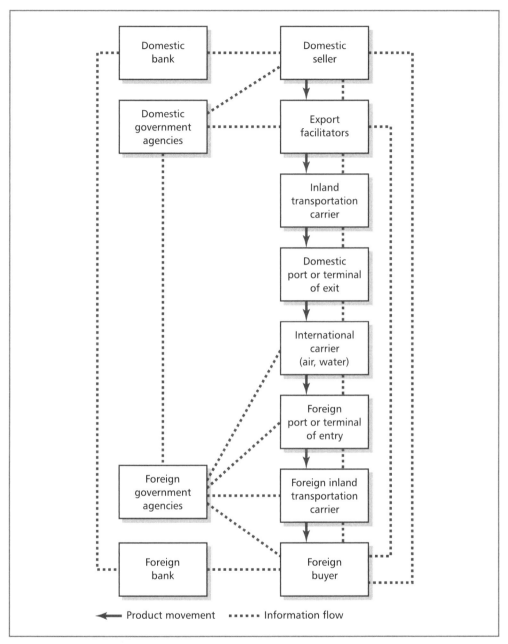

FIGURE 12.1 Major Participants in an International Logistics Transaction

Without direct foreign investment, political uncertainties are less significant because firms are not concerned with the host country nationalizing their operations. Also, it is relatively easy to withdraw if the foreign market does not meet the profit or sales expectations of firms. Exporting is an excellent way to gain experience and test a market before firms expand their own production and marketing operations.

Disadvantages of Exporting

Exporting is not without disadvantages. It is sometimes difficult for firms to compete with firms located in the foreign market. For example, **tariffs** (taxes assessed on goods entering a market), **import quotas** (limitations on the amount of goods that can enter a market) or unfavourable currency exchange rates may adversely affect the price or availability of imported goods.

In addition, domestic firms have little control over the pricing, promotion or distribution of the product they export. Success in international markets depends to a large degree on the capability of **export intermediaries**.

Management must recognize that the export process is not as simple as it first appears. With or without the help of intermediaries, a firm must perform planning, package design, sales negotiation, financial monitoring, banking, insurance and a variety of documentation. Logistics needs to be involved continually in the process from the planning stage. A firm involved in exporting often has to deal with a number of intermediaries who provide a variety of export services.

Licensing

Licensing involves agreements that allow a firm in one country (the licenser) 'to use the manufacturing, processing, trademark, know-how, technical assistance, merchandising knowledge, or some other skill provided by the licenser located in another country'.[3]

Advantages of Licensing

Unlike exporting, licensing allows domestic firms more control over how they distribute their products, because distribution strategy is usually part of preliminary discussions. The specific logistics functions are carried out by the licensees, using the established distribution systems of the foreign country.

Licensing is similar to exporting in that it does not require large capital outlays, thereby reducing risk and increasing flexibility. Licensing is a strategy frequently used by small and medium-sized businesses. It can be an excellent approach if the foreign market has high tariff barriers or strict import quotas. The licenser is usually paid a royalty or a percentage of sales by the licensee.

Disadvantages of Licensing

Although licensing does provide domestic firms with flexibility, it does not mean that licensing agreements can be terminated quickly. While the agreements with licensees may include termination or cancellation provisions, there is usually a time lag between the decision to terminate and the actual date of termination, typically longer than in an exporting situation.

A potentially serious drawback is that licensees can become future competitors. As licensees develop their own know-how and capabilities, they may end the licensing agreement and begin to compete with licensers.

Joint Ventures

Management may wish to exercise more control over the foreign firm than is available in a licensing agreement, but at the same time it may not want to establish a freestanding manufacturing plant or other facility in a foreign market. If so, the **joint venture** offers a compromise.

The risk is higher and the flexibility is lower for a company because an equity position is established in a foreign firm (in that firm's own country). The financial partnership, however, enables a company to provide substantial management input into the channel and distribution strategies of the foreign company. This increased management voice does place additional burdens on the domestic firm – namely, it requires a greater knowledge of the international markets the firm is trying to serve.

The joint venture may be the only method of market entry if management wishes to exercise significant control over the distribution of its products. This would be especially true if wholly owned subsidiaries are prohibited by the foreign government. Such restrictions occur more frequently in developing countries, which often attempt to promote internal industrial or retail development.

Direct Ownership

Complete ownership of a foreign subsidiary offers the domestic firm the highest degree of control over its international marketing and logistics strategies. **Direct ownership** takes place through acquisition or expansion.

Advantages of Direct Ownership

Acquisition of a foreign facility can be advantageous because it minimizes start-up costs: locating and building facilities, hiring employees and establishing distribution channel relationships. Compared with other forms of market entry, ownership of a foreign subsidiary requires the greatest knowledge of a particular international market. The firm is totally responsible for marketing and distributing its product.

Direct ownership in the foreign market allows the company to compete more effectively on a price basis because it can eliminate the transportation costs incurred in shipments from domestic plants to foreign points of entry, customs duties and other import taxes.

Disadvantages of Direct Ownership

Drawbacks include a loss of flexibility because the firm has a long-term commitment to the foreign market. Fixed facilities and equipment cannot be disposed of quickly if sales or profits decline, levels of competition increase or other adversities occur.

The possibility of government nationalization of foreign-owned businesses is another drawback of direct ownership, especially in politically unstable countries. Additionally, exchange rate fluctuations change the relative value of foreign investments because they are valued in local (i.e. foreign) currency instead of the currency of the owner's home country.

Market-entry Strategies

In general, firms follow more than one market-entry strategy. Markets, product lines, economic conditions and political environments change over time, so the optimal market-entry strategy may change. Furthermore, a good market-entry strategy in one country may not be good in another.

A firm considering exporting, licensing, joint venture or ownership should establish a formal procedure for evaluating each alternative. Each market-entry strategy can be evaluated on a set of management-determined criteria. Each functional area of the firm (e.g. accounting, manufacturing, marketing, logistics) must be involved in establishing the criteria and their evaluation. A firm should decide on a method of international involvement only after it has made a complete analysis of each market-entry strategy.

Importing

Many firms will be involved in activities that bring raw materials, parts, components, supplies or finished goods from sources outside the country. This may involve importing, counter-trade and duty drawbacks.

Importing involves the purchase and shipment of goods from an overseas source. Imported items can be used immediately in the production process or sold directly to customers. They can be transported to other ports of entry, stored in bonded warehouses (where goods are stored until import duties are paid) or placed in a free trade zone (where goods are exempt from customs duties until they are removed for use or sale).

Many firms utilize customshouse brokers for importing products into a country. These brokers facilitate the movement of imported products and ensure the accuracy and completeness of import documentation.

> Many . . . brokers help clients choose modes of transportation and appropriate carriers . . . They also provide assistance to importers in assigning shipments the best routes. They handle estimates for landed costs, payments of goods through draft, letters of credit insurance, and redelivery of cargo if there is more than one port of destination.[4]

Counter-trade and Duty Drawbacks[5]

The term **counter-trade** applies to the requirement that a firm import something from a country in which it has sold something else. In essence, counter-trade is any transaction in which part of the payment is made in goods instead of money. The need for counter-trade is driven by the balance of payments problems of a country and by weak demand for the country's products. A likely candidate for counter-trade is a country with a shortage of foreign exchange or a shortage of credit to finance trade flows. Such a country will try to expand its exports or to develop markets for its new products.

Five Forms of Counter-trade

Five basic forms of counter-trade exist: barter, buyback, compensation, counter-purchase and switch.

1 *Barter*, the simplest form of counter-trade, occurs when goods of equal value are exchanged and no money is involved.
2 In *buyback* arrangements, the selling firm provides equipment or an entire plant, and agrees to buy back a certain part of the production. Many developing countries insist on buyback arrangements because they ensure access to western technology and stable markets.
3 A *compensation* arrangement takes place when barter is specified as a percentage of the value of goods being traded to the value of the product being sold.
4 *Counter-purchase* involves transactions with more cash, smaller volumes of goods flowing to the multinational corporation over a shorter period of time, and goods unrelated to the original deal.
5 A *switch* transaction uses at least one third party outside the host country to facilitate the trade. The counter-traded goods or the multinational enterprise's goods are sent through a third country for purchase in hard currency or for distribution.

While counter-trade agreements may be complex, they do offer an opportunity to develop lower-cost sources of supply in the world marketplace. In some cases, they may provide the only means of market entry for the firm.

Duty Drawbacks

Firms that import goods used in manufacturing or export products that contain imported materials can take advantage of drawbacks. A drawback, or **duty drawback**, is a refund of customs duties paid on imported items. Duty drawbacks involve many steps and can be time consuming. However, duty drawbacks can provide significant cost savings for firms that take advantage of them.

Managing Global Logistics

Management of a global distribution system is much more complex than that of a purely domestic network. Managers must properly analyse the international environment, plan for it, and develop the correct control procedures to monitor the success or failure of the foreign distribution system. Figure 12.2 identifies some of the questions the international logistics manager must ask – and answer – about the firm's foreign logistics programmes.

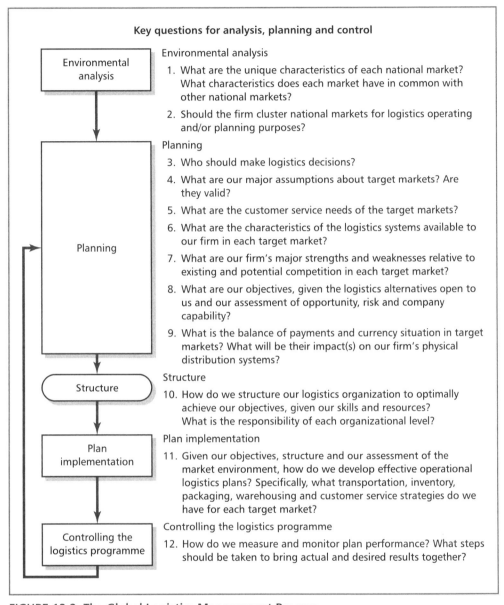

Key questions for analysis, planning and control

Environmental analysis

1. What are the unique characteristics of each national market? What characteristics does each market have in common with other national markets?

2. Should the firm cluster national markets for logistics operating and/or planning purposes?

Planning

3. Who should make logistics decisions?

4. What are our major assumptions about target markets? Are they valid?

5. What are the customer service needs of the target markets?

6. What are the characteristics of the logistics systems available to our firm in each target market?

7. What are our firm's major strengths and weaknesses relative to existing and potential competition in each target market?

8. What are our objectives, given the logistics alternatives open to us and our assessment of opportunity, risk and company capability?

9. What is the balance of payments and currency situation in target markets? What will be their impact(s) on our firm's physical distribution systems?

Structure

10. How do we structure our logistics organization to optimally achieve our objectives, given our skills and resources? What is the responsibility of each organizational level?

Plan implementation

11. Given our objectives, structure and our assessment of the market environment, how do we develop effective operational logistics plans? Specifically, what transportation, inventory, packaging, warehousing and customer service strategies do we have for each target market?

Controlling the logistics programme

12. How do we measure and monitor plan performance? What steps should be taken to bring actual and desired results together?

FIGURE 12.2 The Global Logistics Management Process

Source: adapted from Warren J. Keegan, *Global Marketing Management*, 5th edn, p. 37. Copyright 1995. Reprinted by permission of Prentice Hall, Inc., Englewood Cliffs, NJ.

The questions can be classified into five categories: (1) environmental analysis, (2) planning, (3) structure, (4) plan implementation, and (5) control of the logistics programme.

The overall objective of the process diagrammed in Figure 12.2 is to develop the optimal logistics system for each international target market. This is achieved by examining the various characteristics of the foreign market and developing a set of alternatives or strategies that will fulfil the company's objectives. With a set of objectives or strategies, management can define the proper organizational and channel structures. Once these are established, management implements the distribution system. The final step is to measure and evaluate the performance of the system and to provide feedback to the strategic planning process for purposes of adjustment or modification of the system.

Cost–service Trade-off Analysis

An integral part of the global logistics management process is cost–service trade-off analysis. Whether operations are domestic or international, the ability to properly identify, evaluate and implement the optimal cost–service mix is always important to the firm and its customers. The only difference between domestic and international operations is in the emphasis placed on each cost and service element.

Some particularly important cost and service considerations concern the use of integrated logistics systems to effectively and rapidly manage order completeness, shipping accuracy and shipment condition to any destination economically. Global customers are no longer willing to settle for longer and less reliable order cycle times.

Order completeness is much more important in international logistics, because the costs of back-orders and expedited shipments are substantially higher. Processing and shipping costs must be weighed against the cost of improving order completeness. It is more expensive to ship complete orders all of the time, but this higher service level may be justified in view of the costs of shipping incomplete or partial orders. A similar logic can be used in the case of *shipping accuracy*.

Because of the higher costs linked to shipping errors in international distribution, it is important to maximize the accuracy of shipment routing and the items that make up a shipment. Once the shipment is made, *shipment condition* becomes important because of the time and cost entailed in replacing damaged items.

Guidelines for Developing a Global Logistics Strategy

In developing a global logistics strategy, some general guidelines apply. The following list can be useful to firms in almost any international market.

1 *Logistics planning should be integrated into the company's strategic planning process.* For example, Carrefour, the €91 billion French retailer, has three primary store formats. Hypermarkets offer a wide range of food and non-food products; their shelves stock an average of 70,000 items. Floor areas of hypermarkets range from 5000 square metres to over 20,000 square metres, and their catchment areas are very large. Supermarkets offer a wide selection of mostly food products in outlets featuring floor areas of 1000 to 2000 square metres. Hard discounters stock 800 food products in small stores from 200 to 800 square metres. Carrefour is rapidly expanding these three formats in 30 global markets. Its strategy consists of building group market share in each country in which it does business by expanding the type of retailing best suited to the local market and by taking advantage of the way the three formats complement one another.[6]

2 *Logistics departments need to be guided by a clear vision and must measure output regularly.* The Russian beer brewer Baltic Beverages Holding (BBH) is a joint venture between Scottish & Newcastle (UK) and Carlsberg (Denmark). Russia is Europe's second largest beer market with 74 million hectolitres annual volume. BBH seeks to add value through substantial investment in additional capacity, improved quality and distribution. Scottish & Newcastle and Carlsberg measure results based on return on investment input. For example, their investment in 2003 for process and distribution improvements was about €250 million with the result that five malt houses now provide approximately half of BBH's malt requirements.[7]

3 *Import–export management should try to ensure integrated management of all elements of the logistics supply chain from origin to destination.* This is especially important as major structural

and regulatory changes are under way across the globe. Deregulation in transport in the United States, Mexico, Japan and Europe (including an increase in European cabotage), permits negotiation of creative 'door-to-door' service and price packages with carriers. This allows shippers to design and manage their supply chains, so that delivery can be tailored to customer specifications at a reasonable cost.

4 *Opportunities to integrate domestic and international operations should be pursued to leverage total company volumes with globally orientated carriers.* This usually requires a change in organizational thinking, but major opportunities exist for companies that can move in this direction. A good starting point is to make a list comparing domestic and international logistics activity.

Managers who approach the global logistics process using the above guidelines, good judgement and a determination to succeed are likely to do well. While the international marketplace may be undergoing rapid change, it is certainly manageable and offers exciting opportunities and challenges to firms seeking global markets.

Organizing for Global Logistics

Proper organization and administration of the logistics function is just as important internationally as it is domestically.

> To be competitive, companies must understand all facets of global sourcing and marketing . . . today's international logistics manager must be a well-trained, full-time professional, experienced in managing complex international logistics decisions.[8]

When a company enters the international marketplace, initially through exporting or licensing, the balance of power in the firm will continue to be held by domestic operations. Obviously, that is only proper in the early stages of development, but as foreign operations grow in sales and profits, and thus in importance, the international component of the business must be allowed to make more input into corporate decision-making.

Many companies operating in the global marketplace centralize a large number of logistics activities while decentralizing others. For example, management of customer service tends to work best when it is under local control in the foreign market. On the other hand, material flows into the organization are often centralized, because technology can quickly overcome spatial distances. Most information systems tend to be centralized, which enables decision-making across international boundaries. Also, purchasing economies may occur with centralization.

The Global box discusses how Roche has organized its worldwide distribution system.

'Localized' Global Distribution

Roche, the multinational pharmaceutical firm based in Switzerland, has steered a middle course between centralized and decentralized control of logistics for its vitamins and chemicals unit.

In 1996, the company gave responsibility for distribution and inventories of finished products to area marketing managers. At the same time, the company retained carrier selection, rate negotiation, raw materials supply and intracompany shipments at its Basel headquarters. Roche says the 'localized' global distribution system is a way to stay close to the customer without surrendering the economies of central control.

The company has separate business units (or product divisions) and operates globally, with production facilities in both Europe and in the United States. The vitamins and fine-chemicals business unit has three regional areas: Europe, including Africa and the Middle East; North and South America; and the Far East, including Australia.

Before area managers were given responsibility for distribution and inventory, they already were responsible for marketing.

The headquarters office controls worldwide logistics and transportation of finished goods from production sites to a global distribution centre at Venlo, the Netherlands; from the global centre to area distribution centres within each of the three world areas; and from each area centre to the area's customers.

In Europe, distribution from the area distribution centre to customers is usually by truck and takes only one day at the vitamins and fine-chemicals division. In Asia, however, where ocean transportation is used, distribution can take up to two weeks.

Question: **Can you suggest alternative logistics structures Roche could use for its global operations?**

Source: Philip Damas, '"Localized" Global Distribution', *American Shipper* 39, no. 2 (February 1997), pp. 36, 38.

Financial Aspects of Global Logistics

A firm participating in global logistics faces a financial environment quite different from that of a strictly domestic firm. It has concerns about currency exchange rates, costs of capital, the effects of inflation on logistics decisions and operations, tax structures, and other financial aspects of performing logistics activities in foreign markets.

Working Capital

Global logistics activities require financing for working capital, inventory, credit, investment in buildings and equipment, and accommodation of merchandise adjustments that may be necessary. **Working capital** considerations are extremely important to the international firm owing to time lags caused by distance, border crossing delays and government regulations. Typically, foreign operations require larger amounts of working capital than domestic operations.

Inventories

Inventories are an important aspect of global logistics: in general, higher levels of inventory are needed to service foreign markets because of longer transit times, greater variability in transit times, port delays, customs delays and other factors.

In addition, inventories can have a substantial impact on the international firm because of the rapid inflation that exists in some countries. It is important to use the proper inventory accounting procedure because of the impact of inflation on company profits. The LIFO (last in–first out) method is probably the most appropriate strategy because the cost of sales is valued closer to the current cost of replacement. The FIFO (first in–first out) method gives a larger profit figure than LIFO because old costs are matched with current revenues, although this will increase the tax liability.

In anticipation of higher costs caused by inflation or other factors, management of an international firm must weigh the cost trade-offs involved in the build-up of inventories. The trade-off is between the accumulation of excess inventory and inventory carrying costs on the one hand, and the reduction of carrying costs by holding less inventory on the other, which would require paying higher acquisition costs at a later date.

When management considers direct investment in facilities and logistics networks in the foreign market, the capital budgeting aspects of financial planning become important. As is the case in domestic operations, customers in the international sector do not tender payment to the shipper until the product is delivered. As you have read, many factors might cause the foreign shipment to have a longer delivery time than a comparable domestic shipment. The exporter must be concerned with the exchange rate fluctuations that may occur between the time the product is shipped, delivered to the consignee and paid for by the customer.

Letters of Credit

To ensure that international customers pay for products shipped to them, letters of credit are often issued. A **letter of credit** is a document issued by a bank on behalf of a buyer, which authorizes payment for merchandise received. Payments are made to the seller by the bank instead of the buyer. Figure 12.3 shows how a letter of credit works.

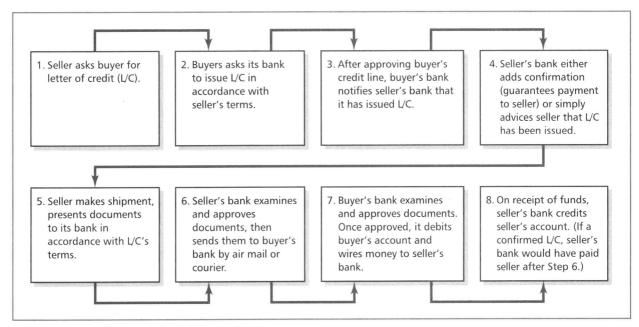

| 1. Seller asks buyer for letter of credit (L/C). | 2. Buyers asks its bank to issue L/C in accordance with seller's terms. | 3. After approving buyer's credit line, buyer's bank notifies seller's bank that it has issued L/C. | 4. Seller's bank either adds confirmation (guarantees payment to seller) or simply advices seller that L/C has been issued. |

| 5. Seller makes shipment, presents documents to its bank in accordance with L/C's terms. | 6. Seller's bank examines and approves documents, then sends them to buyer's bank by air mail or courier. | 7. Buyer's bank examines and approves documents. Once approved, it debits buyer's account and wires money to seller's bank. | 8. On receipt of funds, seller's bank credits seller's account. (If a confirmed L/C, seller's bank would have paid seller after Step 6.) |

FIGURE 12.3 How a Letter of Credit Works

Source: adapted from James Aaron Cooke, 'What You Should Know about Letters of Credit', *Traffic Management* 29, no. 9 (September 1990), pp. 44–5.

Although it might appear to be more the concern of the corporate financial department, it is crucial that logistics managers involved in international trade understand how letters of credit work. If they misinterpret information or fail to diligently follow the shipping instructions contained in the document, it could jeopardize the company's chances of receiving payment for the goods shipped.[9]

The Global Marketplace

All forms of entry into the international marketplace require an awareness of the variables that can affect a firm's distribution system. Some of these factors can be controlled by logistics executives. Others, unhappily, are not subject to control, but must still be dealt with in any international marketing undertaking. Figure 12.4 shows the environment in which the logistics executive operates.

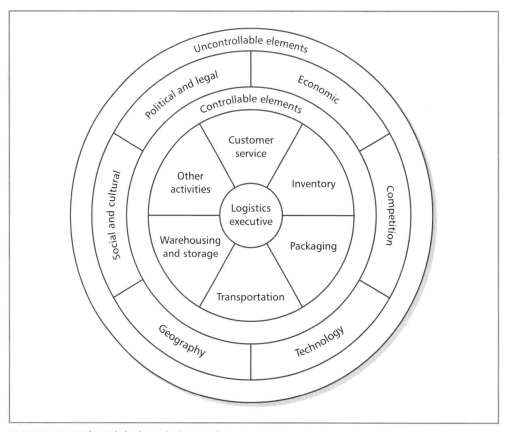

FIGURE 12.4 The Global Logistics Environment

Uncontrollable Elements

Anything that affects the logistics strategy of the international firm, yet is not under the direct control and authority of the logistics manager, is an uncontrollable element. The major uncontrollable elements include:

- political and legal systems of the foreign markets
- economic conditions
- degree of competition in each market
- level of distribution technology available or accessible
- geographic structure of the foreign market
- social and cultural norms of the various target markets.

An uncontrollable environment is characterized by uncertainty, and frequently by volatility. The logistics executive must make decisions within such an environment – for example, cost trade-offs, customer service levels and pricing. To illustrate, shipping component parts instead of finished goods to a foreign market can result in payment of lower duties, although transportation costs may be higher.

Other examples include: paying higher prices for a freight forwarder with lower damage rates, which may allow the firm to reduce packaging costs; using a bonded warehouse instead of a private warehouse, which may result in higher customs supervision costs but lower over-head costs for storage. Such trade-offs are only a few of the many that the logistics executive must evaluate.

It is beyond the scope of this chapter to examine in detail each of the various uncontrollable factors in the global marketplace. A number of international marketing textbooks address these elements. It is sufficient to say that the uncontrollable elements affect the actions of the logistics executive and must be considered in the planning, implementation and control of the firm's global distribution network.

Controllable Elements

When a firm becomes involved in international operations, the scope of the logistics executive's responsibilities often expands to include international distribution activities. Although the logistics executive may have full international responsibility, others within the organization probably have some involvement. A 2004 survey by the former Council of Logistics Management found that approximately 92 per cent of all logistics executives polled had direct or advisory responsibility for their company's global logistics operations.[10]

The management of a firm involved in international distribution must try to administer the logistics components to minimize cost and provide an acceptable level of service to its customers. However, a firm's cost–service mix will vary in international markets.

When all factors are considered, international distribution is generally more expensive than domestic distribution. Increased shipping distances, documentation costs, larger levels of inventory, longer order cycle times, and other factors, combine to increase the expense of international distribution.

Customer Service Strategies

The same consistency of service that a firm provides its domestic customers is not as easily achieved internationally. Because international transportation movements tend to be longer and often require several different types of carrier, multiple transfers and handlings, and the crossing of many international boundaries, time in transit may vary significantly from one shipment to the next. As a result, firms tend to require larger amounts of inventory to meet safety and cycle stock requirements.

Customer service levels may be higher in international markets, such as Japan, where the order cycle time is generally shorter than in Europe. The geographical differences between the two regions, the physical facilities of many wholesalers and retailers, and financial considerations permit the majority of all consumer goods orders in Japan to be delivered in 24 hours or under. For that reason, many international firms operate owned facilities in foreign markets in order to compete effectively on a customer service basis.

The cost of providing a specified level of customer service often varies between countries. A company must examine the service requirements of customers in each foreign market and develop a logistics package that best serves each area. Competition, specific customer needs, and other factors, may cause a firm to incur higher logistics costs, which will result in lower profits. Therefore, top management must make a complete analysis of the situation.

Hewlett-Packard (HP) was able to gain a competitive advantage internationally by improving its level of customer service. HP implemented programmes to tightly link R&D and manufacturing, and has been successful in reducing the 'concept-to-delivery' time. HP's logistics organizations have played a key role in this programme.[11] Improving the order fulfilment process continues to be a key strategic initiative. Objectives have been set to dramatically reduce the cost of taking and processing customer orders, and transporting products to customers. Thus, logistics will continue to play a key role in HP's global strategy.

Inventory Strategies

Inventory control is particularly important to an international company and requires an awareness of the many differences between international and domestic inventory management systems.

> International systems usually have more inventory points at more levels between suppliers and customers; thus multilevel inventory systems are more complex and more common than in domestic systems.
>
> In-transit inventories can be substantially higher than for a domestic operation with similar sales volume. This results from the larger number of locations and levels involved, and longer transportation times.[12]

Depending on the length of transit and the delays that can occur in international product movements, a firm may have to supply its distributors or other foreign intermediaries with higher than normal levels of inventory. A typical domestic firm will have 25 to 30 per cent of its assets in inventory, but firms engaged in international marketing can often have 50 per cent or more of their assets in inventory. For high-value products, the inventory carrying costs as well as the amount of accounts receivable outstanding can be extremely high.

In markets where the firm's products are sold at retail, consumer shopping patterns can be extremely important in determining inventory strategies. Companies in the United States can usually exercise greater control over their inventories because they can influence the amount of product ordered by their customers through discounts. However, this may not be a viable strategy in some international markets.

Since conditions may vary in foreign markets, it is important for the firm to develop inventory policies and control procedures that are appropriate for each market area.

Packaging and Containerization

International shipments require greater protection than domestic shipments, especially when they are not containerized. Other issues to consider include the handling of products, climate, potential for pilferage, communication and language differences, freight rates, customs duties and the customer's requirements. The greater number of handlings of international goods increases the possibility of damage. Generally, the amount of damage and loss in international traffic movements is higher than in domestic movements. Therefore, global shippers must be much more concerned with the protective aspect of the package than their domestic counterparts.

The bottom line in all international packaging decisions is that the item should arrive at its destination undamaged. Logistics executives can help to ensure that goods arrive safely at their international destinations through the proper planning, implementation and control of packaging decisions.

To facilitate product handling and protect the product during movement and storage, many firms have turned to the use of containers. Containers are widely used in international logistics, especially when water movements are part of the transport network. Many companies have adopted standard container sizes (8 feet × 8 feet × 10 feet, 20 feet, 30 feet or longer) that allow for intermodal movements.

Advantages of Containers

The use of standardized materials handling equipment has become commonplace. The advantages of containers are numerous.

- Costs due to loss or damage are reduced because of the protective nature of the container.
- Labour costs in freight handling are reduced because of the increased use of automated materials handling equipment.
- Containers are more easily stored and transported than other types of shipment, which results in lower warehousing and transportation costs.

- Containers are available in a variety of sizes, many of which are standardized for intermodal use.
- Containers are able to serve as temporary storage facilities at ports and terminals with limited warehousing space.

Disadvantages of Containers

On the other hand, containerization is not without disadvantages. The major problem is that ports or terminals with container facilities may not be available in certain parts of the world. Even when these facilities do exist, they may be so overburdened with inbound and outbound cargo that long delays occur.

Another major problem with containerization is that large capital expenditures are required to initiate a container-based transportation network. Significant capital outlays for port and terminal facilities, materials handling equipment, specialized transport equipment and the containers themselves are necessary before a firm can utilize containerization.

Labelling is related to the packaging component of global logistics. From a cost standpoint, labelling is of minor importance in international logistics. However, accurate labelling is essential to the timely and efficient movement of products across international borders. Important issues related to labelling include content, language, colour and location on the package.

Other Activities

Each of the activities or functions of logistics must be performed in the international market. The difference between the domestic and foreign market is not *whether* the logistics activity should be performed, but *how* each activity should be carried out.

One activity where differences occur is in the sourcing of materials. Traditionally, firms obtained raw materials, parts, supplies and components from domestic sources. In recent years there has been an accelerating trend towards international sourcing. For example, Canon uses some local sources of supply for products built outside of Japan; in its California facility, it uses 30 per cent local sources, in Virginia less than 20 per cent, and in Germany 40 per cent.[13]

The concepts of **integrated logistics management**, the systems approach, and **cost trade-off analysis** are very important in international logistics. However, the relative importance of each logistics component may vary from market to market, along with the costs incurred in carrying out each activity. This results in different cost–service equations for each international market.

The best advice for the executive whose company is entering into international logistics for the first time is to obtain as much information as possible about business conditions and operating procedures in each market from as many data sources as possible.

Each of the logistics activities of a company must be performed, although the task may be completed by one or more members of the international channel of distribution. The specific entities involved depend on the channel strategy selected.

Management of the Export Shipment

Many facilitator organizations are involved in the exporting activity. The types of organization utilized most extensively are:

- export distributor
- customshouse broker
- international freight forwarder
- trading company
- non-vessel-operating common carrier (NVOCC).

Other facilitators are used, but to a much lesser degree. These include export brokers, export merchants, foreign purchasing agents and others.

Export Facilitators

A firm involved in exporting for the first time would likely use an export distributor, customshouse broker, international freight forwarder or trading company.

Export Distributor

A company involved in international markets often employs the services of an export distributor (export management company). An **export distributor** is (1) located in the foreign market, (2) buys on its own account, (3) is responsible for the sale of the product, and (4) has a continuing contractual relationship with the domestic firm. The distributor is frequently granted exclusive rights to a specific territory. It may refrain from handling the products of competing manufacturers, or it may sell goods of other manufacturers to the same outlets.

The distributor often performs some of the following functions:

- managing channels of distribution and marketing/sales efforts
- handling customs clearances
- obtaining foreign exchange for payments to suppliers
- maintaining inventories
- providing warehouse facilities
- managing transportation activities
- breaking bulk
- managing credit policies
- gathering market information
- providing after-sale services.

Customshouse Broker

The **customshouse broker** performs two critical functions: (1) facilitating product movement through customs, and (2) handling the necessary documentation that must accompany international shipments.

The task of handling the large number of documents and forms that accompany an international shipment can be overwhelming for many firms. Coupled with differing customs procedures, restrictions and requirements from country to country, the job of facilitating export shipments across international borders requires a specialist: the customshouse broker. In general, if a company is exporting to a number of countries with different import requirements or if the company has a large number of items in its product line (e.g. automotive parts, electronic components, food products), a customshouse broker should be a part of the firm's international distribution network.

International Freight Forwarder

International freight forwarders perform a number of functions to facilitate trade for an international company. They:

- speed the movement of goods from the site of production to the customer's location by using drop shipments, thus eliminating double handling
- receive advanced shipping notices, which speed clearance of customs and rapid preparation of required documentation
- arrange transportation and carrier routings
- coordinate product storage and pick-and-pack operations
- provide full-service logistics to their clients.

Nearly every international company utilizes the services of an international freight forwarder in order to coordinate activities at the destination.

Trading Company

Most trading companies are primarily involved in exporting, but some engage in the import business as well. **Trading companies** match the seller with buyers of goods or services, and manage the export arrangements, paperwork, transportation and foreign government requirements. Examples of trading companies include the United Africa Company (UAC), the largest trading company in Africa, which is part of Unilever, and the Japanese trading house, the Soga Shosha.[14]

Non-vessel-operating Common Carrier (NVOCC)

The NVOCC, sometimes referred to as an NVO, 'consolidates small shipments from different shippers into full container loads . . . and accepts responsibility for all details of the international shipment from the exporter's dock, including paperwork and transportation'.[15] Figure 12.5 shows how these common carriers work.

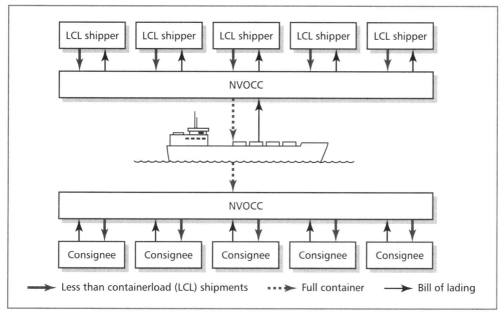

FIGURE 12.5 How NVOCCs Work

Source: Toby B. Estis, 'NVOCCs: A Low-cost Alternative for LCL Shippers', *Traffic Management* 27, no. 6 (June 1988), p. 87.

Documentation

International documentation is much more complex than domestic documentation because each country has its own specifications and requirements. Errors in documentation can create long, costly delays in shipping. Seven of the most widely used documents are listed below.

1 *Air waybill. Issued by*: Airline, consolidator. *Purpose*: Each airline has its own air waybill form, but the format and numbering systems have been standardized by the airline industry to allow computerization. Like the ocean bill of lading, the air waybill serves as contract of carriage between shipper and carrier.

2 *Certificate of origin. Issued by*: Exporter or freight forwarder on exporter's behalf. *Purpose*: Required by some countries to certify the origin of product components. Used for statistical research or for assessing duties, particularly under trade agreements.

3 *Commercial invoice. Issued by*: Seller of goods. *Purpose*: Invoice against which payment is made. Required for clearing goods through customs at destination.

4 *Dock receipt (D/R). Issued by*: Exporter or freight forwarder on exporter's behalf. *Purpose*: No standard form, but the D/R must include shipment description, physical details and shipping information. Used by both shipper and carrier to verify shipment particulars, condition and delivery to carrier.

5 *Ocean bill of lading (B/L). Issued by*: Steamship line. *Purpose*: Each carrier has its own bill of lading form. Serves as the contract of carriage between carrier and shipper, spelling out legal responsibilities and liability limits for all parties to the shipment. The B/L can also be used to transfer title to the goods to a party named in the document. Specifies shipment details, such as number of pieces, weight, destination, and so on.

6 *Packing list. Issued by*: Exporter. *Purpose*: Provides detailed information about contents of each package in the shipment. Customs authorities at destination use this information during clearance and inspection procedures. The packing list is also invaluable when filing claims for damage or shortage.

7 *Sight, time drafts. Issued by*: Exporter or freight forwarder on exporter's behalf. *Purpose*: Request for payment from the foreign buyer. Instructs buyer's bank to collect payment; when collected, the bank releases shipping documents to the buyer. The buyer's bank then remits to seller's bank. Sight drafts are payable on receipt at the buyer's bank. Time drafts extend credit; the foreign bank releases documents immediately, but collects payment later.[16]

Terms of Trade

The **terms of shipment** or **terms of trade** are important information to be included on the actual export documents. The terms of shipment are much more important in international shipping than in domestic shipping because of the uncertainties and problems of control that accompany foreign traffic movements. These terms determine who is responsible for the various stages of delivery, who bears what risks and who pays for the various elements of transportation.

In international trade, 'INCOTERMS' are the equivalent to FOB origin or destination terms, and were developed and defined by the International Chamber of Commerce. The 13 INCOTERMS were last updated in 2000 and are expressed in the terms of sale by a three-letter abbreviation followed by a named location and the designated year of the INCOTERMS definition being used.

1 *EXW* Ex works (. . . named place). Origin should be identified as factory, plant, and so forth. The seller's only responsibility is to make the goods available at the seller's premises, and the seller bears the costs and risks until the buyer is obligated to take delivery. The buyer pays for the documents, must take delivery of the shipment when specified, and must pay for any export taxes.

2 *FCA* Free carrier (. . . named place). This term has been designed to meet the requirements of multimodal transport, such as container or roll-on, roll-off traffic by trailers and ferries. It is based on the same name principle as FOB except the seller fulfils its obligations when the goods are delivered to the custody of the carrier at the named place.

3 *FAS* Free alongside ship (. . . named port of shipment). Similar to FOB vessel, but certain additional port charges for the seller, such as heavy lift, may apply. The buyer is responsible for loss or damage while the goods are on a lighter (small barge) or within reach of the loading device. Loading costs are the responsibility of the buyer.

4 *FOB* Free on board (. . . named port of shipment). The goods are placed on board a ship by the seller at a port of shipment named in the sales agreement. The risk of loss of or damage to the goods is transferred to the buyer when the goods pass 'over the ship's rail' (i.e. off the dock and placed on the ship). The seller pays the cost of loading the goods.

5   *CFR* Cost and freight (. . . named port of destination). The price quoted includes all transportation to the point of destination. The seller pays export taxes and similar fees. The buyer pays the cost of certificates of origin, consular invoices or other documents required for importation into the buyer's country. The seller must prove these, but at the buyer's expense. The buyer is responsible for all insurance from the point of vessel loading.

6   *CIF* Cost, insurance and freight (. . . named port of destination). The price quoted includes the cost of goods, transportation and marine insurance. The seller pays all taxes or fees, as well as marine and war risk insurance. The buyer pays for any certificates or consular documents required for importation. Although the seller pays for insurance, the buyer assumes all risk after the seller has delivered the goods to the carrier.

7   *CPT* Carriage paid to (. . . named port of destination). The seller pays the freight for the carriage of the goods to the named destination. The risk of loss or damage to the goods and any cost increases transfer from the seller to the buyer when the goods have been delivered to the custody of the first carrier, and not at the ship's rail.

8   *CIP* Carriage and insurance paid to (. . . named port of destination). This term is the same as 'freight/carriage paid to (CPT)' but with the additional requirement that the seller has to procure transport insurance against the risk of loss or damage to the goods during the carriage.

9   *DAF* Delivered at frontier (. . . named place). The seller's obligations are fulfilled when the goods have arrived at the frontier but before the customs border of the country named in the sales contract. The term is primarily used when goods are carried by rail or truck. The seller bears the full cost and risk in delivering the goods up to this point, but the buyer must arrange and pay for the goods to clear customs.

10 *DES* Delivered ex ship (. . . named port of destination). The seller will make the goods available to the buyer on board the ship at the place named in the sales contract and bears the full cost and risk involved in bringing the goods there. The cost of unloading the goods and any customs duties must be paid by the buyer.

11 *DEQ* Delivered ex quay (. . . named port of destination). The seller has agreed to make the goods available to the buyer on the quay or the wharf at the place named in the sales contract. The seller bears the full cost and risks in delivering the goods to that point, including unloading over the ship's rail.

12 *DDU* Delivered duty unpaid (. . . named port of destination). The seller's obligations to deliver are fulfilled when the goods are available to the buyer 'uncleared for import' at the point of the named destination. The seller bears all costs and risks involved in bringing the goods to the point or place of named destination.

13 *DDP* Delivered duty paid (. . . named place of destination). DDP represents the seller's maximum obligation and notes that the seller bears all risks and all costs until the goods are delivered. This term can be used irrespective of the mode of transport.

Examples of the correct form in which these terms would be specified are:

EXW, Kuala Lumpur, INCOTERMS 2000
DDU, Frankfurt Schmidt GmbH Warehouse 4, INCOTERMS 2000[17]

Free Trade Zones

Free trade zones (FTZs), sometimes referred to as **foreign trade zones**, are areas where companies may ship products to postpone or reduce customs duties or taxes. Products remaining in the FTZ are not subject to duties or taxes until they are reshipped from the zone into the country of destination. Firms often process, assemble, sort and repackage the product within the FTZ before reshipment. The facilities, services offered and quality of FTZ management vary significantly. Management wishing to utilize an FTZ will have to explore each zone individually to determine its potential usefulness.

Logistics Characteristics of Global Markets[18]

Three major geographic regions account for the bulk of world economic activity and international trade: Europe, including member and non-member states of the EU, North America and the Pacific Rim, which includes China and Japan. These three areas produce over 75 per cent of the world's economic output (see Table 12.2). For this reason, it is necessary to understand the foundations of business and logistics systems in these regions and in other developing areas.

European Region

The European continent has undergone a major political and economic transformation since the early 1990s. This transformation must be clearly understood for businesses to operate successfully in it.

There are two major European regions, each one with clearly differentiated subregions. These regions are the 25 member states of the European Union (EU25) and the seven non-member European states, and the nearby states of the former Soviet Union, Russia, Ukraine, Belarus, Moldova and the Balkan states.

The EU25 states include Austria, Belgium, Cyprus, the Czech Republic, Denmark, Estonia, Finland, France, Germany, Greece, Hungary, Ireland, Italy, Latvia, Lithuania, Luxembourg, Malta, the Netherlands, Poland, Portugal, Slovakia, Slovenia, Spain, Sweden and the United Kingdom. Bulgaria and Romania will join in 2007, while an entry date for Turkey has yet to be agreed. Liechtenstein, Norway, Iceland and Switzerland remain outside the EU. The EU25, with a population of more than 450 million people, has a gross regional product (analogous to the use of the term **gross domestic product**) and per capita income almost equal to the three NAFTA countries: the United States, Canada and Mexico.

Other countries of the former Soviet Union, including Russia and Ukraine, are trying to effect a transition from centrally planned economies to free-market economies. The concurrent political and economic turmoil is not likely to end in the near future. Their economic problems are compounded by political volatility that is disintegrating parts of the region into a conglomerate of smaller, tribal nations. Many of these problems were not envisioned in the early days following the break-up of the Soviet Union.

Eight countries of the former Soviet Union are near to Europe: Russia, Ukraine, Belarus, Moldova, and the four Balkan countries of Albania, Bosnia and Hercegovina, Croatia, and Serbia and Montenegro. These countries have a population of approximately 230 million people, with a per capita income of 26–29 per cent that of the EU25 and NAFTA states, but with enormous reserves of natural resources and a well-educated population. Russia and Ukraine are the economic backbone and encompass most of the population.[19]

From a logistics perspective, these former members of the Soviet Union and those in the EU10 and elsewhere offer special challenges. Most businesspeople travelling to those areas recognize the poor transportation infrastructures resulting from decades of Communist rule. Except in a few rare instances, the concepts of customer satisfaction and customer service are unknown. Sophisticated logistics techniques and the use of computerized order-processing and information systems are impossible in most areas.

Because of the immense opportunities, though, many firms will decide that the benefits outweigh the costs. It will be a slow process, but the firms that penetrate the markets of eastern Europe will gain competitive advantage over those that wait.

Africa, especially North Africa, is also economically important to Europe. It has substantial trade with western Europe and has been the source of massive immigration.

Group	Country	GDP ($US millions)	Per cent
	World	51,480,000	100.0
European Region	European Union (EU25)	11,050,000	
	Turkey	458,200	
	Switzerland	239,300	
	Norway	171,700	
	Romania	155,000	
	Bulgaria	57,130	
	Liechtenstein	825	
	Iceland	8,678	
	Total	12,140,833	23.6
	Russia	1,282,000	2.5
	Ukraine	260,400	0.5
	Balkan States	111,380	0.2
	Belarus	62,560	0.1
	Moldova	7,792	–
NAFTA	United States	10,990,000	
	Canada	958,700	
	Mexico	941,200	
	Total	12,889,900	25.0
Mercosur	Brazil	1,375,000	
	Argentina	435,500	
	Uruguay	43,670	
	Paraguay	28,170	
	Total	1,882,340	3.7
Pacific Rim	China (including Hong Kong)	6,662,000	12.9
	Japan	3,582,000	7.0
ASEAN 10	Indonesia	758,800	
	Thailand	477,500	
	Philippines	390,700	
	Malaysia	207,800	
	Vietnam	203,700	
	Singapore	109,400	
	Myanmar	74,530	
	Cambodia	25,020	
	Laos	10,320	
	Brunei	6,500	
	Total	2,264,270	4.4
	South Korea	857,800	1.6
	Australia and New Zealand	656,740	1.3
	Taiwan	528,600	1.0
Other	India	3,033,000	5.9

TABLE 12.2 Comparison of Gross Domestic Product Among Global Regions, 2003

Source: Central Intelligence Agency, 'The World Factbook', http://www.cia.gov/ (updated 10 February 2005).

Economic Issues for the New Century

Into the twenty-first century, the EU25 countries will be the locomotive of the European economy. The following business-related issues are of major importance to companies operating in Europe.

- *Common currency.* The conversion of all 25 members to a common currency, the **euro (€)**, while slow in coming, will eliminate the costs and work of converting Europe's many currencies.
- *Tax equalization.* The eventual equalization of personal and corporate taxes throughout the region will bring logistic costs to the forefront as the critical ones necessary to establish facility locations.
- *Political homogenization.* The creation of uniform political institutions among all members will compound the benefits of tax equalization.
- *Standards homogenization.* In January 1993, the EU implemented some 1500 safety, health, environmental and quality standards. The ultimate goal is to establish some 10,000 standards in Europe. The new rules are being drafted by the European Committee for Standardization, better known by its French acronym, CEN, and the International Standards Organization (ISO).

Logistics Issues Facing Europe

The developments affecting these issues will have major consequences for business in Europe, especially for logistics decisions concerning the location of plants and warehouses, and the determination of their missions and areas of convergence.

However, one fact is clear from the development of optimum logistics strategies on a European scale: The optimum number and location of plants and warehouses is substantially lower than that required to optimize the logistic networks of each individual country. Typical cost reductions obtained by optimum continental systems are 15 to 20 per cent compared with optimum national systems. Thus, companies that streamline their logistics networks on a continental basis before their competitors will gain major competitive advantages. If those advantages are retained for a few years, they may become insurmountable to their competitors.

The current economic heart of Europe is the 'golden triangle' with vertexes in the English Midlands, eastern Germany and southern France. Most production and consumption takes place within that area and is therefore of major logistics importance. This area contains major distribution centres to supply European customers or secondary warehouses. However, this triangle will undoubtedly shift with the addition of the 10 accession, or EU10, countries of the Czech Republic, Cyprus, Estonia, Hungary, Latvia, Lithuania, Malta, Poland, Slovakia and Slovenia in May 2004.

Other major considerations governing west European logistics are as follows.

Customs and transit procedures

The EU has in theory eliminated customs and transit procedures among its members; thus there should be a free flow of goods between its members and the four members of the European Free Trade Association (EFTA): Iceland, Liechtenstein, Norway and Switzerland. However, there have been practical problems regarding implementation of these procedures and the single market is 'only 43 per cent operational' and is 'still an imperfect reality with numerous hurdles and obstacles that hinder intra-Community trade'.[20]

Transportation deregulation

The deregulation of the transportation industry has provided liberalization along with standardization of vehicle specifications, trailer lengths and weights, and vehicle emissions. However, challenges remain such as the 'vignette' charge for commercial traffic travelling in Austria and Switzerland, and the road ban for commercial vehicles on Sundays in France, Germany and

Austria. These measures are ostensibly for environmental reasons but are considered by many to offer protection for national hauliers.[21]

Transportation modes

Trucks and pipelines account for the majority of tonnage transported in Europe. They continue to gain tonnage at the expense of rail and coastal and inland water transportation although, for environmental reasons, the national railroads are waging a campaign to force more traffic to use rail. The result is likely to be an increase in intermodal transportation.

Subcontracting of services

Logistics services are increasingly being subcontracted to logistic service companies, rather than performed in-house. This trend is likely to accelerate as the average shipment size continues to decrease and the average shipping frequency to increase; consolidation provides substantial economies. But while third-party logistics service providers (3PLSP) have grown, 'there is still some way to go before the largest [3PLSPs] can claim to offer homogeneous levels of service across all European markets'.[22]

Eurotunnel

The opening of the Eurotunnel (the 'Chunnel') under the English Channel improved service and reduced transportation costs between the United Kingdom and continental Europe. This has affected facility location decisions.

Palletization

Palletized freight is increasingly using the ISO standard pallet sizes.

Five Major Areas of Change in the European Union

Changes in the EU are expected to have a substantial impact on the marketing and manufacturing strategies of firms selling products in Europe. The anticipated changes will occur in five major areas: manufacturing, transportation, distribution channels, administration and organization.

Manufacturing

Many firms have traditionally established plants in each country, but with the economic unification of Europe, companies are establishing plants that can service multiple countries. With larger plants, greater economies of scale can be realized, which will have a variety of effects on logistics activities carried out in the channel of distribution due to the larger amounts of goods being produced. If firms maintain several plant locations, the focused factory concept could be implemented, with each plant specializing in one or a few product lines.

Transportation

Accompanying the unification of Europe on 1 January 1993 was the deregulation of transportation, particularly motor transport. As happened in the United States after the deregulation of transportation, customers in Europe are demanding more services. Transportation providers have become more competitive in pricing and more innovative in service. Just as in the United States, a shakeout of the industry was anticipated, and has begun with the merger of Exel and Tibbett & Britten and the sale of Hays (now ACR) to a US investment bank in late 2004.[23]

The transport industry should experience financial improvement after a shakeout occurs. Freight consolidation opportunities should increase because of many factors, including deregulation, more optimal routing and scheduling opportunities, and the development of pan-European (across Europe) services. In addition, the 'Chunnel' links the UK with the rest of Europe and will facilitate additional freight transport between England and France.

Distribution Channels

Within the European channel of distribution, there may be a decline in the use of channel intermediaries. With a single European market, separate channels will not be required for each country. Instead, larger wholesalers and distributors will develop. Vertical integration will occur, and direct marketing and distribution will increase. The result will be less inventory within the system and a lowering of channel-wide logistics costs. How Xerox has achieved logistics success in Europe is discussed in the Creative Solutions box.

CREATIVE SOLUTIONS

Xerox's European Logistics Success

In Europe, Xerox meets its customers' needs for new product, suppliers and maintenance through a central distribution centre in Venray, the Netherlands. (Interestingly, the company's sophisticated logistics modelling analysis had located the theoretically best site for its European logistics operations only 20 miles from this existing facility.) Including older machines, Xerox estimates it must maintain a 100,000-item catalogue to meet customer-support requirements. The company uses air transport for priority shipments. To meet the needs of a more unified Europe, Xerox has entered into long-term alliances with surface carriers that transport copiers and replacement parts. These contracts include provisions for a common tracking system for inventory management. Xerox estimates that 80 per cent of all repairs and parts deliveries can be performed within one day, 98 per cent by the second day and 100 per cent by the third day. The company's goal is to improve its performance to 98 per cent within one day.

Xerox sees its European logistics systems as a model for those in the United States and, eventually, in all its markets around the globe. The company aims to establish a new manufacturing plant near its Venray distribution centre. Its benchmarking approach focuses on qualitative measurements of logistics and overall corporate performance, including responsiveness to customers, profitability and return on assets. Plans call for establishing a build-to-order/direct delivery system for manufacturing and transportation, and linking this electronically with Xerox management, suppliers, shippers and customers. The company estimates that its progress towards this goal has reduced its inventory as a percentage of revenue from 25 per cent to 14 per cent.

As a large and complex multinational firm, Xerox has shown the value of combining integrated logistics, advanced information technology and customer-responsive management to prosper in a fast-changing global marketplace. The company appears well positioned to meet the challenges of a unified Europe.

Question: **What are the cost and customer service trade-offs involved in using air freight for priority shipments?**

Source: William W. Goldsborough and David L. Anderson, 'The International Logistics Environment', in *The Logistics Handbook*, ed. James F. Robeson and William C. Copacino. New York: Free Press, 1994, p. 667.

Administration

The removal of customs procedures will result in greater efficiencies in transportation, packaging and labelling. Technology improvements can be implemented throughout Europe instead of within individual countries. More centralization of order processing, inventory control, warehousing and computer technology can occur with a unified Europe.

Organization

Significant organizational changes are occurring in Europe. The centralization trend mentioned previously, and the organizational structures that recognize national boundaries to a lesser extent will result in pervasive changes in all industries.

In sum, the competitive situation in Europe will intensify. The penalties for poor performance will be greater, but the rewards much higher for those firms that can effectively implement optimal manufacturing, marketing and logistics strategies in a unified European economy.

North and South America

The North American Free Trade Association (NAFTA), launched in January 1994, brought together the economies of Canada, the United States and Mexico with a resultant population of 431 million people. In 2004, the flow of trade among NAFTA members was $725 billion. Imports from Mexico to the United States were $156 billion, while exports from the United States to Mexico were $111 billion. Imports from Canada to the United States were $256 billion, while exports from the United States to Canada were $190 billion. Two-way trade between Canada and Mexico amounted to only about $13 billion.[24]

This area has a larger domestic economic product than the EU25 and European Free Trade Association (EFTA) combined. South of the NAFTA countries is the rest of Latin America, a potential addition to NAFTA with a population of some 350 million. There is already a customer union among four South American countries. **Mercosur** was created by Argentina, Brazil, Paraguay and Uruguay in March 1991 with the signing of the Treaty of Asuncion. The combined gross domestic products of the Mercosur nations accounted for about 4 per cent of the world's economic output in 2003 (see Table 12.2).

The provisions of NAFTA are significant in that they directly impact a variety of logistics activities and affect how supply chains are structured when Canadian, US and Mexican companies are involved. Some of the more substantive changes brought about by NAFTA are that it:
- eliminates many tariff and non-tariff barriers
- enhances carriers' ability to operate across borders, especially between the United States and Mexico, where cross-border movements occur much more easily
- liberalizes foreign investment (allows US and Canadian companies the right to establish firms in Mexico or to acquire existing Mexican companies)
- standardizes customs initiatives, local content rules, and packaging and labelling requirements.[25]

Canada and the United States have the most advanced logistics infrastructure and systems in the world. North America offers a wide choice of suppliers in all transportation modes and very good, competitively priced warehousing facilities and ancillary services throughout the continent. Thus, the development of logistics strategies and operations are seldom limited by the physical facilities available.

In North America, it is possible to find common, contract and private carriers offering transportation services by air, highway, railroad, pipeline and water. In the United States and Canada, prices are negotiable in most cases, depending on freight type (e.g. hazardous materials, refrigerated goods) and product characteristics (e.g. annual volumes, seasonality, shipment size, type of product).

Unitization, in the form of pallets and slip sheets, and freight containerization, have been commonplace for decades. Pallets have been standardized mostly along industry lines (e.g. grocery manufacturers). Containers, most 40 feet and longer in length, with some 20-foot containers, have served as the basis for the International Standards Organization's standards.

The use of electronic data interchange (EDI) to support logistics operations was pioneered in North America several decades ago, and the continent is the largest user of that technology in the world today (see the Technology box).

TECHNOLOGY

NAFTA and Technology Combine to Cut Customs Delays

Today, the documentation for more than 1.5 million motor carrier shipments into Canada is processed by customs before the freight even reaches the border, shortening the time spent releasing the freight when it arrives.

In 1990 going into Mexico meant clearances that could take three or four days. Today, 90 per cent of all goods from the United States are cleared by Mexican Customs at the border in 20 seconds or less.

The ultimate goal of [the United States, Canada and Mexico] is to completely automate the customs clearance process. Canadian Customs is implementing Customs 2000 initiatives with the goal of making its customs system paperless by the year 2000. One initiative that has boosted the automation efforts in Canada has been implementation of the Pre-Arrival Review System (PARS) and the Inland Pre-Arrival Review System (INPARS) . . . Through PARS and INPARS, information on a shipment is sent from the carrier to the broker or designate who prepares the release documentation and forwards it to Canadian Customs for review and processing. These systems allow brokers to prepare release documentation for customs review prior to the freight's arrival . . . This allows paperwork errors to be corrected while the freight is en route instead of delaying the shipment after it arrives at the border.

Automation has helped Mexican Customs . . . From 1990–1995, traffic volume through Mexican Customs has increased 300 per cent while the agency was able to reduce its staff by 60 per cent through automation.

Question: **What could the European Union take from the NAFTA experience to move towards the free movement of goods?**

Source: Robert B. Carr, 'Don't Let Borders Be Barriers', *Transportation and Distribution* 36, no. 3 (March 1995), pp. 65–70.

Many North American manufacturers, retailers, and logistics service providers are taking advantage of NAFTA opportunities. These 'leading edge,' or 'best practice,' companies have adopted one or more of the following strategies.

- *Customer service*. Best-practice companies manage key accounts in a consistent, coordinated manner in all three countries. They are working to create more uniform service levels across an integrated North American marketplace.
- *Manufacturing*. Best-practice companies modify their product development and manufacturing approaches to take advantage of the market and tariff advantages inherent in NAFTA.
- *Channel design*. Best-practice companies establish market research groups in each North American country to facilitate distribution channel designs.

- *Sourcing.* Best-practice companies regularly revisit and revise their sourcing strategies. Increasingly, this means moving away from offshore vendors to suppliers in North America.
- *Distribution.* Best-practice companies establish core carrier programmes with their major North American carriers. They develop cross-border shipping programmes that include innovative freight consolidation and 'double stack' train approaches, where two containers are stacked together on rail flatbeds. Finally, they establish a strong border presence to expedite cross-border product flow.
- *Sales and marketing.* Best-practice companies develop sales and marketing strategies targeted to specific markets and customers within North America.
- *Organization.* Best-practice companies create internal NAFTA units dedicated to managing business in the NAFTA trading area. They train their own people and their vendors in NAFTA rules and regulations.[26]

The Pacific Rim: China

The Pacific Rim was a $14.6 trillion economy in 2003 (see Table 12.2). It is important to note for our discussion that Pacific Rim refers to countries on the eastern Pacific – China, Japan, the ASEAN 10 countries, South Korea, Australia and New Zealand, and Taiwan – because the United States, Canada, Mexico and Chile, for example, are also countries on the Pacific Rim. The two largest players in the Pacific Rim are China and Japan, and they will be the focus of our discussions in these next two sections. However, India, which is in the Asian subcontinent, also deserves discussion due to its size and economic output, and this follows the next two sections.

Firms that import from or export to the Pacific Rim of Asia, those outsourcing materials from the region and those interested in entering markets there recognize that differences in economics, politics and culture greatly influence business activities in specific Asian countries. Ranging from the affluence of Japan to the poverty of Indonesia and some parts of China, the region offers a myriad of immense problems and significant opportunities for companies.

Firms marketing products to Japan, South Korea, Hong Kong and other industrialized areas of the Pacific Rim will find logistics environments similar to those found in North America and the European Union. While cultures and politics are different, transportation infrastructures are developed, a variety of warehousing options exist, the use of automated systems is widespread, and customer service concepts are understood and accepted by logistics service providers.

This cannot be said about the developing countries of the Pacific Rim, such as China. In China, most logistics activities are still handled by the government. The economy is characterized by some materials shortages, planned distribution by the government, and inefficient logistics systems that include many intermediaries and much double handling.

Firms penetrating the Chinese market will find some logistics difficulties. The process will be slow, requiring great patience both managerially and financially. However, conditions are changing for the better. The China Communications and Transport Association (CCTA) has developed a link with the United Kingdom's Chartered Institute of Logistics and Transport (CILT) to deliver CILT qualifications to Chinese logisticians. This is important for the 29th Olympic Games, which will be held in Beijing in 2008. It is estimated that logistics services worth approximately Chinese Yuan 41.7 billion, or $5 billion, will be demanded for the Games. Several large multinational logistics services providers will be competing to meet this demand.[27] However, two firms, DHL and FedEx, are already well established in China, as discussed in Box 12.1.

Business transactions in countries like China may not show significant payback for many years, and firms will not be able to utilize the same financial criteria in evaluating Chinese logistics efficiencies as they use in other parts of the world.

Box 12.1 The Growth of Logistics Service Providers in China

The booming Chinese logistics market is attracting many of the world's leading logistics companies.

DHL, the express and logistics company, has announced that its local joint venture, DHL-Sinotrans, will invest an additional $200 million during 2004–08 as part of its Chinese expansion plan. The investment will boost DHL-Sinotrans' capacity further and support the anticipated high growth of the express delivery service in the country. DHL-Sinotrans will use the money to expand and enhance its existing gateways in Beijing, Shanghai, Guangzhou and Shenzhen, add 14 new regional branches, buy 1200 new vehicles and recruit more staff – creating up to 2100 new positions for DHL in China.

DHL's new expansion programme follows recent initiatives to cement its leading position in China, which included the acquisition of a 5 per cent stake in Sinotrans, its

Chinese partner, and the offering of overnight freight services between Hong Kong and Shanghai four times a week in cooperation with Dragonair, a Hong Kong-based carrier.

The logistics sector in China, fuelled by the growth of manufacturing industries and trade, has grown dramatically in the past few years. DHL has experienced an annual growth of between 35 per cent and 45 per cent during the past three years in China.

FedEx Corp, another global logistics service provider, recently announced its expansion plan for China. FedEx now operates in 224 cities in China, up from 97 cities in 1998, and expects to increase that number by 10 over the next five years. The company set up regional headquarters in Shanghai earlier this year in a bid to capture a bigger slice of the booming air freight market. FedEx's Chinese sales were up to 40 per cent in the first quarter of 2004.

Source: 'Expansion of DHL and FedEx in China: Good Examples of Opportunity in a Growing Economy', *Logistics & Transport Focus* 6, no. 5 June 2004, p. 63.

The Pacific Rim: Japan[28]

Japan remains an economic powerhouse of the Asian side of the Pacific Rim despite some slowing in its economic growth rate. The Japanese distribution system is by far the most complex and inefficient of the industrialized countries, although that may be changing, as described in Box 12.2. This is a consequence of historical preferences given by the government to small business enterprises.

Most aspects of goods distribution in Japan are tightly controlled by the government. The recent affluence of Japan, coupled with complaints from foreign governments and companies that consider the system a major impediment to entering the Japanese market, have provoked political actions. These have resulted in the beginnings of major liberalization of the distribution system in Japan. However, distribution costs are so high that retail prices are several times their respective wholesale prices.

Although Japan is an archipelago comprising more than 5000 islands, the bulk of its population lives on the four major islands of Hokkaido, Honshu, Kyushu and Shikoku. Of these, the island of Honshu contains all the major cities and most of the population of Japan.

Some of the major characteristics of Japanese logistics are as follows.

- *Transportation modes.* Ninety per cent of domestic tonnage is transported by truck, and this is likely to continue. Truck transportation requires licensing from the Ministry of Transport. Licenses distinguish between:
  - long-distance trucks, which carry loads between major regions; for example, from plants to distribution centres
  - short-distance trucks, which carry loads within a region; for example, between a wholesaler and a retailer
  - district trucks, which can carry loads anywhere, but whose routes must originate and terminate within a designated district and can carry goods only for a single shipper
  - route trucks, which can carry loads along their licensed route for multiple shippers.

Box 12.2 Is Japan Beginning to Adopt Western Supply Chain Management Concepts in its Retail Distribution Systems?

A survey performed by Japan's Ministry of International Trade and Industry (MITI) in 1989 found that Japan averaged 2.21 wholesale steps between the producer and retailer. This compared to .73 in France, .90 in the former West Germany and 1.0 in the United States. This heavy 'middleman' structure has several effects: it increases prices to Japanese consumers, promotes inefficiency and makes it extremely difficult for anyone new, such as foreign competitors, to break into the market. They have to establish a relationship with wholesalers in order to have a point of entry.

These long distribution channels are part of Japan's Keiretsu system, where manufacturers own full or partial shares in many of the retailers and wholesalers that are part of their channel. These channels are tightly controlled through interlocking directorates and secret meetings among channel members to plan the long-term strategy of the channel. Interpersonal relationships are very close and stable. Information flow is excellent, and transaction costs are relatively low due to the familiar, established network.

Things began to change in the early 1990s when Japan suffered an economic slump, making increased efficiency more important. As in the United States and Europe, power is shifting away from manufacturers to retailers. Large retailers don't want to deal with long channels that add costs to their processes. Direct sales and mail order have been growing faster than retail sales, causing retailers to want to become more competitive. As a result, innovative Japanese manufacturers have begun to respond, reducing the length of their channels. An example of this is KOA Soap. The firm used three levels of wholesalers in the 1960s and two levels, including its own exclusive wholesaler, in the 1970s and 1980s. In the 1990s, KOA Soap was using only its own exclusive wholesaler.

This could be a trend that further shakes up the Japanese retail sector. As distribution becomes simplified, it may be easier for foreign competitors to enter Japanese markets, and bring their own logistics innovations, Japanese consumers should come out the clear winners.

Sources: John Fahy and Fuyuki Taguchi, 'Reassessing the Japanese Distribution System', *Sloan Management Review*, Winter 1995, pp. 49–61; and Lisa M. Ellram and Martha C. Cooper, 'The Relationship between Supply Chain Management and Keiretsu', *The International Journal of Logistics Management* 4, no. 1 (1993), pp. 1–12.

■ *Logistics heartland*. The main area for production in Japan is the triangle bound by the cities of Tokyo, Nagoya and Osaka on the island of Honshu. It is about 500 kilometres from Tokyo to Osaka. The triangle includes the metropolitan area around Tokyo known as Kanto (including Yokohama and Kawasaki) and that around Osaka known as Kansai (including Kobe and Kyoto).

■ *Traffic congestion*. Traffic congestion on roads and highways is a critical problem in the triangle, especially in and around the major cities where traffic speed averages less than 15 kilometres per hour. For this reason, just-in-time systems require many small facilities or substantial fleets of small vehicles to meet customer requirements quickly, reliably and economically.

■ *Distribution systems*. Distribution systems for varying products are usually diverse because of traditional differences in trade practices and channels of distribution.

■ *Distribution channels*. Non-traditional distribution channels, especially non-store channels, are booming. They often represent the best way to introduce new products into the Japanese market. These channels include mail order, catalogue sales, door-to-door sales, teleshopping and vending machines.

■ *Shared distribution*. Shared distribution is common. Competitors delivering to the same stores share delivery facilities and trucks.

■ *Palletization*. Large companies tend to use ISO standard pallet sizes. These are not mandatory, and a proliferation of different pallet sizes complicates logistics operations significantly.

■ *Warehousing.* Business warehouses are supervised by the Ministry of Transport and regional Transport Bureaus. These distinguish between private, agricultural, cooperative and public warehouses. Public warehouses are further classified into general-purpose, cold storage, open-air, storage tanks, floating storage (e.g. for logs) and dangerous goods warehouses. These are treated differently by the Ministry of Transport, which issues them permits.

India[29]

India enjoyed a GDP of over $3 trillion in 2003 (see Table 12.2). Indian industry spends 14 per cent of its GDP on logistics. The Indian logistics environment comprises road transport companies, railways, air freight companies, intermodal transport providers, ports and shipping companies, as well as 3PL companies. Their performance is critically dependent on the state of the Indian logistics infrastructure.

India has a road network totalling 2.7 million kilometres of road length. However, express and national highways constitute only 1.4 per cent of the total road length but carry nearly 40 per cent of all road freight. The overall quality of roads is poor, resulting in slow transport speeds, increased wear and tear of vehicles and high accident rates. The Indian government has initiated the National Highway Development Programme to build four-lane highways connecting the four metropolitan areas of India and the north–south and east–west corridor. Completion of this project by 2007 is expected to have a major impact on transportation times and costs.

The Indian Railway network is a government monopoly in India and is fraught with hidden inefficiencies. It is the second largest railroad system in the world covering a route length of 62,809 kilometres and facilitates 450 million tonnes of freight every year. The cost of using the rail network is high due to handling requirements and the time and cost of arranging pick-up and drop of consignments to and from railway facilities. This results in the slow average speed of freight movement and low average wagon turnaround time, which are major concerns for Indian logisticians.

There are 11 major seaports that handle the total foreign trade of the country, amounting to almost 272 million tonnes. The facility and infrastructure of Indian ports are rated low in terms of global standards primarily on account of lack of storage space and outdated handling equipment. Most Indian seaports are inefficient in loading and unloading operations, with the result that ships spend more time there, which increases costs for a shipper by 10–20 per cent.

The six international and 87 domestic airports handle 220,000 tonnes of domestic cargo and 468,000 tonnes of international cargo, which is low in terms of world standards. To make air cargo more attractive and efficient, the Indian government has initiated the introduction of an 'open sky' policy, integrated cargo management systems at the four metropolitan airports and provision of centres for perishable cargo.

All the factors related to transport infrastructure above have adversely affected the logistics network in the country both in terms of lead-time and costs. However, the noted policy changes currently under way are expected to bring about a positive change in the Indian transportation environment. This provides vast opportunities for companies offering logistics services in the country, so Indian organizations will be able to reduce logistics costs by using third-party logistics services for enhanced supply chain efficiencies.

Logistics practices in the rest of the Pacific Rim of Asia present significant national differences, although many countries in the region look to Japanese practices as a model.

SUMMARY

More companies are expanding their operations into the international sector. As firms locate and service markets in various countries, they must establish logistics systems to provide the products and services that customers demand. While the components of a global logistics system may be similar to those in a domestic system, the management and administration of the international network can be vastly different.

To be a global company, management must be able to coordinate a complex set of activities – marketing, production, financing, procurement – so that least total cost logistics is realized. This will allow a firm to achieve maximum market impact and competitive advantage in its international target markets.

In this chapter, we examined some of the reasons firms expand into global markets. Companies that do so can become involved in exporting, licensing, joint ventures, direct ownership, importing or counter-trade. As part of the exporting process, we described the specific roles of the export distributor, customshouse broker, international freight forwarder and trading company. In addition, we looked at the importance of documentation and the use of free trade zones.

The international logistics manager must administer the various logistics components in a marketplace characterized by a number of uncontrollable elements: political and legal, economic, competitive, technological, geographical, and social and cultural. Within the uncontrollable environment the manager attempts to optimize the firm's cost–service mix. A number of differences exist between countries in administering logistics activities.

We examined the financial aspects of global logistics. Since logistics management is concerned with the costs of supplying a given level of service to foreign customers, it is important to recognize the factors that influence the costs of carrying out the process.

Some global market opportunities for companies in various regions of the world – Europe, North America and the Pacific Rim (including India) – were identified and described. In addition to identifying some of the opportunities within each of these regions, we presented the many challenges or disadvantages facing firms attempting to penetrate these markets.

With the first 12 chapters as background, we are now ready to develop an overview of a firm's implementation strategy for logistics. This is the topic of Chapter 13.

KEY TERMS

A full Glossary can be found at the back of the book.

QUESTIONS AND PROBLEMS

1 An increasing number of firms are engaging in international marketing and distribution. What factors would influence a company to enter international markets?

2 Companies that enter global markets have four main channel strategies available: (a) exporting, (b) licensing, (c) joint ventures, and (d) ownership. Briefly discuss each strategy and identify the advantages and disadvantages of each.

3 Explain the role of each of the following exporting organizations in global logistics:
■ export distributor
■ customshouse broker
■ international freight forwarder
■ trading company.

4 Explain why it is usually more difficult for a firm to provide the same level of customer service in its international markets that it provides in its domestic markets. Under what circumstances might an organization be able to provide better customer service to international markets than to domestic markets?

5 Discuss the relative importance of inventories in domestic and global logistics. In your response, consider the financial impact of inventory decisions on the strategic position of the firm.

6 Discuss how 'letters of credit' are used in international business transactions. Why are they important?

7 Briefly identify the opportunities and challenges facing firms seeking to market products in the following regions:
■ Europe
■ North America
■ Pacific Rim.

8 What are the logistics implications of trading blocs such as NAFTA and ASEAN?

THE LOGISTICS CHALLENGE!

PROBLEM: THE BLAME GAME

Frank Havlat is a freight forwarder and customs broker with DFM International, Inc. His problem is related to shipment security, for which he sees no easy solution.

Let's say that DFM is the Spanish forwarder of a shipment of high-value computer parts headed to South America via Miami. The cartons are contained in half a dozen shrinkwrapped pallets, and the shipper's instructions are explicit: don't break down the pallets. As a result, DFM is able to perform only a cursory inspection of the outside.

The shipment reaches Miami where an agent receives it, sees nothing wrong and sends it on. But when the shipment gets to its destination, several of the cartons deep inside the pallets are empty. Havlat starts to worry.

The question is one of liability. Havlat may well have signed a liability release based on the shipper's instructions not to disturb the contents of the pallet, but that doesn't mean the consignee won't try to include him in a lawsuit for damages.

Havlat says that shippers often are ignorant of the basics of international moves, including what must be done to protect their shipments. For some, the first impulse is to blame the middleman.

What advice do you have for Havlat, a relatively small forwarder, that would protect him from lawsuits while helping him to ensure the security of shipments moving under his control? And when inspections are possible, how can he monitor the hundreds, if not thousands, of individual cartons that are generated by his global customers on a daily basis? 'If we had to do this for every single shipment,' he says, 'we wouldn't move anything.'

What Is Your Solution?

Source: 'Distribution: The Challenge', *Distribution* 96, no. 3 (March 1997), p. 86.

SUGGESTED READING

BOOKS

Anderson, David L. and Dennis Colard, 'The International Logistics Environment', in *The Logistics Handbook*. James F. Robeson and William C. Copacino (eds). New York: Free Press, 1994.

Augustin, Siegfried, Peter G. Klaus, Ernst W. Krog and Ulrich Mueller-Steinfahrt, 'The Evolution of Logistics in Large Industrial Organizations in Europe', *Proceedings of the Annual Conference of the Council of Logistics Management*, 20–23 October 1996, pp. 535–53.

Bender, Paul S., 'International Logistics', in *The Distribution Management Handbook*. James A. Tompkins and Dale Harmelink (eds). New York: McGraw-Hill, 1994.

Carpenter, Susan, *Special Corporations and the Bureaucracy: Why Japan Can't Reform*. London: Palgrave Macmillan, 2003.

Fishman, Ted C., *China Inc*. London: Simon & Schuster, 2005.

Goldsborough, William W. and David L. Anderson, 'Import/Export Management', in *The Logistics Handbook*. James F. Robeson and William C. Copacino (eds). New York: Free Press, 1994.

Paliwoda, Stanley and Michael Thomas, *International Marketing*, 3rd edn. London: Butterworth-Heinemann, 1998.

Schary, Philip and Tage Skjøtt-Larsen, *Managing the Global Supply Chain*, 2nd edn. Copenhagen: Copenhagen Business School Press, 2001.

Stone, Marilyn A. and J.B. McCall, *International Strategic Marketing: A European Perspective*. London: Routledge, 2004.

JOURNALS

Aurik, Jonan C. and Jan Van De Dord, 'New Priorities in Logistics Services in Europe', *Transportation and Distribution* 35, no. 2 (February 1995), pp. 43–8.

Copacino, William C. and Frank F. Britt, 'Perspectives on Global Logistics', *The International Journal of Logistics Management* 2, no. 1 (1991), pp. 35–41.

Fernie, John, 'Quick Response: An International Perspective', *International Journal of Physical Distribution and Logistics Management* 24, no. 6 (1994), pp. 38–46.

MacDonald, Mitchell E., 'Who Does What in International Shipping', *Traffic Management* 30, no. 9 (September 1991), pp. 38–40.

Min, Hokey and Sean B. Eom, 'An Integrated Decision Support System for Global Logistics', *International Journal of Physical Distribution and Materials Management* 24, no. 1 (1994), pp. 29–39.

Rinehart, Lloyd M., 'Global Logistics Partnership Negotiation', *International Journal of Physical Distribution and Logistics Management* 22, no. 1 (1992), pp. 27–34.

Roberts, John H., 'Formulating and Implementing a Global Logistics Strategy', *The International Journal of Logistics Management* 1, no. 2 (1990), pp. 53–8.

Van der Ven and A.M.A. Ribbers, 'International Logistics: A Diagnostic Method for the Allocation of Production and Distribution Facilities', *The International Journal of Logistics Management* 4, no. 1 (1993), pp. 67–83.

Vantine, José G. and Claudirceu Marra, 'Logistics Challenges and Opportunities within MERCO-SUR', *The International Journal of Logistics Management* 8, no. 1 (1997), pp. 55–66.

Zinn, Walter and Robert E. Groose, 'Barriers to Globalization: Is Global Distribution Possible?', *The International Journal of Logistics Management* 1, no. 1 (1990), pp. 13–18.

REFERENCES

[1] Stanley Paliwoda and Michael Thomas, *International Marketing*, 3rd edn. Oxford: Butterworth-Heinemann, 1998, pp. 392–441; and Marilyn A. Stone and J.B. McCall, *International Strategic Marketing: A European Perspective*. London: Routledge, 2004, pp. 48–82.

[2] Paliwoda and Thomas, *International Marketing*, pp. 15–17.

[3] David L. Anderson and Dennis Colard, 'The International Logistics Environment', in *The Logistics Handbook*, ed. James F. Robeson and William C. Copacino. New York: Free Press, 1994, pp. 658–9.

[4] R. Neil Southern, *Transportation and Logistics Basics*. Memphis, TN: Continental Traffic Service, 1997, p. 295.

[5] This section is taken from Lisa M. Ellram and Laura Birou, *Purchasing for Bottom Line Impact*. Burr Ridge, IL: Irwin Professional Publishing, 1995, p. 61.

[6] Carrefour, http://www.carrefour.com/ (2005).

[7] Scottish & Newcastle, http://www.scottish-newcastle.com/ (2005).

[8] William W. Goldsborough and David L. Anderson, 'Import/Export Management', in *The Logistics Handbook*, ed. James F. Robeson and William C. Copacino. New York: Free Press, 1994, p. 674.

[9] James Aaron Cooke, 'What You Should Know about Letters of Credit', *Traffic Management* 29, no. 9 (September 1990), pp. 44–5.

[10] Bernard J. La Londe and James Ginther, 'The Ohio State University 2004 Survey of Career Patterns in Logistics', http://www.cscmp.org/ (2005), p. 9.

[11] Anderson and Colard, 'International Logistics Environment', pp. 669–70.

[12] Paul S. Bender, 'The International Dimension of Physical Distribution Management', in *The Distribution Handbook*, ed. James F. Robeson and Robert G. House. New York: Free Press, 1985, pp. 785–6.

[13] Anderson and Colard, 'International Logistics Environment', p. 666.

[14] Paliwoda and Thomas, *International Marketing*, p. 130.

[15] Southern, *Transportation and Logistics Basics*, p. 297.

[16] 'Ten Key Trade Documents', *Traffic Management* 29, no. 9 (September 1990), pp. 53, 55.

[17] Thomas A. Foster, 'Anatomy of an Export', *Distribution* 79, no. 10 (October 1980), pp. 76–7.

[18] This section was adapted from Paliwoda and Thomas, *International Marketing*, pp. 392–495; and James H. Bookbinder and Chris S. Tan, 'Comparison of Asian and European Logistics Systems', *International Journal of Physical Distribution and Logistics Management* 33, no. 1 (2003), pp. 36–58.

[19] Central Intelligence Agency, 'The World Factbook', http://www.cia.gov/ (updated 10 February 2005).

[20] Paliwoda and Thomas, *International Marketing*, p. 407.

[21] Stone and McCall, *International Strategic Marketing: A European Perspective*, pp. 66–7, 72–3.

[22] John Manners-Bell, 'Europe's Top Ten', *Logistics Europe* 12, no. 7 (September 2004) p. 31.

[23] Ibid., pp. 30–6.

[24] US Census Bureau, http://www.census.gov/ (2005).

[25] David G. Waller, Robert L. D'Avanzo and Douglas M. Lambert, *Supply Chain Directions for a New North America*. Oak Brook, IL: Council of Logistics Management, 1995, pp. 2–3.

[26] Ibid., p. 8.

[27] 'Co-operating with China', *Logistics & Transport Focus* 6, no. 5 (June 2004), pp. 61–4.

[28] This section is adapted from Paul S. Bender, 'International Logistics', in *The Distribution Management Handbook*, ed. James A. Tompkins and Dale Harmelink. New York: McGraw-Hill, 1994, pp. 8.18–8.19.

[29] This section is adapted from B.S. Sahay and Ramneesh Mohan, 'Third Party Logistics Practices: An Indian Perspective', *POMS 2004 2nd World Conference on Production and Operations Management*, Cancun, Mexico (http://www.poms.org/, 2005).

CHAPTER 13
LOGISTICS STRATEGY

384

CHAPTER OBJECTIVES

■ To develop an understanding of the concept of strategy and strategic planning

■ To illustrate how logistics can contribute specifically to the strategic success of an organization, and the importance of logistics participation in the strategic planning process

■ To show the key steps and issues in developing a logistics strategic plan

■ To discuss trends and issues that will present challenges for logistics professionals in the future

Logistics is well positioned to be a full participant and valuable contributor to an organization's strategy and strategic planning process. The chapter opens with an introduction to the concepts of mission, strategy and the hierarchical nature of planning. This is followed by a presentation of the key elements of the strategic planning process and, more specifically, how to develop a logistics strategic plan. Critical issues to consider in the strategic planning process are developed. These issues are based on the key challenges facing logistics professionals now and in the future.

The chapter closes with a description of the many opportunities and challenges that logistics professionals will face in the twenty-first century.

What are Strategy and Strategic Planning?

Because the strategic planning process exists to support an organization's strategy, it is important to develop an understanding of the overall concept of organizational strategy before discussing logistics strategy. **Strategy** is defined as 'the direction and scope of an organization over the long term which achieves advantage for the organization through its configuration of resources within a changing environment, to meet the needs of markets and fulfill stakeholder expectations'.[1] Thus strategy represents the overall actions or approach to be taken to achieve the firm's goals and objectives.

Mission Statements

The corporate mission statement is the overriding objective of the organization, which serves to guide the organization's strategy, activities and goals. It describes the organization's business. Every organization should have a mission to guide management's strategy development and implementation. These should be communicated clearly to all employees, so that they understand and can support the organization's overall direction. It is critical that all functional areas understand the general and specific strategies that an organization is pursuing, so that they can properly support those strategies and formulate strategies of their own. Therefore, each functional area should have a mission to guide its actions in supporting the corporate mission.

Why Strategy is Important to Logistics

If logistics managers do not understand corporate strategy, they will not be able to make decisions that are in the best interests of the organization. Even if logistics managers use the systems approach to make decisions and analyse trade-offs, they will still not be able to make the best decisions without a good understanding of the corporate strategy and the corresponding logistics strategy. Without this knowledge, logistics personnel will not know how to value various alternatives in making trade-offs.

For example, if the goal is to achieve differentiation by offering fast, reliable deliveries (i.e. it is making the trade-off of cost versus delivery reliability), management would choose trucking over rail. If low cost is the primary objective, management might choose rail to deliver products to customers.

It is clear that if logistics managers do not understand corporate strategy, they will be unable to make decisions that are consistently in the best interests of the company as a whole. Thus, a plan should be developed in order to execute strategy and monitor progress.

Why Plan?

There are many reasons to plan. As one popular expression states: 'If you don't know where you're going, how can you expect to get there?' When planning, management should consider the overall mission of the organization and develop specific action plans and activities to move the organization in the desired direction. In today's rapidly changing business environment, it is essential for managers to anticipate changes and prepare their organizations to best incorporate, respond to and take advantage of such change. Without taking a proactive approach, managers will be constantly reacting in crisis mode, and they will not be able to move forward in achieving their firm's mission.

The Hierarchy of Planning

Planning within an organization exists at many levels, as well as in many functional areas. At a minimum, most organizations *formally* update their plans on a yearly basis. However, planning is ideally an ongoing process. In addition, it is important to tie all of the functional plans together to ensure that they mesh and support the overall corporate plan and objectives. It is also important to have plans for different time frames, and that these time-phased plans fit together to support the long-range plan. The various types of planning are shown in Table 13.1 and are described in the following sections.

Type	Time frame	Focus	Level of detail	Level of integration
Operational	Day to day < 1 year	Efficiency	Heavy financial orientation	Functional
Tactical	> 1 to 5 years	Event	Somewhat financially orientated	Integrated-functional
Strategic	5 to 10 years or more	Competition, resources, stakeholders	Few financials, more goal orientated	Integrated-corporate and supply chain

TABLE 13.1 Characteristics of Planning Types

Source: adapted from Martha C. Cooper, Daniel E. Innis and Peter R. Dickson, *Strategic Planning for Logistics*. Oak Brook, IL: Council of Logistics Management, 1992, p. 28.

Strategic Plan

Organizations use a variety of terms to explain the various planning levels. At the highest level, which extends the furthest in time, is the **strategic plan**. Most European organizations tend to extend their planning horizon about five to ten years. The strategic plan for Japanese firms may look forward 50 or more years. The further into the future a plan extends, the less detail it will need. This is true because it is extremely difficult to anticipate the changes that may occur in the environment and the organization that will affect the organization's mission and its strategy.

The strategic plan considers an organization's objectives, overall service requirements and how management intends to achieve the corporate vision. The plans are very general and usually include projected revenues and expenses, lines of business, anticipated relative share of business within the market, and sales and profits from existing business lines compared with new lines of business.

Tactical Plan

At an intermediate level, generally one to five years into the future, an organization may have a medium-range plan, often called a **tactical plan**. Tactical plans are often more specific than strategic plans in terms of product lines, and may be broken down into detailed quarterly revenues and expenses. Nevertheless, such plans tend to show only a 'top' line, without much detail about sales by stockkeeping unit (SKU).

Tactical plans usually include a capital expenditure plan that indicates how much the organization will invest each year in new plant, equipment and other capital expenditure items. Issues like building warehouses, purchasing transportation or materials handling equipment, and other major expenditures to support the logistics infrastructure should be addressed as part of the capital expenditure plan.

Operating Plan

The most detailed level of plan is called the **operating plan** or the annual plan. It breaks out revenues, expenses, and associated cash flows and activity by month for a one-year period. The detailed operating plan is prepared to guide the activities for the following year. Actual performance is monitored and compared to planned performance in order to anticipate problems and respond accordingly, and to communicate results.

Production scheduling and materials purchases may be based on the operating plan. A firm can use this plan to anticipate its logistics needs from warehouse space to shipping. This allows logistics to anticipate its labour needs and to negotiate contracts with third-party providers. As the year unfolds and actual results occur, the plan may be adjusted for actual activity levels and revised to reflect expected performance.

Linking Logistics Strategy with Corporate Strategy

Logistics strategic planning can be defined as:

> A unified, comprehensive, and integrated planning process to achieve competitive advantage through increased value and customer service, which results in superior customer satisfaction (where we want to be), by anticipating future demand for logistics services and managing the resources of the entire supply chain (how to get there). This planning is done within the context of the overall corporate goals and plan.[2]

Logistics strategic planning is a complex process that requires an understanding of how the different elements and activities of logistics interact in terms of trade-offs and the total cost to the organization. Only by understanding the corporate strategy can logistics best formulate its own strategy. Recent studies by A.T. Kearney noted an increase in the complexity of logistics and supply chain environments that necessitates a better appreciation of strategic planning by logistics professionals. They defined four types of complexity in such environments: (1) market-facing with regard to product development and channel selection; (2) internal operating decisions and practices; (3) external factors such as competitors and government; and (4) organizational factors such as corporate governance, IT and cross-functional capabilities.[3]

Another study sponsored by the former Council of Logistics Management reported that a majority of logistics professionals surveyed indicated that their company's executives believed that the logistics plan was critical to the corporation's strategic plan, while less than a third disagreed and the remainder were neutral.[4] Another finding was that strategic planning in logistics is less common than strategic planning in marketing or manufacturing.

A.T. Kearney believes that organizations need to take a proactive role in the strategic logistics planning process in their companies, and differentiate their activities from a uniform and 'predictable' model to more responsive models in order to handle increasing complexity. Figure 13.1 is an example of how organizations can do so. The Type 1 model focuses on a lean and efficient

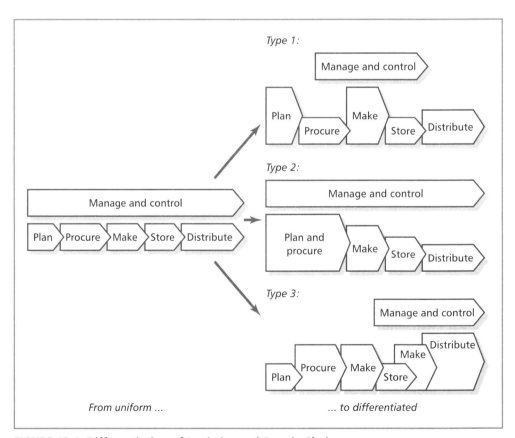

FIGURE 13.1 Differentiation of Logistics and Supply Chains

Source: adapted from European Logistics Association and A.T. Kearney, *Differentiation for Performance: Excellence in Logistics 2004*. Hamburg: Deutcher Verkehrs-Verlag GmbH, 2004, p. 29.

operation that is dominated by making products. The Type 2 model focuses on supplying complex products to specific requirements, with long lead-times, which requires collaborative planning with supply chain partners. The Type 3 model focuses on maximizing efficiency to meet customer demands in terms of volume and mix, thus requiring flexibility and late configuration of finished goods.[5]

Logistics can contribute to and support an organization's strategic planning process in a number of ways. Intel has identified six specific ways that show how logistics supports corporate strategy (see Table 13.2). The benefits of this participation in strategic planning include operating improvements (e.g. lower inventory and shorter lead-times), which can lead to strategic advantages (e.g. lower total cost and improved customer service). Before discussing the logistics planning process, we present an overview of the organizational planning process.

- ■ Increased planning capability and reduced inventory as a result of reliable delivery time

- ■ Increased margin and improved customer service

- ■ Reduced inventory levels through shorter cycle times

- ■ Increased marketing advantage from consistent, shorter order cycles

- ■ Uninterrupted supply of inbound material

- ■ Reduced total cost by incorporating logistics into the corporate planning process

---

TABLE 13.2 The Value Logistics Adds to the Corporation

Source: adapted from Lisa Ellram and L. Wayne Riley, 'Purchasing/Logistics Strategic Planning: Value to the Corporation', *Proceedings of the Annual Conference of the Council of Logistics Management*. Oak Brook, IL: Council of Logistics Management, 1993, p. 461.

The Organizational Planning Process

The logistics plan is dependent upon and takes direction from corporate strategic planning, which requires that consideration be given to the following environments:

- ■ legal and political
- ■ technological
- ■ economic and social
- ■ overall competitive.

Some of the key issues to consider in each environment are illustrated in Figure 13.2, which shows how these four environments interact with one another. The effects of each environment upon the others greatly complicates the environmental assessment process. It is not unusual for an organization to have economists on staff to help forecast trends and identify external data sources to use as a basis for plan assumptions. Government-forecast data of projected inflation and economic growth rates are frequently used as a starting point. Major steps in the corporate strategic planning process are shown in Figure 13.3.

The decisions made to support the steps in corporate strategic planning have a strong influence on the cost trade-offs required in marketing and logistics. For example, the decisions related to the evaluation of the potential consumers, and the identification, evaluation and selection of the target markets will have a profound impact on the type or types of logistics channels, intermediaries, facility locations, and so on. Thus the 'place' decision is highly dependent on the

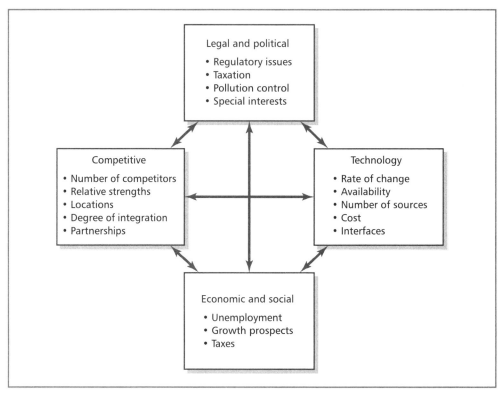

FIGURE 13.2 Environmental Influences in Planning

target customer. Steps 1–4 in Figure 13.3 were presented in more detail in Chapter 1 and Chapter 2. Steps 5–7 are described in greater detail below.

Formulation of Channel Objectives and Strategy

Formulation of the channel objectives and strategy can begin only after target customers and target markets have been selected. It is critical that the logistics function participate in the strategic planning process from this point onwards.

The logistics function plays a major role in the channel of distribution as both a performer of many activities and an interface with channel members who support the organization's channel of distribution. Most logistics executives agree that the logistics plan needs to mesh well with the plans of other departments. With the channel strategy determined, the next step is to identify channel alternatives.

Identifying Channel Alternatives

With greater interest and participation in partnering, outsourcing and supply chain management, the complexity of identifying channel alternatives has increased. By limiting itself only to traditional options and structures, management may weaken an organization's competitiveness.

Issues to consider in formulating channel alternatives include the desired consistency and speed of delivery, information flows, the degree of control desired and the cost of service. Activities may be performed internally or externally (outsourced). External service suppliers may

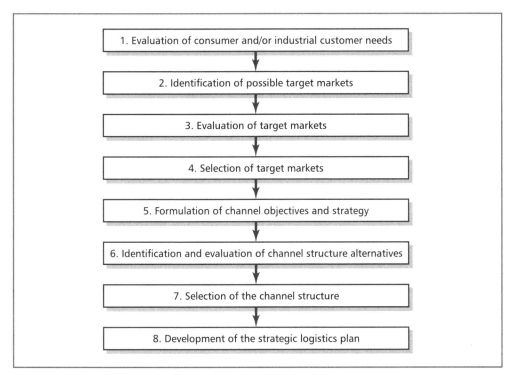

1. Evaluation of consumer and/or industrial customer needs

2. Identification of possible target markets

3. Evaluation of target markets

4. Selection of target markets

5. Formulation of channel objectives and strategy

6. Identification and evaluation of channel structure alternatives

7. Selection of the channel structure

8. Development of the strategic logistics plan

FIGURE 13.3 Major Steps in the Corporate Strategic Planning Process

provide full or limited service. If the current channel structure is meeting the organization's objectives, a less comprehensive review of channel alternatives may be sufficient.

Even if current objectives are being met, it is good policy to review the channel structure to identify opportunities for meeting objectives more effectively and efficiently, and to remain aware of new opportunities in channel design. The Creative Solutions box discusses how Tesco changed the way it viewed its distribution channel in order to improve its performance.

CREATIVE SOLUTIONS

Tesco Understands the Strategic Importance of Logistics

Tesco plc is the UK's leading grocery chain and recorded profits of £1.7 billion on £30.8 billion of worldwide sales in 2004. Tesco's share of the UK retail market is about 12.5 per cent; thus it accounts for £1 of every £8 (€11.60) spent in UK retail stores.

Tesco's growth over the past 25 years has been based largely on its strategy to take control of its supply chain and improve its logistics functions. As a result of various strategic initiatives, Tesco has been able to pursue a volume-led domestic strategy to reduce operating costs and cut prices for consumers and, in 1997, established a four-part growth strategy: core UK business, non-food, retailing services and international.

The logistics and supply chain initiatives that have fostered this success are centralization, composite distribution and vertical collaboration. In the mid-1970s Tesco operated direct-to-store delivery, where suppliers and manufacturers delivered 'as and when' they chose, making product volumes and quality inconsistent. Further, store managers developed their own relationships with suppliers and manufacturers that made central control and standardization difficult.

Tesco adopted a centrally controlled and physically centralized distribution service in 1980 that provides delivery to stores within a 48-hour lead-time. Integral to this service was massive investment in new primary consolidation and regional distribution centres (RDCs), information technology (IT), handling systems and working practices that allowed faster inventory turnover to achieve such decreased lead-times. Centralization allowed Tesco to increase its number of stores in the UK from a little over 500, with an average size of less then 929 square metres, in 1980 to 1878 stores with an average size of about 2165 square metres today.

Composite distribution is an extension of centralization that enables temperature-controlled products (ambient or fresh, chilled and frozen) to be distributed through one system of multi-temperature RDCs and vehicles. Benefits include lower inventory levels from daily deliveries of composite product groups, improvement in quality delivered to the stores with less wastage, and increased productivity through economies of scale and enhanced equipment use.

Tesco's vertical collaboration includes information sharing, electronic trading and collaborative improvements with suppliers. Tesco improved store scanning of products and introduced sales-based ordering with its suppliers over the Tesco Information Exchange – TIE, an Internet-based data exchange system. Tesco also adopted category management and introduced a 'continuous replenishment' system in the late 1990s.

Coupled with continuous replenishment is 'primary distribution', whereby Tesco orders from all suppliers more than once a day and takes responsibility for the delivery of orders to either a primary consolidation centre or RDC. Store orders are assembled as the stock arrives rather than being held pending batched orders and 'flowed through' by cross-docking.

Primary distribution is considered by Tesco to be a strategic change in goods flow, and about achieving continuous and efficient flows and not a pricing policy. However cost reduction is a key driver as primary distribution, also referred to as 'factory gate pricing', separates out transportation costs from the purchase price and puts it under Tesco's control. Tesco spends over £2.5 billion a year on its supply chain costs from the suppliers' factories through to on-shelf, in-store. About £500 million a year or 20 per cent of these costs are attributable to primary distribution.

In November 2004, Tesco began the rollout of its Secure Supply Chain initiative, tracking high-value and high-shrink products through its supply chain into stores using radio frequency identification (RFID) tags to drive improvements in on-shelf availability and help reduce shrink. Suppliers will tag product cases at source, giving Tesco the ability to track them throughout the supply chain from origin to shelf. The rollout will reach a total of 1400 stores and 30 distribution centres by the end of 2005.

The strategic changes Tesco has introduced to its logistics functions and supply chain management over the last quarter-century have enabled it to develop one of the most sophisticated and efficient distribution systems in the world, affording it a key advantage over its rivals in terms of growth and profitability.

Question: **How will suppliers have to strategically react to retail initiatives such as primary distribution, efficient consumer response (ECR) and quick response (QR)?**

Sources: adapted from David Smith and Leigh Sparks, 'Logistics in Tesco: Past, Present and Future', pp. 101–20 and 'Temperature-controlled Supply Chains', pp. 121–37, both in John Fernie and Leigh Sparks (eds), *Logistics and Retail Management*, 2nd edn. London: Kogan Page, 2004; and the Institute of Grocery Distribution, *Retail Logistics*. Watford, UK: Institute of Grocery Distribution, 2004, pp. 211–29.

Selection of the Channel Structure

The channel structures that appear to hold the most promise in terms of meeting organizational objectives should be analysed in depth. Factors to consider are operating costs, investment required, degree of control, flexibility and the ability to meet channel objectives. It is not unusual for an organization to use multiple channels to meet the needs of different customers.

For example, a paint manufacturer may sell directly to contractors, own some retail stores to sell directly to consumers, and have a dealer network to sell to other retail stores. The channel must be matched with the firm's objectives. Once the channel structure has been determined, the strategic logistics plan can be formulated.

The Strategic Logistics Plan

The strategic logistics plan is not developed in isolation. It depends on a number of inputs from various functional areas, each of which will be described below.

1. *Marketing* provides the key inputs to the logistics plan because of the close interrelationship between marketing and logistics discussed throughout this text, and illustrated in Figure 13.4. Marketing provides information about product or service offerings, pricing and promotion for each channel. This includes planned sales volume by month, type of customer, and regional area; product introductions and deletions; and customer service policies for various types of customer and geographical area. In certain instances, there may be specific policies related to key customers, which the logistics function must be aware of and be able to support.

Customer service policies are very important to logistics strategy. Logistics should be involved in setting such policies to make sure that they are feasible and economical. In addition, participation and input by logistics increases its understanding of and commitment to those policies. Customer service policies should include information by type of customer and region, covering issues such as:

- order placement methods
- order entry
- target cycle time
- order cycle variability
- desired fill-rate or in-stock levels.

Customer service policies should also relate to product substitution, expediting, transshipment and customer pick-up.

Unilever recently changed its IT system organization to forge more efficient links with its trading partners: customers, suppliers and others (see the Technology box). This example shows some of the operational, day-to-day elements of the role of logistics in supporting corporate strategy.

TECHNOLOGY

Global E-commerce Strategy for Unilever

Unilever is one of the world's largest consumer products companies, producing an extensive range of foods, home products and personal care items. It sells its products in around 150 countries across the globe, with 150 million people purchasing its well-known brands such as Knorr, Lipton, Surf laundry detergents, Dove and Cif to feed their families and clean their homes everyday. Historically, Unilever has grown through a series of acquisitions, which means that the

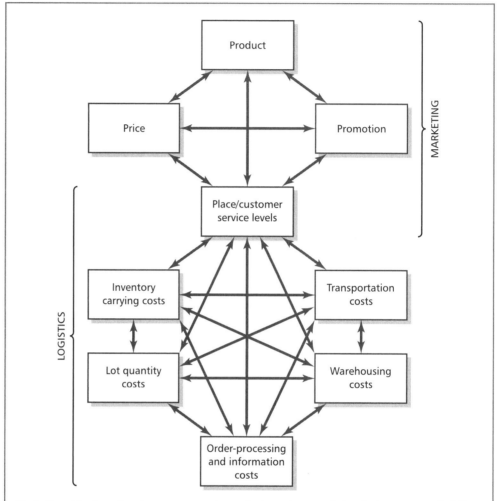

Marketing objective: allocate resources to the marketing mix to maximize the long-run profitability of the firm.

Logistics objectives: minimize total costs given the customer service objective share: Total costs = transportation costs + warehousing costs + order-processing and information costs + lot quantity costs + inventory carrying costs.

FIGURE 13.4 Cost Trade-offs Required in Marketing and Logistics

Source: adapted from Douglas M. Lambert, *The Development of an Inventory Costing Methodology: A Study of the Cost Associated with Holding Inventory* (Chicago: National Council of Physical Distribution Management, 1976), p. 7.

company has inherited a large number of disparate business and technology systems all over the world. The problem it faced was knitting these systems together in order to facilitate communications along the company's many supply chains.

Operating in around 100 countries across five continents using various electronic data interchange (EDI) standards and value-added networks (VANs), the company found that it was both complex and expensive to run so many different systems in diverse locations. Unilever realized it had to become more effective at dealing with its trading partners and that installing an electronic exchange would make perfect sense for the company.

In order to increase revenue and maximize competitive advantage Unilever wanted to achieve much closer collaboration with all sections of its value chain, consisting of customers, suppliers and trading exchanges. This strategy would enable Unilever to manage internal data transfer at a regional and global level across its entire product portfolio, and would also provide an external route or gateway to its customers across the world.

Unilever looked at a broad range of suppliers before deciding to work with IBM for this project. Working closely together the two companies developed and implemented a powerful cross-enterprise web-based collaboration platform: the 'Unilever Private Exchange'. The centralized platform addresses two major areas: first, it provides a common data infrastructure for Unilever's global operations to communicate internally; second, it enhances the company's relationships with all trading partners that make up the supply chain.

The Unilever Private Exchange enables the company to improve efficiency, reach new markets and enhance the traditional EDI approach. The new technology gives Unilever much more visibility across the enterprise and enables far greater collaboration with suppliers.

Central to the collaboration platform is IBM's Websphere Integration suite, in particular the trading gateway. The global hub, or switch, is based in North America, with five regional hubs across the world. A key feature here is Websphere's ability to enable the exchange to deal with different communication protocols from the disparate systems run by the various Unilever companies.

The Unilever Private Exchange not only facilitates the company's internal applications, such as SAP, but is also expected to help in transforming and simplifying Unilever's business by enabling the company to get products to market faster and react quickly to changing market conditions. Unilever expects to make significant operational savings, grow revenue and raise productivity levels.

The Unilever Private Exchange is live in several parts of Unilever's business, mainly in North America, and the company is now in the process of integrating its key customers and suppliers. Unilever is confident that all of its largest partners, including those responsible for generating 80 per cent of the company's annual €48 billion revenues, will transact business through the hub as it is rolled out. North America and Europe will be linked up to the Exchange first, with other regions following.

Research has shown that 15 per cent of all *Fortune* 2000 companies have set up private exchanges, with a further 28 per cent expected to follow suit by the end of 2003. During the next three years, private exchanges are expected to garner up to 90 per cent of new investment in marketplace infrastructure. As an early adopter of a private exchange, Unilever is taking the lead on a best practice example for collaboration with both partners and customers.

Question: **What impact do you think Internet-based information and business flows will have on future logistics activities?**

Source: 'A Private Function', *Logistics Europe* 11, no. 6 (July 2003), pp. 26–8.

2. *Manufacturing* provides information important to the logistics strategic plan, such as locations of current and planned production facilities, and planned volume and product mix for each site. When the same product is produced at multiple locations, logistics can determine how to serve each market most efficiently.

3. *Finance/accounting* provides cost forecasts related to inflation rates and growth assumptions that need to be built into the planning process to project future costs. Finance/accounting provides the cost data required to perform the cost trade-off analysis. It also is responsible for capital budgeting, which determines the availability of capital to finance expenditures to improve logistics equipment and infrastructure.

4. *Logistics* provides data and analysis related to the existing logistics network to the other functions, including current storage and distribution facilities owned and rented, both at manufacturing locations and in the field; equipment capacity and capabilities at each location; and current transportation arrangements between various channel members. Logistics must identify the costs associated with these activities and the various channels used and proposed. Evaluation and selection of channel members are critical aspects of this process.

Evaluation and Selection of Channel Members

Management needs to put the logistics plan into operation through the channel members it chooses. Thus alternatives related to the choice of carriers, warehousers and other logistics service providers need to be developed. Channel members should be judged and selected according to predetermined criteria designed to meet logistics objectives, such as reliability, consistency, geographical coverage, variety of service offerings, use of information technology and cost.

Ongoing Channel Evaluation and Improvement

Keeping performance on track requires regular monitoring and reporting of actual performance results. It is important to report expected levels of performance as a basis for comparison. Some of the relevant measures of performance of both the logistics function and external channel members were described in previous chapters. These include, but are not limited to, on-time delivery, response speed for emergencies, delivery time variability and response to customer enquiries.

Additional examples of such measures are given in the Creative Solutions box. If management determines that channel performance is unsatisfactory, it must decide whether to continue to work with existing channel members towards improvement, replace the intermediary with one providing similar services or restructure the channel.

Developing a Strategic Logistics Plan

The development of a strategic logistics plan requires the following:

- a thorough grasp and support of corporate strategy and supporting marketing plans in order to optimize cost–service trade-offs
- a thorough understanding of how customers view the importance of various customer service elements, and the performance of the firm compared with its competitors
- a knowledge of the cost and profitability of channel alternatives.[6]

It is clear that the strategic planning process as it relates to logistics should focus on determining required customer service levels. Information can be obtained utilizing internal and external market research (see Chapter 2). Firms often obtain this information as part of a logistics audit.

Logistics Audit

The logistics audit should be conducted routinely and is a review of how logistics is performing versus its objectives (see Figure 13.5). The audit complements and supports the strategic planning process by:

- linking logistics strategies and objectives to corporate strategies and objectives
- identifying key measures of logistics performance
- comparing customer perceptions of logistics performance with key measurements and objectives
- analysing 'gaps' in actual performance and comparing them with desired performance
- analysing trade-offs among desired performance levels on key results areas
- designing systems to reflect key goals and objectives
- identifying expected performance and continuously monitoring results.

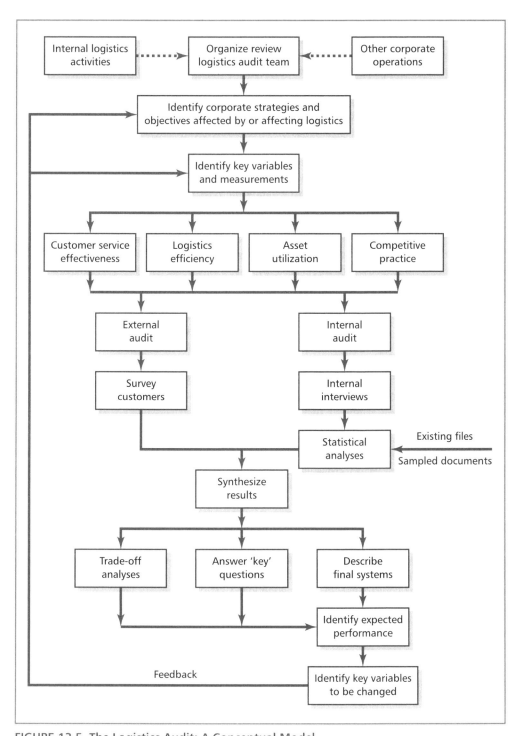

FIGURE 13.5 The Logistics Audit: A Conceptual Model

Source: Jay U. Sterling and Douglas M. Lambert, 'A Methodology for Assessing Logistics Operating Systems', *International Journal of Physical Distribution and Materials Management* 15, no. 6 (1985), p. 13.

A formal logistics audit may be an annual event to help align the efforts of the logistics function and to focus the strategic planning process.

A logistics audit is performed most effectively by a team of representatives that includes logistics personnel and other company personnel. Customer input to this group is important. A mix of people is required to provide a balanced view of logistics performance and critical issues. Each of the elements of a logistics audit will be described. The logistics audit process begins with the formulation of a logistics audit team.

Logistics Audit Team

The logistics audit team should include those involved in managing or performing logistics activities, such as warehousing, customer service and traffic, and those who use logistics services, such as other corporate functions, particularly MIS and finance/accounting and even external customers. The team may include representatives from sales, manufacturing and key customers.

Inviting participation from multiple functions helps to improve cooperation and support, and gives logistics more visibility and a broader organizational perspective. This should ease the acceptance and implementation of the team's recommendations. The logistics audit team begins with a review of the organization's strategy.

Review of Organizational Strategy

The corporate mission, goals and strategies must be reviewed early in the logistics audit to provide an overall direction and perspective. In addition, the strategies of both marketing and production/operations must be reviewed to ensure that logistics objectives and activities are supporting these key business functions. In more advanced organizations, management solicits logistics input when it develops corporate strategy and strategies for other functional areas. This review process helps to identify:

■ critical areas for review and focus during the logistics audit
■ important logistics performance measures in the eyes of the customers
■ alternative strategies to achieve desired objectives.

A review of corporate strategies helps the logistics audit team focus on the areas and issues to be investigated in the audit process.

Development of Key Issues to Investigate

To develop a comprehensive assessment of logistics performance on key activities, the logistics audit team needs to develop a list of questions to ask customers, as well as key operating personnel within the company. The questions should be broad enough to identify potential opportunities and barriers to logistics contributing to a competitive advantage for the organization. Examples of possible questions include the following.

■ What is the organization's overall customer service strategy and what should it be?
■ How should the overall customer service strategy differ by market or customer segment?
■ What approaches to logistics are the competition using, and what are their strengths and weaknesses?
■ What are the best opportunities for cost reduction in the organization's logistics system?
■ Are there benefits to outsourcing all or part of our logistics activities?
■ Are there any opportunities for consolidating any logistics facilities and efforts, either within or between strategic business units?
■ What type of order-processing flexibility and response time must the organization meet to be the industry leader in customers' minds?
■ How can the organization improve logistics productivity?

Identifying Critical Measures and Variables

Once key questions have been identified, the logistics audit team can determine *how* it will judge whether logistics is meeting the organization's objectives. This requires identification of specific variables and measurements in each of four broad categories: customer service effectiveness, logistics efficiency, asset utilization and analysis of competitor performance.

Customer Service Effectiveness

This category addresses issues such as order cycle time and order cycle consistency, fill rate, response times to customer enquiries, and flexibility to adapt to changes in order quantities, delivery dates, production schedules and product substitutions.

Logistics Efficiency

This category focuses on the costs of meeting customer service objectives, such as the cost of transportation, warehousing, inventory management, purchasing, order entry and the scheduling of shipments.

Asset Utilization

This category considers the efficiency and effectiveness of asset use, focusing on inventory, storage facilities and transportation equipment.

Analysis of Competitor Performance

It is important to measure how the organization's performance of logistics activities compares with that of other organizations. This activity is often termed **benchmarking**, and may consider the performance of both competing and non-competing organizations concerning important customer service activities and asset utilization. Proper investigation of these four categories means that both an external and an internal audit must be conducted.

External Audit

An external audit of customers may involve a survey, interviews or focus groups with various customer constituencies (see Chapter 2). It is important to reach a representative mix of the firm's customers in terms of their size, rate of growth and product demand. The external audit considers the following questions.

- How are current logistics systems performing in customers' eyes? How would customers like the system to perform?
- How do these perceptions compare with customers' perceptions of the performance of the competition's logistics systems? Does the firm's performance meet the customers' requirements?
- What changes in customer requirements are anticipated in the future?
- What changes in logistics customer service might the competition offer in the future?

Management will often employ outside organizations to conduct the logistics audit in order to encourage unbiased feedback from their customers and the customers of competitors. In addition, outside researchers often have specialized training and experience that allow them to conduct such studies more effectively.

Internal Audit

The internal audit involves collecting data from the firm's own operations, and examining external documentation and flows within the organization's logistics processes. To properly and consistently address the relevant logistics interfaces with other functions, a formal interview guide should be prepared and should cover areas such as the following.

- What is the activity level in the area?
- What are the key performance parameters?
- How is actual performance measured? How often is it measured? How are the results reported?
- Who is empowered to make decisions?
- How does that area interface with other functions?

A copy of the interview guide should be provided in advance, so that individuals can be prepared and have the data readily available.

For the more quantitative phase of the logistics audit, many internal documents and data sources can be relied upon, including:

- order history
- bills of lading
- freight bills
- private fleet trip logs
- warehouse time cards
- inbound and outbound shipment contents
- destination and trailer utilization
- warehouse labour hours for various activities
- order cycle and fill rate information.

Most of this information will include a great deal of detail related to customers, locations, items ordered, and so on. This detail makes it possible to perform a wide range of analysis. If the organization does not have a user-friendly decision support system to allow easy analyses and interface with the organization's database, the data sources listed previously can be 'coded' or put into a standard format, and input into a file for manipulation and analysis with a standard statistical package. Many reports can be developed.

Analysis of Results

On completion of the internal and external audits, the logistics audit team is ready to analyse the identified trade-off opportunities and to review the key questions originally developed to ensure that they have been addressed. Based on this analysis, the team should be able to develop and recommend a strategy and the supporting measurements needed to monitor the progress made in pursuing that strategy.

The team should be able to use the data gathered to analyse the historical performance of the firm and to predict how changes in service levels will affect costs. It is desirable to analyse the impact of proposed strategy on order cycle times, fill rates, various customer segments, product segments and channel segments. This will help to ensure that there are few unforeseen consequences.

The Logistics Plan

Now that a logistics strategy has been formulated, a logistics plan needs to be developed to support that strategy. The plan includes the specific activities that the logistics function will undertake to achieve its objectives. Thus, logistics decisions are made in a hierarchical manner.

The highest level is the strategic level, which considers issues such as business objectives and customer service requirements. The next level is the tactical level, which considers decisions such as the number, size and location of distribution centres; transportation modes preferred; and the type of inventory control system. Finally, the operating level focuses on day-to-day decisions, such as expediting policies, vehicle routing and scheduling. Because all of these decisions are related, they must be made in an iterative fashion (see Figure 13.6).

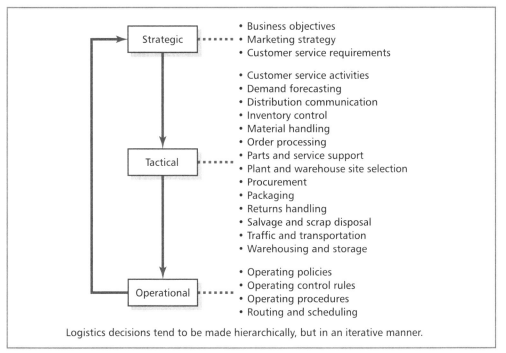

Strategic ········
- Business objectives
- Marketing strategy
- Customer service requirements

Tactical ········
- Customer service activities
- Demand forecasting
- Distribution communication
- Inventory control
- Material handling
- Order processing
- Parts and service support
- Plant and warehouse site selection
- Procurement
- Packaging
- Returns handling
- Salvage and scrap disposal
- Traffic and transportation
- Warehousing and storage

Operational ········
- Operating policies
- Operating control rules
- Operating procedures
- Routing and scheduling

Logistics decisions tend to be made hierarchically, but in an iterative manner.

FIGURE 13.6 Making Logistics Decisions

Source: adapted from William C. Copacino, Andersen Consulting, from a presentation at the International Logistics Management and Strategy Seminar, University of North Florida, 9–11 March 1992. All rights reserved by the author.

The logistics plan itself covers a variety of issues and requires inputs from representatives who participate in each of the logistics activities. A sample outline of a logistics strategic plan is shown in Table 13.3. This outline illustrates that the plan moves from the general to the specific. It begins with the mission because all logistics actions and activities should be driven by the logistics mission.

The key issues and objectives, review of past performance, and internal and external analysis are all completed as part of the logistics audit. The five-year vision represents the performance level that logistics would like to be moving towards over the next five years, based on the logistics audit results. The next step is the development of specific action plans to meet those goals, as well as how progress towards the vision will be measured. This part of the plan needs to be broken down into detailed activity on a year-by-year basis, with the greatest detail shown for the next year. Implementation issues, such as funds needed, personnel to be hired or redeployed, and other potential resource requirements need to be identified and addressed along with other critical issues.

Future Challenges and Critical Issues in the Strategic Planning Process

Several major issues are emerging in logistics; these should be described here, since many of them are critical considerations and some of them have not been dealt with in earlier chapters. The focus of this section is on supply chain management and integrated channel management, including quick response and efficient consumer response, total quality management, just-in-time, information systems, reengineering, time-based competition,

Logistics mission

Key issues and objectives

- Customer service performance
- Information systems
- Human resource management
- Supplier relationships
- Outsourcing

Comparison with previous plan performance

Internal analysis (current position)

- Organization
- Human resources
- Transportation
- Relations with internal customers
- Quality
- Service

External/situational analysis

- Competitor logistics performance
- Trends
- Public, private and contract warehouses
- Public, private and contract carriage

Five-year vision

Action plans to achieve vision

Implementation issues

Other critical issues

TABLE 13.3 Example Outline of Logistics Plan

Source: adapted from Martha C. Cooper, Daniel E. Innis, and Peter R. Dickson, *Strategic Planning for Logistics*. Oak Brook, IL: Council of Logistics Management, 1992, pp. 72–9.

environmental issues and reverse logistics. Management should integrate these issues into the strategic planning process and consider them in terms of how logistics can be used to gain a sustainable competitive advantage.

Supply Chain Management

Supply chain management, described in detail in Chapter 1, reflects the notion that the channel should be viewed and analysed as a whole, from a systems perspective. Whichever party within the channel can most efficiently and effectively perform a task – from holding inventory to adding value through product differentiation – should perform that task. Thus, an organization should consider supply chain management during the strategic planning process when management is deciding to what extent it should manage the firm's supply chain, and what activities should be performed, where, and by whom, within the supply chain. The Global box presents one consultant's views on how firms may re-shape their global supply chain strategies to improve their operations.

Shrinking Global Supply Chain Strategies

'Expect to see companies moving production back to Europe,' says Charles Davis, vice-president at A.T. Kearney.

Perversely, as press and politicians across the developed world obsess about the outsourcing of jobs and functions to 'faraway countries of which we know little', Davis identifies onshoring as a likely development. 'Companies have done loads of offshoring', he says, 'but they never understand the supply chain complexity they are creating. They haven't budgeted for the emergency air freight to sort out problems in China or wherever, and they don't understand how it increases the cost of changeover to a new product and how that becomes impossible to manage. More importantly, they have shifted production even when the cost benefit isn't really there.

'Take printed circuit boards; 90–95 per cent of the cost is in materials, the prices of which are essentially the same around the world. What's the benefit of making these in Asia? There is a totally uncertain supply chain, long communications – how do I collaborate when they are not even close to my time zone? I expect to see many companies moving back – at least to east or central Europe.'

Davis is equally trenchant on the lean/agile debate. 'The lean idea is overplayed,' he says. 'It can properly be applied in some areas but not where you need flexibility. If you start with the wrong supply chain model you are going to draw the wrong conclusions.

'And I worry about the emphasis on KPIs (key performance indicators). This stuff is all historical, it's measuring how the supply chain has performed in the past. We must switch to key capability indicators (KCIs) – measures of how our supply chains are fitted to adapt and change. We must look forward, not back.

'Chief executives don't generally see issues of supply chain vulnerability or resilience until it's too late. This is a bit of a broader question than risk management. For example, the high-tech industries are all about flexibility in theory. But the practical response tends to be, when the market goes up you throw in inventory and capital, when the market goes down you write off billions. And the problem is getting worse in semiconductors, for example: people used to plan on a seven-year business cycle but now the cycle is shortening and the problem of lousy capital allocation is spreading. And that's because there are no forward-looking KCIs. Decision-makers just look at their KPIs and allocate money on historical data.'

Another related area that Davis sees as key is that of supply chain control. 'There is lots of outsourcing and contract manufacturing – and firms have given everything away, They are now beginning to say, 'I want to retain control of the supply chain but I don't necessarily want to do the execution. I don't need to do manufacturing or packing or whatever, but I increasingly do need to control them.'

'But in reality, companies confuse ownership with control. They've outsourced not just the operations, but also the control. They are misaligned with their suppliers; have introduced more rather than fewer supply chain companies into the process and, as a consequence, aren't creating 'integrated' supply chains, but rather the reverse.

'So we have 4PLs. But no one knows what a 4PL is. There's scope for companies to fill that space if they really have superior IT, although it's got to be pretty good to beat what you can get through web-based providers. Control of physical assets, and so on – that's really just advanced 3PL. There is a space for firms that can really help the client control the supply chain, and that has to be a growth area: how you can retain control without performing the execution.'

Question: **How much of their manufacturing and logistics activities do you think European firms will relocate to the new accession (EU10) countries?**

Source: Charles Davis, 'Face to Face', *Logistics Europe* 12, no. 9 (November/December 2004), p. 60.

The supply chain management concept does not dictate a particular channel arrangement. It may involve forming partnerships with a few channel members in an effort to integrate most of the channel's activities, or making a conscious decision to let other channel members manage as they have been, perhaps because they have been doing an excellent job.

A supply chain management approach focuses on making a conscious, coordinated effort to improve channel management and channel efficiency through increased information sharing, efficient inventory placement and coordinated decision-making across all key business processes. Supply chain management has been referred to as 'integrated channel management'.

Integrated Channel Management

This is the concept of integrating all channel members' programmes and activities in order to achieve a higher level of customer satisfaction. Benetton, an Italian clothing manufacturer and retailer, has focused its efforts on this. Benetton's channel structure is discussed in Box 13.1. The company is at the centre of this channel structure and exercises a great deal of control over all aspects of it.

How to Carry Out Integrated Channel Management

One of the challenges of the integrated channel management philosophy is how to execute the strategy. Four major options are identified: implement channel integration strategies, assume channel leadership, form alliances with the channel leader, and improving selected, high leverage activities. Each of these approaches is described opposite.

Box 13.1 Benetton: A Supply Chain Leader

Benetton is a global fashion manufacturer and retailer with headquarters near Venice, Italy. The company was founded in 1965 by the three Benetton brothers and a sister. It began as a manufacturer before opening three retail stores of its own in 1968. Benetton experienced rapid growth throughout the 1980s. It experienced difficulty when it attempted to diversify into financial services, but it has since refocused its efforts and left that arena.

Benetton follows a global strategy of selling the same garments throughout the world in similar small speciality shops. This creates a uniform look, which is supported by strict corporate merchandising guidelines and media support.

Because Benetton is in the fashion industry, it must deal with a highly competitive, time-sensitive market with extremely short product life cycles. The short product life cycles – often a complete change of product line 10 times a year – help hold consumer interest. With a rapid changeover in products and the need to get the merchandise to market quickly to meet the immediate demand of consumers, Benetton needs a highly responsive logistics system. To remain competitive, Benetton's system must also be fast and flexible, yet efficient.

To fully leverage its marketing and logistics prowess, Benetton has taken a clear leadership role in its supply chain (see figure). Benetton uses a network of subcontractors, most of whom are exclusive manufacturers for Benetton, to produce its product. Benetton has established relationships with raw materials suppliers and provides the subcontractors with a schedule data, raw materials, technical assistance and financial assistance to buy or lease production equipment.

Benetton has agents that coordinate stores' demand information, ordering and marketing, product mix and financial management. These agents use EDI to transmit orders directly to Benetton. Benetton also receives other demand information, which it uses to determine its production. It does all the cutting and dyeing of garments based on actual demand patterns. Relevant demand data are forwarded to subcontractors. Undyed pieces are sent to subcontractors who sew them and return them to Benetton. These 'grey goods' are held undyed until the consumer's preferred colours are determined based on actual sales. At that time, Benetton dyes the product and sends it to its $50 million highly automated distribution centre located in Treviso, 20 km north of Venice. This operation supplies all garments

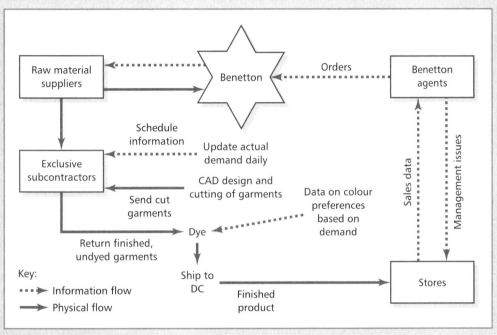

Benetton's Central Supply Chain Position

to more than 4000 Benetton stores in 83 countries, using only six operators per shift for three shifts.

In this way, Benetton has taken control of all aspects of its logistic operations, and has chosen to play the role of channel leader. By doing this, the firm is able to leverage economies of scale with in-house and outsourced expertise. It relies heavily on information technology, such as CAD/CAM (computer-aided design/computer-aided manufacturing) in cutting and dyeing materials, and EDI and bar coding for gathering point-of-sale data. It leverages its volume by performing centralized strategic and capital-intensive activities. Benetton takes advantage of the flexibility and relatively low cost of small subcontractors in its assembly operations.

Sources: adapted from Joseph R. D'Cruz and Alan M. Rugman, 'Developing the Five Partners Model', *Business Quarterly*, Winter 1993, pp. 60–71; and Peter Dapiran, 'Benetton – Global Logistics in Action', *International Journal of Physical Distribution and Logistics Management* 22, no. 6 (1992), pp. 7–11.

Implement Channel Integration Strategies

This approach involves utilizing some or all of the following activities:

- joint inventory management by means of channel-wide distribution resources planning
- joint management of transportation activities
- coordination of promotional effort
- channel-wide electronic data interchange (EDI)
- advance information sharing (POS data, inventory levels and positions, promotion response)
- automatic replenishment of inventory
- co-location of suppliers at manufacturer's site to place orders, manage inventory and solve problems
- channel-wide performance measurement.[7]

Channel integration strategies are most likely to be appropriate when large firms, such as distributors, retailers or manufacturers, deal with other large firms, although firms of unequal size can work together. The keys to success are mutual dependence and not having one firm dominate the other. These strategies can be maintained by firms of varying size, but more typically occur with firms who are comparable in size or power.

Assume Channel Leadership

Assumption of channel leadership is appropriate when a large firm is dealing with smaller trading partners, as in the Benetton example (see Box 13.1). Such an approach allows the channel leader and its trading partners to improve their efficiency and to lower costs by sharing information and focusing on channel productivity improvements. The central role assumed by the channel leader further reinforces and enhances its leadership role.

Form Alliance with Channel Leader

This is a strategy for smaller or less powerful firms to pursue. The goal is to maintain good relations with the channel leader in order to influence the actions taken in the channel, and to ensure that the interests of smaller firms are considered when the leader sets channel policy.

Focus on Selected, High-leverage Activities

When small or medium-sized organizations deal with other organizations of similar size they often have limited resources and find themselves competing against much larger firms. They may need to focus their improvement efforts on a few key areas instead of on the whole channel, by implementing automated ordering, sharing long-term plans, and focusing on time compression to speed up the time to market and to reduce channel investment in inventory and associated assets.

Formation of Partnerships

As described in Chapter 4, one of the issues management must consider is the **formation of partnerships**. 'A partnership is a tailored business relationship based on mutual trust, openness, shared risk and shared rewards that yields a competitive advantage, resulting in business performance greater than would be achieved by the firms individually.'[8] This type of arrangement can work well in supporting logistics objectives such as improved order cycle consistency or reduced lead-time by working closely with key suppliers. Partnerships are not appropriate with all customers and suppliers because of the time and effort to successfully implement them. Therefore, they should be reserved for those key firms where the outcomes justify the effort.

Retail Quick Response Programmes

As described throughout this text, *quick response (QR)* and *efficient consumer response (ECR)* programmes are ways for retailers and manufacturers to integrate their information systems, giving them the ability to better coordinate production and inventory levels, reordering and product stocking.

Quality Management

Quality management and the operations management concept of total quality management (TQM) should be kept at the forefront of decision-making throughout the strategic planning process and in the implementation of the strategies chosen. Quality is one performance issue that could be 'traded off' as the organization attempts to reduce costs or cycle time to market. Yet, quality remains a key aspect of customer service, so any reduction in the quality of service might spur an increase in complaints and/or fewer sales, which could more than offset any cost savings. Customers expect quality in service.

The impact of any changes in the level of quality must be explicitly evaluated. Quality management must focus on the customer's view of quality, and what changes in performance quality mean to the customer.[9] This should be explicitly investigated as part of the logistics external audit.

Just-in-time

The use of a just-in-time (JIT) inventory philosophy has a profound affect on logistics activity at all levels. JIT is a philosophy aimed at reduction of waste, redundancy and inefficiency throughout the entire production system. JIT has been extended to the supply chain to eliminate waste and redundancy among trading partners beyond the immediate organization.

For organizations using or considering the use of JIT, the implications should be considered throughout the processes of corporate strategic planning and annual logistics planning. For example, management should consider:

- which parts of the organization will implement JIT
- since the number of carriers and suppliers are generally reduced in JIT systems, how many should be utilized and how should they be selected
- what kind of information system linkages are required to ensure visibility of production schedules and inventory levels
- how logistics will interface with manufacturing to coordinate shipments.

These are only a few of the many issues to be considered in planning for JIT.

Information Systems

Logistics activities are transaction intensive. Transactions, such as receiving, stocking, order filling and shipment, generate large quantities of data. The data can be useful in assessing not only actual logistics performance but in identifying attractive areas for improving performance. As shown in Chapter 3, the sheer volume of data makes it virtually impossible to manage without an adequate information system.

Information systems are much more than receptacles for storing data. The manner in which the data are stored and the types of database management system available within an organization play a large role in whether that data can be readily utilized for analysis of logistics performance. These data are valuable in the strategic planning process.

Information systems enable logistics to communicate and interface with customers and others within the organization. Information technology makes QR and ECR possible. Examples of key information systems technology for sharing information include EDI, point-of-sale (POS) information gathering, electronic funds transfer, access to DRP/MRP files through information linkages, and bar coding. Without such technologies, vendor-managed inventories and QR systems would be extremely difficult, if not impossible, to manage. Logistics executives must ensure that these technologies are getting the proper visibility and funding commitment as part of the information systems plan and capital budgeting process of the strategic planning process.

Reengineering

Reengineering focuses on an important means of improving an organization's efficiency and effectiveness. The term reengineering is used to describe the elimination of old methods of operation and the creation of new, better approaches from scratch. More than simply modifying existing systems, reengineering challenges the organization to reinvent its operations without considering the way it currently does business. This creates the potential for major breakthroughs instead of only small incremental improvements.

The strategic planning process represents an excellent opportunity to evaluate making such bold moves. Reengineering works particularly well in people-intensive, transaction-intensive processes such as accounts payable or order processing. It can be used to identify ways to eliminate unnecessary or redundant activities, and to automate routine, mundane activities. This frees up the productive resources of the organization to provide greater value added. In describing reengineering, the following analogy has been used:

- the optimist says the glass is half full
- the pessimist says the glass is half empty
- the reengineer says the glass is twice as large as is needed.[10]

Time-based Competition

Time-based competition refers to ways of 'taking time out' of operations. It could entail reducing the order cycle time, speeding up order placement or introducing new products to market more quickly. Time-based competition is receiving a great deal of attention as organizations have discovered that time really is money. Longer processes can: create inefficiencies; require higher inventory levels, greater handling and more monitoring; incur a greater possibility for error and obsolescence; decrease the efficiency of the whole supply chain.[11]

Logistics is in an excellent position to help reduce the organization's cycle time by working with carriers, suppliers and customers to share more real-time information, improve information accuracy and identify current inefficiencies. This represents a significant opportunity for logistics to provide a competitive advantage for the organization.

The importance of lead-time reduction to customers should be investigated as part of the logistics audit. Lead-time reduction must be linked to customer requirements and the firm's marketing efforts to have a positive impact on the organization's competitiveness. Shorter lead-times may result in lower inventories for the customer, depending on the volatility of sales and the degree of difficulty in forecasting.

Lean Versus Agile Logistics and Supply Chains[12]

Two different logistics and supply chain paradigms emerged during the late 1990s: 'lean' and 'agile'. In the United Kingdom the lean paradigm is prevalent at Cardiff Business School's Lean Research Centre, while the agile paradigm has stemmed from the Cranfield University School of Management's Centre for Logistics and Supply Chain Management.

The lean paradigm is based on the principles of lean production in the automotive sector, detailed in the book *The Machine that Changed the World*. The idea behind this is the development of a value stream to eliminate all waste, including time, and ensure a level production system.

Conversely, the agile paradigm has its origins in principles of postponement that were discussed in Chapter 1. Being agile means using market knowledge and a virtual corporation to exploit profitable opportunities in a volatile marketplace.

The lean approach seeks to minimize inventory of components and work-in-progress, and to move towards a 'just-in-time' environment wherever possible. Firms using an agile approach are meant to respond in shorter time frames to changes in both volume and variety demanded by customers. Thus, lean works best in high volume, low variety and predictable environments, while agility is needed in less predictable environments where the demand for variety in high.[13]

While the paradigms appear dichotomous, in reality most organizations probably have a need for both lean and agile logistics, and supply chain solutions, suggesting a hybrid strategy. Such a strategy has also been called 'leagile'; Figure 13.7 illustrates such a hybrid solution. The materials 'decoupling point' represents a change from a lean, or 'push', production strategy to an agile, or 'pull', production strategy. An example is Benetton's use of delaying the final colour dyeing of garments until market demand information has been received (see Box 13.1). The information 'decoupling point' represents the point where market sales information can assist forecasting efforts within the lean approach of this hybrid solution.

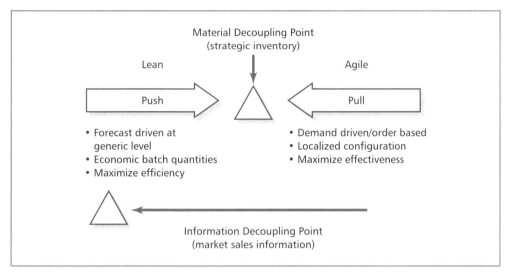

FIGURE 13.7 A Hybrid Lean and Agile Strategy

Sources: adapted from Martin Christopher, *Logistics and Supply Chain Management: Creating Value-Adding Networks*, 3rd edn. Harlow, UK: FT Prentice Hall, 2005, pp. 115–42; Daniel T. Jones, Peter Hines and Nick Rich, 'Lean Logistics', *International Journal of Physical Distribution and Logistics Management* 27, no. 3/4 (1997), pp. 153–73; and J. Ben Naylor, Mohamed M. Naim and Danny Berry, 'Leagility: Integrating the Lean and Agile Manufacturing Paradigms in the Total Supply Chain', *International Journal of Production Economics* 62 (1999), pp. 107–18.

Environmental Issues/Reverse Logistics

Environmental issues have received a great deal of attention in recent years. Not only has there been an increase in legislation on a domestic and global scale, but there has been an increasing demand by consumers for 'green', or environmentally friendly, products and practices. Recycling, proper management and disposal of hazardous waste, reusable packaging materials, use of renewable resources and energy conservation have become visible political and emotional issues. As such, consideration of environmental factors should be ongoing throughout the strategic planning process and during its implementation.

Reverse Logistics

Reverse logistics is concerned with issues such as reducing the amount of raw materials or energy used, recycling, substitution, reusable packaging and disposal. However, logistics cannot deal effectively with these issues in isolation. It must interface with manufacturing, marketing, purchasing and packaging engineering. Decisions made in each of these areas have an impact on the ability of logistics to conserve resources and achieve green goals.[14]

Hazardous Materials

Specific legislation covers the movement and disposal of *hazardous materials*. Personal liability can be created if materials are knowingly mishandled or mishandled through negligence. It is important for logistics professionals to become familiar with the laws governing any hazardous materials that their organization uses, and to ensure proper handling, transportation, storage and disposal. Hazardous materials are generally disposed of in approved dump sites, incinerated or otherwise destroyed. The best approach is to avoid the creation of waste through careful selection of materials and recycling, reclamation and revision of materials on-site.[15]

Life Cycle Analysis

Implied in the process of reverse logistics is that organizations use **life cycle analysis** in evaluating product or packaging decisions. Life cycle analysis involves viewing the purchased item from cradle to grave to determine its impact on the environment and on the total cost of doing business. Looking at it from an environmental perspective, this approach is useful because it examines the impact of the item on the environment at all stages of its life cycle, from product development up to and including ultimate disposal. Life cycle analysis should consider all available methods for waste reduction and management, including source reduction, recycling, substitution and disposal.

Legislation

Italy, Denmark, Norway and Germany have led the way in environmental and recycling legislation in Europe. There are a number of bans on plastic foam, polyethylene, corrugated containers and similar materials from landfills and incineration. Logistics and packaging engineering need to work on reducing and reusing packaging, and developing more creative packaging materials, such as the foam 'peanuts' that melt in water and actual 'popcorn', which is biodegradable.

In Europe, Hewlett-Packard switched from shipping printers in individual units to packaging them in bulk. This not only significantly reduced the amount of packaging materials, but reduced the shipping weight and allowed more efficient space utilization during shipment.[16]

Many countries have tried to minimize the creation of waste through the use of tax incentives and the provision of public information and technical assistance. Waste minimization policies tend to be favoured as the most efficient strategy. As the trend towards greater environmental awareness continues, logistics professionals must remain attuned to legislation and public sentiment, and be ready to suggest and support green strategies.

Such strategies represent a potentially attractive marketing opportunity because they create a favourable public image. The correct corporate response could create a distinct advantage for an organization, allowing it to command higher prices for its products and services. As you have read throughout these chapters, many other opportunities exist for logistics to contribute to an organization's competitive advantage.

Logistics as a Source of Distinctive Competitive Advantage

How can logistics provide the firm with a distinct competitive advantage? This should be one of the key focuses of logistics during the strategic planning process and the implementation of strategy. There are many opportunities for logistics' services and activities to give the firm an important advantage: supply chain management, quality of service, outstanding information systems and effective time-based competition. Organizations such as Tesco and Carrefour have effectively used information technology to give them a distinct advantage in terms of improved customer service and in-stock availability while lowering inventory.

An example of the strategic use of logistics is Coca-Cola's operation in Japan. Coca-Cola has a variety of types of customer in Japan, all of which demand different types of service. Coca-Cola's major Japanese customers – supermarkets and convenience store chains – want predictable deliveries and to have displays set up for them. Operators of vending machines want assurance that their machines are full to avoid lost sales. Coca-Cola has adapted its logistics services to meet the needs of each customer segment or channel.

Coca-Cola focuses on delivery, timing, frequency, in-store display and merchandising for supermarkets and retailers because that is what these customers want. In small, family-owned stores, Coca-Cola focuses on helping with paperwork, setting up displays and cleaning the storeroom.

On the other hand, vending machine owners do not care about such services; they want a fully stocked machine. To meet their needs, Coca-Cola has installed sophisticated information systems in the vending machine to monitor inventory levels. This allows the delivery vehicle to stock the right mix and level of products, and to deliver in a timely manner before the inventory is depleted. As a result, Coca-Cola does not dispatch trucks when machines do not need replenishment and it has the correct mix of product, which reduces costs.

By monitoring demand, Coca-Cola avoids carrying products that do not sell.[17] Coca-Cola in Japan uses logistics strategically, differentiating its service offerings to provide it with a competitive advantage. This has been a successful strategy for Coca-Cola, which holds a 34 per cent market share in Japan, compared with Pepsi's 5 per cent.[18]

On the other hand, 'traditional services' that logistics has provided should not be overlooked. Tailoring the logistics system to meet customer needs better than the competition can be an important source of competitive advantage. If logistics is to continue to grow in stature and be recognized as an important player in corporate strategy, it is incumbent upon logistics professionals to recognize and seize the opportunities to contribute to the overall success and distinctive competence of the organization.

SUMMARY

While logistics has been recognized as an important business function worthy of study only in the past 45 years, much has been gained during that time. Logistics has grown from a transaction-orientated, tactical function to a process-orientated, strategic function. The challenges and opportunities for logistics professionals to participate actively in setting strategy and to contribute to the success of the organization have never been greater. The rewards for recognizing and accepting these challenges in a creative and proactive manner should prove to be substantial. It is the hope and sincere desire of the authors that we have presented the material in this book in such a way as to encourage bright, ambitious young men and women to seek careers within the logistics profession.

KEY TERMS

A full Glossary can be found at the back of the book.

QUESTIONS AND PROBLEMS

1 Why is planning likely to become an increasingly important activity for logistics managers?
2 How does the logistics strategic planning process interface with the marketing strategic planning process?
3 What are the various approaches to integrated channel management? How should managers choose the appropriate channel management strategy?
4 Why is it important to perform both an internal and an external analysis in developing the logistics strategic plan?
5 What are the critical elements of the internal and external logistics audit?
6 Why is it advisable to use a team in performing a logistics audit? What functional areas should be represented on the team for a retailer? For a manufacturer?
7 Which of the many challenges facing the logistics profession in the years ahead do you believe is the most significant? Discuss why.
8 Give some examples of how increased environmental concern has affected logistics.
9 How are new operating techniques such as JIT, ECR and QR affecting logistics operations?

THE LOGISTICS CHALLENGE!

PROBLEM: AN UN-MERRY CHRISTMAS

Sony faced a major logistics problem in getting stocks of its best-selling mini-PlayStation 2 (PS2) games console into stores for Christmas 2004. The firm hired giant cargo planes to airlift PS2s directly from China to the UK, having failed to secure enough by sea, since many stores had sold out of that year's 'must have' toy.

Sony spokesman David Wilson said that supplies were likely to be back on track by mid-December, though most retailers were refusing to take orders for the model. The problem began when a Russian-owned oil tanker got stuck in the Suez Canal for three days in November 2004. The tanker held up 100 ships, including Sony's Christmas shipment which was bound from China with thousands of PS2s destined for the UK market.

The closure of the 120-mile-long waterway was the first for more than a day in 37 years. Neil Smith, marine manager at Lloyds Market Association, which represents Lloyds insurance underwriters, said 'A ship will carry thousands of containers carrying all sorts of items and tracking one down on a particular voyage is not feasible. You can't just turn around a ship if you get stuck halfway down the Suez Canal to drop off one container at one port'.

The cargo planes were landing at least twice a week at Stanstead and Gatwick airports near London. Each aircraft has capacity for about 40,000 PS2s. The times of the flights and the deliveries were a tightly guarded secret for fear that lorries could be hijacked by crooks eager to cash in on the Europe-wide shortage.

The new PS2 was selling at a rate of 70,000 a week in Britain until the end of November when stocks depleted and there were only 20,000 sold. During the first week of December the figure was down to 6000.

You are a consultant hired by Sony to examine its supply chain to prevent this situation from happening in the future. How would you face this challenge?

What is your solution?

Source: adapted from Valerie Elliot, 'Merry Christmas, Your PlayStation 2 is Stuck in Suez', *The Times*, 9 December 2004, p. 9.

SUGGESTED READING

BOOKS

Camp, Robert C., *Benchmarking: The Search for Industry Best Practices that Lead to Superior Performance*. Milwaukee, WI: ASQC Quality Press, 1989.

Christopher, Martin, *Logistics and Supply Chain Management: Creating Value-Adding Networks*, 3rd edn. Harlow, UK: FT Prentice Hall, 2005.

Cooper, Martha C., Daniel E. Innis and Peter R. Dickson, *Strategic Planning for Logistics*. Oak Brook, IL: Council of Logistics Management, 1992.

European Logistics Association and A.T. Kearney, *Differentiation for Performance: Excellence in Logistics 2004*, Hamburg: Deutcher Verkehrs-Verlag GmbH, 2004.

Hamel, Gary and C.K. Prahalad, *Competing for the Future*. Cambridge, MA: Harvard Business School Press, 1994.

Hammer, Michael and James Champy, *Reengineering the Corporation*. New York: HarperCollins, 1993.

Johnson, Gerry and Kevan Scholes, *Exploring Corporate Strategy*, 5th edn. Harlow, UK: FT Prentice Hall, 1999.

Porter, Michael E., *Competitive Strategy*. New York: Free Press, 1980.

Porter, Michael E., *Competitive Advantage: Creating and Sustaining Superior Performance*. New York: Free Press, 1985.

Stock, James R., *Development and Implementation of Reverse Logistics Programs*. Oak Brook, IL: Council of Logistics Management, 1998.

Womack, James P., Daniel T. Jones and Daniel Roos, *The Machine that Changed the World*. New York: Macmillan Publishing Company, 1990.

JOURNALS

Bowersox, Donald J., 'The Strategic Benefits of Logistics Alliances', *Harvard Business Review* 68, no. 4 (July–August 1990), pp. 36–45.

Fuller, Joseph B., James O'Conor and Richard Rawlinson, 'Tailored Logistics: The Next Advantage', *Harvard Business Review* 71, no. 3 (May–June 1993), pp. 87–98.

Jones, Daniel T., Peter Hines and Nick Rich, 'Lean Logistics', *International Journal of Physical Distribution and Logistics Management* 27, no. 3/4 (1997), pp. 153–73.

Lambert, Douglas M., Margaret A. Emmelhainz and John T. Gardner 'Developing and Implementing Supply Chain Partnerships', *The International Journal of Logistics Management* 7, no. 2 (1996), pp. 1–12.

McGinnis, Michael A. and Jonathan Kohn, 'A Factor Analytic Study of Logistics Strategy', *Journal of Business Logistics* 11, no. 2 (1990), pp. 41–63.

McGinnis, Michael A., 'Logistics Strategy, Organizational Environment and Time Competitiveness', *Journal of Business Logistics* 14, no. 2 (1993), pp. 1–24.

Naylor, J. Ben, Mohamed M. Naim and Danny Berry, 'Leagility: Integrating the Lean and Agile Manufacturing Paradigms in the Total Supply Chain', *International Journal of Production Economics* 62 (1999), pp. 107–18.

REFERENCES

[1] Gerry Johnson and Kevan Scholes, *Exploring Corporate Strategy*, 5th edn. Harlow, UK: FT Prentice Hall, 1999, p. 10.

[2] Martha C. Cooper, Daniel E. Innis and Peter R. Dickson, *Strategic Planning for Logistics*. Oak Brook, IL: Council of Logistics Management, 1992, pp. 4–5.

[3] A.T. Kearney, 'The Challenge: Successfully Managing Complexity', presentation to the 2004 Council of Logistics Management Conference, Philadelphia, 4 October; and European Logistics Association and A.T. Kearney, *Differentiation for Performance: Excellence in Logistics 2004*. Hamburg: Deutcher Verkehrs-Verlag GmbH, 2004, pp. 14–30.

[4] Cooper, Innis and Dickson, *Strategic Planning for Logistics*, p. 43.

[5] European Logistics Association and A.T. Kearney, *Differentiation for Performance*, pp. 28–30.

[6] Louis W. Stern and Fredrick D. Sturdivant, 'Customer-driven Distribution Systems', *Harvard Business Review* 64, no. 4 (July–August 1986), pp. 34–41.

[7] This material is adapted from William C. Copacino and Douglas M. Lambert, 'Integrated Channel Management', unpublished manuscript, 1992. All rights reserved by the authors.

[8] Douglas M. Lambert, Margaret A. Emmelhainz and John T. Gardner, 'Developing and Implementing Supply Chain Partnerships', *The International Journal of Logistics Management* 7, no. 2 (1996), p. 2.

[9] Nigel Slack, Stuart Chambers and Robert Johnson, *Operations Management*, 4th edn. Harlow, UK: FT Prentice Hall, 2004, pp. 717–47.

[10] Thomas A. Stewart, 'Evidence that Reengineering has Lost its Buzz', *Fortune*, 13 November 1995, p. 60.

[11] Thomas E. Hendrick, *Time Based Competition*. Tempe, AZ: Center for Advanced Purchasing Studies, 1994, p. 4.

[12] This section was adapted from Martin Christopher, *Logistics and Supply Chain Management: Creating Value-adding Networks*, 3rd edn. Harlow, UK: FT Prentice Hall, 2005, pp. 115–42; Daniel T. Jones, Peter Hines and Nick Rich, 'Lean Logistics', *International Journal of Physical Distribution and Logistics Management* 27, no. 3/4 (1997), pp. 153–73; and J. Ben Naylor, Mohamed M. Naim and Danny Berry, 'Leagility: Integrating the Lean and Agile Manufacturing Paradigms in the Total Supply Chain', *International Journal of Production Economics* 62 (1999), pp. 107–18.

[13] Christopher, *Logistics and Supply Chain Management*, pp. 117–18.

[14] Terrence L. Pohlen and M. Theordore Farris II, 'Reverse Logistics in Plastics Recycling', *International Journal of Physical Distribution and Logistics Management* 22, no. 7 (1992), p. 35.

[15] James R. Stock, *Development and Implementation of Reverse Logistics Programs*. Oak Brook, IL: Council of Logistics Management, 1998.

[16] Frances Cairncross, 'How Europe's Companies Reposition to Recycle', *Harvard Business Review* 70, no. 2 (March–April 1992), pp. 35–45.

[17] Joseph B. Fuller, James O'Conor and Richard Rawlinson, 'Tailored Logistics: The Next Advantage', *Harvard Business Review* 93, no. 3 (May–June 1993), pp. 87–98.

[18] Patricia Sellers, 'How Coke is Kicking Pepsi's Can', *Fortune*, 28 October 1996, p. 78.

GLOSSARY

360-degree evaluation: A qualitative management evaluation technique where anonymous input is sought from a manager's boss, workers, peers and subordinates.

ABC analysis: A method of categorizing products or customers where A is the most important, B is the next important, and so on. See also Pareto's law or 80/20.

Absorption costing: In cost management, an approach to inventory valuation in which variable costs and a portion of fixed costs are assigned to each unit of production. The fixed costs are usually allocated to units of output on the basis of direct labour hours, machine hours or material costs.

Activity-based costing (ABC)/activity-based management (ABM): A discipline focusing on the management of activities within business processes as the route to continuously improve both the value received by customers and the profit earned in providing that value. ABC analysis in this context is a methodology that measures the cost and performance of cost objects, activities and resources.

Ad hoc committees: Companies will form various committees on an ad hoc or impromptu basis that are concerned with one specific purpose, case or situation at hand. See standing committees.

Allowances: In sales and accounting, a company may give an allowance or price reduction, for example an allowance granted in exchange for used merchandise.

Artificial intelligence (AI): The understanding and computerization of the human thought process.

Automatic guided vehicle systems (AGVSs): A transportation network that automatically routes one or more material handling devices, such as carts or pallet trucks, and positions them at predetermined destinations without operator intervention.

Bar code: A symbol consisting of a series of printed bars representing values. A system of optical character reading, scanning and tracking of units by reading a series of printed bars for translation into a numeric or alphanumeric identification code. A popular example is the UPC code used on retail packaging.

Benchmarking: A process of comparing performance against the practices of other leading companies for the purpose of improving performance. Companies also benchmark internally by tracking and comparing current performance with past performance.

Blanket orders: A long-term commitment to a supplier for material against which short-term releases will be generated to satisfy requirements. Often blanket orders cover only one item with predetermined delivery dates.

Block trains: Block trains are complete freight trains as long as 700 metres and with a gross weight of up to 5,500 tonnes. They generally run for one company transporting large quantities.

Bonded warehouses: Warehouses where goods are stored until import duties or tariffs are paid.

Cabotage: Is where a carrier from one European Union (EU) country is allowed to perform domestic transportation transport in another EU country.

Capital budgets: A budget that contains budgets for capital expenditures on long-term assets or investments such as plant, property and equipment.

Centre-of-gravity approach: A supply chain planning methodology for locating distribution centres at approximately the location representing the minimum transportation costs between the plants, the distribution centres and the markets.

Channel of distribution: This means by which products are moved from producer to ultimate consumer. See also channels.

Channel power: The amount of business power a member of a channel of distribution can hold or exert over other channel members.

Channels: A method whereby a company dispenses its product, such as a retail or distribution channel, call centre or web based electronic storefront. See also channel of distribution.

Choice process: The third stage in the transportation carrier selection process where a company selects a carrier.

Collaborative planning and forecasting (CPFR): A collaboration process whereby trading partners can jointly plan key logistics and supply chain activities from production, and delivery of raw materials to production and delivery of final products to end customers. Collaboration encompasses business planning, sales forecasting, and all operations required to replenish raw materials and finished goods.

Common carrier: A carrier that provides transportation to the public and does not provide special treatment to any one party, and is regulated as to the rates charged, the liability assumed and the service provided.

Compatibility: Refers to whether goods stored in a warehouse are compatible for storage purposes.

Complementarity: Refers to whether goods stored in a warehouse are ordered together and are there complementary for storage purposes.

Components: In a purchasing or procurement partnership model, components are the joint activities and processes used to build and sustain the partnership.

Concurrent engineering: A cross-functional, team-based approach in which the product and the manufacturing process are designed and configured within the same time frame, rather than sequentially.

Continuous replenishment (CR): Continuous replenishment is the practice of partnering between distribution channel members that changes the traditional replenishment process from distributor-generated purchase orders, based on economic order quantities, to the replenishment of products based on actual and forecasted product demand.

Contract carrier: A carrier that does not serve the general public, but provides transportation for hire for one or a limited number of shippers under a specific contract.

Contract warehousing: A variation of public warehousing where the warehouse provider and customer have an arrangement that might see customized services and facilities within the warehouse on a shared cost and risk basis.

Control chart: A statistical approach used in statistical process control that permits examination of process behaviour against upper and lower limits

Corporate culture: See organizational climate.

Cost of lost sales: The forgone profit or cost associated with a product stockout.

Cost trade-off analysis: The interrelationship among system variables indicates that a change in one variable has cost impact upon other variables. A cost reduction in one variable may be at the expense of increased cost for another.

Cost-of-service pricing: Where the price paid to a transportation carrier covers fixed and variable costs and allows a margin for profit.

Cost-to-sales ratio: A company performance measure comparing total costs to actual sales.

Counter-trade: A requirement where a company must import goods from a country where it has exported or sold its own goods.

Cross-docking: A distribution system in which merchandise received at the warehouse or distribution centre is not put away, but instead is readied for shipment to retail stores. Cross-docking requires close synchronization of all inbound and outbound shipment movements.

Cumulative quantity discounts: Discounts for transportation services based on cumulative orders over a fixed period of time.

Customer order cycle: The elapsed time between when a customer places an order until the customer receives the order.

Customer satisfaction: Represents a customer's overall assessment of all elements in the marketing and sales process when purchasing goods or services.

Customer service: Activities between the buyer and seller that enhance or facilitate the sale or use of the seller's products or services.

Customshouse broker: A business firm that oversees the movement of international shipments through customs and ensures that the documentation accompanying a shipment is complete and accurate.

Data analysis: Advanced analysis of data by mathematical or statistical means to provide management with information for strategic and operational decision-making.

Data processing: The ability to transform raw data into a more useful form by relatively straightforward conversion.

Data retrieval: The ability to recall data in its raw form conveniently and quickly.

Decision support systems (DSSs): Software that speeds access and simplifies data analysis, queries, etc. within a database management system.

Decision-making unit: In purchasing or procurement, a group of decision-makers and decision-influencers who combine to make acquisition decisions.

Dedicated storage: A warehouse storage technique where goods are placed in a permanent location within a warehouse. Although this method often requires more storage space than a randomized storage method, it is used in manual labour situations where employee performance improves as they learn each product's location.

Delivered pricing system: A system that provides the actual price of a good or product delivered to the customer's desired location. See INCOTERMS.

Deregulation: In a logistics context the revisions or complete elimination of economic regulations controlling transportation.

Direct costing: A cost that can be directly traced to a cost object since a direct or repeatable cause-and-effect relationship exists. A direct cost uses a direct assignment or cost causal relationship to transfer costs.

Direct ownership: Corporate ownership of a foreign business subsidiary.

Direct store delivery: Process of shipping direct from a manufacturer's plant or distribution centre to the customer's retail store, thus bypassing the customer's distribution centre.

Direct trains: A form of combined transport by members of the International Union of combined Road-Rail transport (UIRR) where trains operate non-stop between two points.

Discrepancy of assortment: There is a discrepancy between what assortment of goods a customer requires and the assortment a manufacturer has on-hand or in production. A channel intermediary will adjust this discrepancy by performing various functions such as sorting out, accumulating, allocating and assorting.

Distribution centre (DC): The warehouse facility which holds inventory from manufacturing pending distribution to the appropriate stores.

Distribution requirements planning (DRP I): A system of determining demands for inventory at distribution centres and consolidating demand information in reverse as input to the production and materials system.

Distribution resource planning (DRP II): The extension of distribution requirements planning into the planning of the key resources contained in a distribution system: warehouse space, workforce, money, trucks, freight cars, etc.

Drivers: In a purchasing or procurement partnership model, drivers are the compelling reasons to partner.

Duty drawback: A refund or drawback of customs duties paid to bring goods into a country from another country.

Early supplier involvement: The process of involving suppliers early in the product design activity and drawing on their expertise, insights and knowledge to generate better designs in less time and designs that are easier to manufacture with high quality.

Economic order quantity (EOQ): An inventory model that determines how much to order by determining the amount that will meet customer service levels while minimizing total ordering and holding costs.

Efficient consumer response (ECR): A demand-driven replenishment system designed to link all parties in the grocery logistics channel to create a massive flow-through distribution network. Replenishment is based upon consumer demand and point-of-sale information.

Electronic data interchange (EDI): Intercompany, computer-to-computer transmission of business information in a standard format. An EDI transmission consists only of business data, not any accompanying verbiage or free-form messages.

Electronic mail (e-mail): The computer-to-computer exchange of messages. E-mail is usually unstructured (free-form) rather than in a structured format. X.400 has become the standard for e-mail exchange.

Enterprise resource planning (ERP): A class of software for planning and managing 'enterprise-wide' the resources needed to take customer orders, ship them, account for them and replenish all needed goods according to customer orders and forecasts. Often includes electronic commerce with suppliers.

Euro (€): The currency of the European Union. Not all member nations are using the euro.

Expert systems (ES): A computer program that mimics a human expert.

Export distributor: A company that acquires goods from producers for resale in another country.

Export intermediaries: Companies that assist other companies in conducting international business.

Exporting: A term used to describe products produced in one country that are sold in another.

Facilitators: In a purchasing or procurement partnership model, facilitators are the supportive corporate environmental factors that enhance partnership growth and development.

Field warehouse: A warehouse on the property of the owner of the goods that stores goods that are under the custody of a bona fide public warehouse manager.

Fill rate: The percentage of order items that the picking operation actually fills within a given period of time.

Finished goods inventory: Products completely manufactured, packaged, stored and ready for distribution.

Fixed order interval: Within the economic order quantity model, a point in time where the EOQ is ordered based on the level of inventory on-hand

Fixed order point: Within a company's inventory policy, a point in time that is a fixed interval, such as one or two weeks, where an amount of inventory is ordered based on an upper level of inventory the company wants to maintain.

Fixed-slot storage: See dedicated storage.

Floating slot storage: See randomized storage.

Focused factories: Factories that are smaller, simpler and totally focused on one or two key manufacturing activities.

Forecasting: An estimate of future demand. A forecast can be constructed using quantitative methods, qualitative methods or a combination of methods, and it can be based on extrinsic (external) or intrinsic (internal) factors. Various forecasting techniques attempt to predict one or more of the four components of demand: cyclical, random, seasonal and trend.

Foreign trade zone: See free trade zone.

For-hire carrier: A carrier that provides transportation service to the public on a fee basis.

Form utility: A value created in a product by the process of creating it in the right form that customers require. Manufacturing provides form utility.

Formation of partnerships: The initiation of business relationships where companies are partners and share costs and profits from the venture.

Forward buying: When a customer buys more of a product than required for current use in order to avoid possible supply restrictions or inflationary markets.

Framework agreement: A purchasing agreement between European Union (EU) governments and public sector suppliers to establish terms and conditions for all public-sector purchasing in the EU.

Free port: A provision in some countries where a company may hold inventory tax-free for up to a year.

Free trade zone (FTZ): An area in a country where goods may enter without attracting any duties or taxes until they are removed for use or sale.

Free-on-board (FOB): Contractual terms between a buyer and a seller that define where title transfer takes place.

Freight forwarders: Companies that provide logistics services as an intermediary between the shipper and the carrier, typically on international shipments.

Functional silos: A view of an organization where each department or functional group is operated independently of other groups within the organization. Each group is referred to as a silo.

Green logistics: The concept of introducing an environmental or ecological approach to the usual economic approach in logistics management. An example would be an initiative to reduce vehicle emissions in freight transport.

Gross domestic product: The total market value of all final goods and services produced in a country in a given year, equal to total consumer, investment and government spending, plus the value of exports, minus the value of imports.

Hollow corporation: A term describing a company that outsources most of its operations including logistics and exists as a small organization of managers who look after all the outsourced activities. Sometimes called a network.

Hoover's model: A warehouse location approach devised by Edgar Hoover in the 1940s that identified three different strategies: market positioned, production positioned and intermediately positioned.

Import quotas: Limitations on the amount of goods that can enter a country.

Importing: Bringing goods into one country that are produced in another country.

INCOTERMS: International terms of sale developed by the International Chamber of Commerce to define sellers' and buyers' responsibilities.

Independent versus dependent demand: An independent demand item is one that can be produced independently of the demand for its raw materials, for example finished goods. Conversely, dependent demand items are the raw materials or components used in a finished good that are dependant on customer demand for the finished good to be required or introduced in the production process.

Integrated logistics management: A comprehensive, system-wide view of the entire supply chain as a single process, from raw materials supply through to finished goods distribution. All functions that make up the supply chain are managed as a single entity, rather than managing individual functions separately.

Integrated supply: Where a customer combines purchases of all like items, such as office or laboratory supplies, from one supplier to reduce purchase and administrative costs.

Intermediately positioned strategy: A warehouse location strategy whereby a warehouse is located between where goods and products are manufactured and the market it serves.

Intermodal marketing companies (IMCs): Intermediaries that sell intermodal services to shippers.

International freight forwarders: See freight forwarders.

Interorganizational or interfunctional teams: See work teams.

Inventory carrying costs: One of the elements comprising a company's total supply-chain management costs. These costs consist of opportunity or capital costs, inventory service costs, storage space costs and inventory risk costs.

Inventory risk costs: The costs of obsolescence, pilferage or shrinkage, relocation within the inventory system and damage to inventory.

Inventory service costs: The costs of insuring inventories and taxes associated with the holding of inventory.

Inventory turnover: The cost of goods sold divided by the average level of inventory on hand. This ratio measures how many times a company's inventory has been sold during a period of time. Operationally, inventory turns are measured as total throughput divided by average level of inventory for a given period; how many times a year the average inventory for a firm changes over, or is sold.

Job performance: An employee's responsibilities and outputs regarding his or her job in a company.

Joint venture: A joint partnership between two companies in different countries to conduct business in one of the countries.

Judgemental sampling: A method of forecasting where a company solicits opinions of salespeople or experts in the field. This method is quick and reasonably inexpensive but may be subject to personal biases of those persons consulted.

Just-in-time (JIT): An inventory control system that controls material flow into assembly and manufacturing plants by coordinating demand and supply to the point where desired materials arrive just in time for use. Developed by the automobile industry, it refers to shipping goods in smaller, more frequent lots. See also Kanban and Toyota production system.

Just-in-time II (JIT II): Vendor-managed operations taking place within a customer's facility. The supplier representatives place orders to their own companies, relieving the customer's buyers from this task. Many also become involved at a deeper level, such as participating in new product development projects and manufacturing planning (concurrent planning).

Kanban: Japanese word for visible record, loosely translated means card, billboard or sign. Popularized by Toyota Corporation, it uses standard containers or lot sizes to deliver needed parts to assembly lines just in time for use. See also just-in-time and Toyota production system.

Landbridge: The movement of containers across one country or landbridge en route to another country, for example from Finland to China via Russia.

Less-than-truckload (LTL): Trucking companies may consolidate and transport smaller (less-than-truckload) shipments of freight by utilizing a network of terminals and relay points.

Letter of credit: An international business document that assures the seller that payment will be made by the bank issuing the letter of credit upon fulfilment of the sales agreement.

Licensing: Licensing involves agreements that allow a firm in one country (the licensee) to use the manufacturing, processing, trademark, know-how, technical assistance, merchandising knowledge or some other skill provided by the licenser located in another country.

Life cycle analysis: A financial and performance analysis of a product's life cycle, from production and sale to final disposal.

Linear programming: Linear programming is a mathematical technique for finding optimal or minimal solutions for an objective function, subject to various constraints.

Local area network (LAN): A data communications network spanning a limited geographical area, usually a few miles at most, providing communications between computers and peripheral devices.

Logistics mission statement: A mission statement is an overriding objective of a company as to what business it is in. A logistics mission statement is an overriding objective of a company's logistics strategy, for example it will outsource all of its logistics activities.

Logistics strategic planning: The process of planning logistics strategy within a company.

Lot quantity costs: Costs associated with manufacturing or producing lots of finished goods or products.

Management style: A pattern of management behaviour in companies. For example, some managers may go around the company to see how things are done, this is known as management by walking around.

Manufacturing resource planning (MRP II): A method for the effective planning of all resources of a manufacturing company. It is made up of a variety of processes, each linked together: business planning, production planning (sales and operations planning), master production scheduling, material requirements planning, capacity requirements planning, and the execution support systems for capacity and material.

Marketing concept: A concept developed in the early 1960s that argues companies should create products or service to meet customers' needs, but at a profit.

Market-positioned strategy: A warehouse location strategy whereby a warehouse is located close to the market it serves.

Materials management: The movement and management of materials and products from procurement from suppliers (inbound logistics) through the production process.

Materials requirements planning (MRP I): A decision-making methodology used to determine the timing and quantities of materials to purchase.

Matrix organizations: An organizational structure in which two (or more) channels of command, budget responsibility and performance measurement exist simultaneously.

Mercosur: A trading association created in 1991 by Argentina, Brazil, Paraguay and Uruguay.

Modelling: The process of developing a symbolic representation of a real, total system for managerial purposes.

Natural accounts: A cost accounting technique where similar costs across a company are grouped together in natural accounts, for example all salary payments are grouped into a salaries account.

Network: See hollow corporation.

Non-cumulative quantity discounts: Discounts for transportation services based on individual orders.

Operating plan: An annual plan that details operations, anticipated revenues and associated costs for the forthcoming year.

Opportunity cost of capital: Is the return that a company could make on money it has tied up in inventory or other asset investments.

Organizational attachment: The extent to which employees identify with their employer.

Organizational climate: The internal factors in a company that affect the effectiveness of employees and the company. Also known as corporate culture.

Outsourcing: To utilize a third-party provider to perform logistics services previously performed in-house.

Own account carrier: A company that transports its own goods and supplies in its own equipment.

Pareto's law or 80/20 rule: The law suggests that most effects come from relatively few causes; that is, 80 per cent of the effects (or sales or costs) come from 20 per cent of the possible causes (or items). See also ABC analysis.

Place utility: A value created in a product by changing its location. Transportation creates place utility.

Plant warehouse costs: Costs related to storage and handling at a warehouse located at a company's plant or production facility.

Point-of-sale (POS): The time and place at which a sale occurs, such as a cash register in a retail operation, or the order confirmation screen in an online session. Supply chain partners are interested in capturing data at the POS, because it is a true record of the sale rather than being derived from other information such as inventory movement.

Point-to-point service: Transportation service that operates from the point where a product is picked up to where a product is delivered to its final destination.

Popularity: Refers to the order popularity of goods stored in a warehouse, which may see such goods stored near shipping and receiving docks for cost and handling efficiency.

Possession utility: The value created by marketing's effort to increase the desire to possess a good or benefit from a service.

Postponement: A strategy to delay final activities such as assembly, production or packaging until the latest possible time to eliminate excess inventory in the form of finished goods which may be packaged in a variety of configurations. This is the opposite of speculation.

Post-choice evaluation: The fourth stage in the transportation carrier selection process where a company institutes an evaluation process to determine the performance level of the chosen carrier.

Private warehouses: Warehouses owned and operated by a company for its own goods and products.

Problem recognition: The first stage in the transportation carrier selection process where a company recognizes a need for different or improved transportation services.

Procurement: The business functions of procurement planning, purchasing, inventory control, traffic, receiving, incoming inspection and salvage operations.

Production-positioned strategy: A warehouse location strategy whereby a warehouse is located close to where the goods or products are produced.

Proprietary systems: An EDI system that is owned, managed and maintained by a single company.

Public warehouse costs: Costs to a company to rent or lease a warehouse or storage space owned by a third party.

Pull versus push systems: If a company produces products to a forecast in anticipation of customer demand then it is 'pushing' its inventory. Conversely, if a company waits to produce its products until a customer demands them, the customer is 'pulling' the inventory.

Purchasing: The functions associated with the actual buying of goods and services required by a company.

Quick response (QR): A strategy widely adopted by general merchandise retailers and manufacturers to reduce retail out-of-stocks, forced markdowns and operating expenses. These goals are accomplished through shipping accuracy and reduced response time. QR is a partnership strategy in which suppliers and retailers work together to respond more rapidly to the consumer by sharing point-of-sale scan data, enabling both to forecast replenishment needs.

Randomized storage: A warehouse storage technique in which goods are placed in any space that is empty when they arrive at the warehouse. Although this method requires the use of a locator file to identify part locations, it often requires less storage space than a dedicated storage method.

Reengineering: A fundamental rethinking and radical redesign of business processes to achieve dramatic improvements in performance. Also called business process reengineering (BPR).

Report generation: The ability to generate various management reports from data processing and analysis in a logistics information system.

Return on investment (ROI): A financial performance measure where profit is divided by total assets or investment in a company. ROI represents the ratio of profit provided by all assets in the company.

Reverse logistics: A specialized segment of logistics focusing on the movement and management of products and resources after the sale and after delivery to the customer. Includes product returns for repair and/or credit.

Search process: The second stage in the transportation carrier selection process where a company searches information for different or improved transportation services.

Self-directed teams: See work teams.

Shuttle trains: A form of combined transport by members of the International Union of combined Road-Rail transport (UIRR) where trains have a fixed composition of wagon loads.

Simulation: A mathematical technique for testing the performance of a system due to uncertain inputs and/or uncertain system configuration options.

Speculation: A strategy of producing finished goods well in advance of their introduction in the marketplace in order to reduce uncertainty over demand. This is the opposite of postponement.

Speculative buying: When a customer purchases goods to resell at a later date for profit as opposed to using them in the company's business processes.

Standard costs: A budgeting technique where standard costs represent actual costs if the company is operating as planned.

Standing committees: Companies will form various committees on an indefinite basis that are concerned with one or various purposes or cases. See ad hoc committees.

Statistical process control (SPC): A visual means of measuring and plotting process and product variation. Results are used to adjust variables and maintain product quality.

Stock spotting: A process of shipping a consolidated truckload of inventory to a public warehouse just prior to a period of maximum seasonal sales in order to ensure demand is met.

Stockkeeping unit (SKU): A category of unit with unique combination of form, fit and function (i.e. unique components held in stock).

Stockless purchasing: A practice whereby the buyer negotiates a price for the purchases of annual requirements of MRO items and the seller holds inventory until the buyer places an order for individual items.

Storage space costs: Warehouse space-related costs that change with the level of inventory. These costs do not include usual warehousing costs such as capital costs, usual labour, etc.

Strategic alliances: Business relationships in which two or more independent organizations cooperate and willingly modify their business objectives and practices to help achieve long-term goals and objectives.

Strategic plan: Looking one to five years into the future and designing a logistical system (or systems) to meet the needs of the various businesses in which a company is involved.

Strategy: A specific action to achieve an objective.

Supplier certification: Certification procedures verifying that a supplier operates, maintains, improves and documents effective procedures that relate to the customer's requirements. Such requirements can include cost, quality, delivery, flexibility, maintenance, safety, and ISO quality and environmental standards.

Supplier development: Where a supplier and customer make dedicated investments in their relationship, and create various bonds where the customer takes the lead in setting performance improvement targets for the supplier.

Supply chain: The supply chain links many companies together, starting with unprocessed raw materials and ending with the final customer using the finished goods, and consists of the material and information interchanges in the chain for a particular product or service.

Supply chain management (SCM): As defined by CSCMP, supply chain management encompasses the planning and management of all activities involved in sourcing and procurement, conversion, and all logistics management activities, and includes coordination and collaboration with channel partners, which can be suppliers, intermediaries, third-party service providers and customers.

Systems approach: A concept that considers all functions or activities in a system need to be understood in terms of how they affect or are affected by other elements they interact with in the system.

Tactical plan: The process of developing a set of tactical plans (e.g. production plan, sales plan, marketing plan, and so on). Two approaches to tactical planning exist for linking tactical plans to strategic plans – production planning and sales and operations planning.

Tariffs: Taxes levied on goods entering a country.

Telemarketing: Marketing by telephone, where a salesperson will contact a customer to discuss new orders or provide customer service information.

Terms of shipment: Terms and conditions for two or more companies to ship or transport goods or products. See INCOTERMS.

Terms of trade: Terms and conditions for two or more companies to do business. See INCOTERMS.

Throughput: A measure of warehousing output volume (weight, number of units). Also, the total amount of units received plus the total amount of units shipped, divided by two.

Time utility: A value created in a product by having the product available at the time desired. Transportation and warehousing create time utility.

Time-based competition: A process for taking time out of operations and logistics in order to reduce costs and give a company a competitive advantage.

Time-in-transit: The time taken for a product to move from one point to another.

Total cost analysis: A decision-making approach that considers minimization of total costs and recognizes the interrelationship among system variables such as transportation, warehousing, inventory and customer service.

Total cost concept: A concept that considers the costs of all logistics activities needs to be considered in total to effectively manage logistics processes. Total cost analysis is the tool used in this concept.

Total quality management (TQM): A management approach where managers constantly communicate with organizational stakeholders to emphasize the importance of continuous quality improvement.

Toyota production system (TPS): A system of automotive production developed by Toyota to speed up the production process, improve quality and reduce waste. See also Kanban and just-in-time.

Trading companies: Trading companies match buyers and sellers in different countries and manage the documentation and export arrangements.

Traffic management: The management and controlling of transportation modes, carriers and services.

Transportation brokers: Are companies that provide services to both shippers and carriers by arranging and coordinating transportation of products.

Unitization: In warehousing, the consolidation of several units into larger units for fewer handlings.

Value analysis: A method to determine how features of a product or service relate to cost, functionality, appeal and utility to a customer (i.e. engineering value analysis).

Value-added networks (VANs): A company that acts as a clearing-house for electronic transactions between trading partners that receives EDI transmissions from sending trading partners and holds them in a 'mailbox' until retrieved by the receiving partners.

Value-of-service pricing: Pricing according to the value of the product being transported; third-degree price discrimination; demand-orientated pricing; charging what the traffic will bear.

Vehicle telematics: Vehicle telematics systems integrate telecommunications and informatics and are used for a number of purposes, including collecting road tolls, managing road usage, tracking fleet vehicle locations, recovering stolen vehicles, providing automatic collision notification and providing location-driven driver information services.

Very large crude carriers (VLCCs): Are large crude oil supertankers of between 200,000 and 300,000 dead weight tons (DWT) in size.

Tankers that exceed 300,000 DWT are known as ultra large crude carriers (ULCCs).

Virtual corporation: A variant of the hollow corporation and a logical extension of outsourcing. In a virtual corporation the capabilities and systems of the company are merged with those of the suppliers, resulting in a new type of corporation where boundaries between the suppliers' systems and those of the company seem to disappear.

Volume contracts: When a customer establishes a large volume purchase contract with a supplier for several company business units or departments in order to take advantage of volume discount pricing.

Warehouse: Storage place for products. Principal warehouse activities include receipt of product, storage, shipment and order picking.

Warehousing: The storing (holding) of goods in a warehouse.

Work teams: A small number of people committed to certain tasks in the workplace that are ongoing in nature and have specific, continuing goals. These people are drawn from cross-functional groups in the company and are also known as self-directed teams.

Working capital: Working capital measures how much in liquid assets, i.e. current assets minus current liabilities, a company has available to operate its business.

Work-in-process inventory: Parts and sub-assemblies in the process of becoming completed finished goods. Work in process generally includes all of the material, labour and overhead charged against a production order which has not been absorbed back into inventory through receipt of completed products.

Zone pricing: The constant price of a product at all locations within a particular geographic zone.

Index